CAMBRIDGE STUDIES IN ADVANCED MATHEMATICS 102

# RANDOM FRAGMENTATION AND COAGULATION PROCESSES

Fragmentation and coagulation are two natural phenomena that can be observed in many sciences and at a great variety of scales. This book is the first comprehensive theoretical account of mathematical models for situations where either phenomenon occurs randomly and repeatedly as time passes.

The fragmentation and coalescent processes considered in this text describe the evolution of particle systems, where particles are characterized by their sizes. In a fragmentation process, each particle splits at a rate which depends on its size, and independently of the other particles. In a coalescent process, coagulations occur at rates which depend only on the particles involved in the merging, and not on the other particles in the system. The book starts by developing the theory of fragmentation chains, that is processes in which each fragment remains stable for some random time and then splits; it then turns to the general situation where each fragment may split instantaneously, using Kingman's theory of exchangeable random partitions. Then, two quite different types of coagulation process are considered: "exchangeable" coalescents, where rates of coagulation do not depend on the masses in the system and coagulations may occur simultaneously and involve an arbitrary number of components, and "stochastic" coalescents, where only binary coagulations are permitted and the rate of such coagulation may depend on the two fragments involved.

This self-contained treatment develops the models in a way that makes recent developments in the field accessible to readers with a solid background in probability. Each chapter ends with a comments section in which important aspects not discussed in the main part of the text (often because the discussion would have been too technical and/or lengthy) are addressed and precise references are given.

Cambridge Studies in Advanced Mathematics



All the titles listed below can be obtained from good booksellers or from Cambridge University Press. For a complete series listing visit: http://www.cambridge.org/mathematics

*Already published*

68 K. Sato *Lévy processes and infinitely divisible distributions*
69 H. Hida *Modular forms and Galois cohomology*
70 R. Iorio & V. Iorio *Fourier analysis and partial differential equations*
71 R. Blei *Analysis in integer and fractional dimensions*
72 F. Borceaux & G. Janelidze *Galois theories*
73 B. Bollobás *Random graphs*
74 R.M. Dudley *Real analysis and probability*
75 T. Sheil-Small *Complex polynomials*
76 C. Voisin *Hodge theory and complex algebraic geometry, I*
77 C. Voisin *Hodge theory and complex algebraic geometry, II*
78 V. Paulsen *Completely bounded maps and operator algebras*
79 F. Gesztesy & H. Holden *Soliton Equations and Their Algebro-Geometric Solutions, I*
81 S. Mukai *An Introduction to Invariants and Moduli*
82 G. Tourlakis *Lectures in Logic and Set Theory, I*
83 G. Tourlakis *Lectures in Logic and Set Theory, II*
84 R.A. Bailey *Association Schemes*
85 J. Carlson, S. Müller-Stach & C. Peters *Period Mappings and Period Domains*
86 J.J. Duistermaat & J.A.C. Kolk *Multidimensional Real Analysis I*
87 J.J. Duistermaat & J.A.C. Kolk *Multidimensional Real Analysis II*
89 M. Golumbic & A. Trenk *Tolerance Graphs*
90 L. Harper *Global Methods for Combinatorial Isoperimetric Problems*
91 I. Moerdijk & J. Mrcun *Introduction to Foliations and Lie Groupoids*
92 J. Kollar, K.E. Smith & A. Corti *Rational and Nearly Rational Varieties*
93 D. Applebaum *Lévy Processes and Stochastic Calculus*
94 B. Conrad *Modular Forms and the Ramanujan Conjecture*
95 M. Schechter *An Introduction to Nonlinear Analysis*
96 R. Carter *Lie Algebras of Finite and Affine Type*
97 H.L. Montgomery, R.C. Vaughan & M. Schechter *Multiplicative Number Theory I*
98 I. Chavel *Riemannian Geometry*
99 D. Goldfeld *Automorphic Forms and L-Functions for the Group GL(n, R)*
100 M.B. Marcus & J. Rosen *Markov Processes, Gaussian Processes, and Local Times*
101 P. Gille & T. Szamuely *Central Simple Algebras and Galois Cohomology*

# RANDOM FRAGMENTATION AND COAGULATION PROCESSES

JEAN BERTOIN
*Laboratoire de Probabilités et Modèles Aléatoires*
*Université Pierre et Marie Curie*

CAMBRIDGE
UNIVERSITY PRESS

CAMBRIDGE
UNIVERSITY PRESS

University Printing House, Cambridge CB2 8BS, United Kingdom

Cambridge University Press is part of the University of Cambridge.

It furthers the University's mission by disseminating knowledge in the pursuit of education, learning and research at the highest international levels of excellence.

www.cambridge.org
Information on this title: www.cambridge.org/9780521867283



First published 2006

*A catalogue record for this publication is available from the British Library*

ISBN 978-0-521-86728-3 Hardback

# Contents

# Introduction

Fragmentation and coagulation are two natural phenomena that can be observed in many sciences, at a great variety of scales. To give just a few examples, let us simply mention first for fragmentation, the studies of stellar fragments in astrophysics, fractures and earthquakes in geophysics, breaking of crystals in crystallography, degradation of large polymer chains in chemistry, DNA fragmentation in biology, fission of atoms in nuclear physics, fragmentation of a hard drive in computer science, ... For coagulation, we mention the formation of the large structures (galaxies) in the universe and of planets by accretion in astrophysics, of polymer chains in chemistry, of droplets of liquids in aerosols or clouds, coalescence of ancestral lineages in genealogy of populations in genetics, ...

The main purpose of this monograph is to develop mathematical models which may be used in situations where either phenomenon occurs randomly and repeatedly as time passes. For instance, in the case of fragmentation, we can think of the evolution of blocks of mineral in a crusher. The text is intended for readers having a solid background in probability theory.[1] I aimed at providing a rather concise and self-contained presentation of random fragmentation and coagulation processes; I endeavored to make accessible some recent developments in this field, but did not try to be exhaustive. Each chapter ends with a 'Comments' section in which some important aspects that have not been discussed in the main part of the text (often because the discussion would have been too technical and/or lengthy) are addressed and precise references are given.

[1] Preliminaries on classical properties of some fundamental stochastic processes (including Markov chains in continuous time, Poisson random measures, ...) will be provided in the text; however, the reader is expected to be already acquainted with basic notions in this area.

Let us first briefly discuss fragmentation. In order to deal with models that can be studied mathematically, we are led to make hypotheses which may look at first sight somewhat stringent, but that are however, commonly assumed in applications. First, we suppose that the system has a memoryless evolution, that is its future only depends on its present state and not on its past. In particular, this excludes the possibility that an object might be more fragile (i.e. more likely to split) due to former shocks. Second, we assume that each fragment can be characterized by a real number that should be thought of as its size. This stops us from considering the spatial position of a fragment or further geometrical properties like its shape; physicists call such models *mean field*. Finally, we shall always suppose that the evolution of a given fragment does not depend on its environment, in the sense that fragments split independently of each other, or in other words, that the branching property is fulfilled.

Similarly, the coalescent processes that we shall consider here are particle systems with a Markovian (i.e. memoryless) evolution, where particles are characterized by their sizes. Roughly, the key assumption is that coagulations occur at rates which only depend on the particles involved in the merging, and not on the other particles in the system. Very loosely speaking, this hypothesis plays a role similar to the branching property for fragmentation processes.

Naively, fragmentation and coagulation are dual notions, in the sense that a simple time-reversal changes one into the other. However, due to the fundamental requirements we impose (these are needed to get notions that can be investigated mathematically), this naive duality relation fails for the fragmentation and coalescent processes that are considered in this text. This means that in general, time-reversal does not transform a fragmentation process into a coalescent process, and vice-versa.[2] Therefore we will have to develop distinct theories. However, they share similarities, not just because they deal with random processes taking values in the same state space but more significantly because the same techniques are often relevant to investigate both notions.

The plan of this text is as follows. In the first chapter, we develop the theory of fragmentation chains, that is fragmentation processes in which each fragment remains stable for some random time and then splits. We focus on processes having a natural property of self-similarity. The heart of the study lies in the underlying genealogical structure of the fragmentation chain as a randomly marked tree, and the branching property. We investigate asymptotic properties of the empirical measure of the fragments as time tends to infinity;

[2] Nonetheless, there are a few striking and important examples in which the duality holds, that will be emphasized in this text.

the so-called intrinsic martingale induced by the genealogical tree plays a fundamental role in the study. When the index of self-similarity is negative, we point at unexpected phenomena of extinction and formation of dust.

Plainly, this discrete approach does not enable us to consider situations where splitting can occur continuously, that is when each fragment may split immediately. We shall make a key step to solve this fundamental difficulty in the second chapter. There, we will start by discussing various notions of partitions, which provide natural frameworks for fragmentation and coalescent processes. One obvious notion is that of a partition of a unit mass, that is a sequence of non-negative real numbers which add up to at most 1. An important class of random mass-partitions can be constructed from Poisson random measures, and we shall develop some material in this field, including the classical Poisson-Dirichlet distributions and its two-parameter generalization by Pitman and Yor. A probably less intuitive notion, which is due to Kingman, is that of an exchangeable random partition of $\mathbb{N}$, the set of natural integers. Roughly, the latter arises by sampling points at random in the object which splits, and provides a powerful technique of spatial discretization for mass-partitions.

Exchangeable random partitions play a key role in Chapter 3 for the construction and study of general self-similar fragmentation processes, in which each fragment may split immediately. We shall characterize their dynamics in terms of an erosion coefficient, a rate of sudden dislocation, and the index of self-similarity. The evolution of a fragment containing a point tagged at random can then be described in terms of some subordinator (i.e. an increasing process with independent and stationary increments) whose characteristics are expressed in terms of the erosion coefficient, the dislocation rate, and the index of self-similarity. Statistical properties of the tagged fragment yield extensions of results proved for fragmentation chains in the first chapter to this more general setting.

Coalescent processes are considered in the last two chapters; we shall focus on two quite different types of dynamics. The first are the so-called exchangeable coalescents for which the rates of coagulation do not depend on the masses in the system. In general, coagulations may occur simultaneously, and each coagulation may involve an arbitrary number of components. The second type comprises the so-called stochastic coalescents, which are often used as simple models for random coagulation in physics. There, only binary coagulations are permitted, but now the rate of such a coagulation may depend on the two fragments which are involved.

Exchangeable coalescents are studied in Chapter 4. We first introduce the celebrated coalescent of Kingman (the essentially unique exchangeable coalescent having only binary coagulations), which plays an important role in the study of the genealogy of large populations. Then, we shall consider general exchangeable coalescents, which have been introduced by Pitman, Möhle, Sagitov and Schweinsberg. The analyses often rely on the same techniques and ideas as those which have been used for homogeneous fragmentations, even though time-reversing a homogeneous fragmentation does not produce an exchangeable coalescent as one might expect naively. Finally, we develop connections with certain stochastic flows, following a recent work by Bertoin and Le Gall, and investigate the special case of the so-called Bolthausen-Sznitman coalescent which was motivated by considerations in statistical physics.

Finally, Chapter 5 is devoted to the stochastic coalescents of Marcus and Lushnikov, and their asymptotics. We shall pay special attention to the problem of the regularity of the transition probabilities as a function of the initial conditions, following a beautiful approach due to Fournier. We shall also develop the connection with Smoluchowski's coagulation equations, by presenting some results about the hydrodynamic behavior of certain stochastic coalescents. In this direction, we shall first investigate the multiplicative coalescent via its representation in terms of the Erdős-Rényi random graph model, and then extend the analysis to sub-multiplicative kernels by coupling. The ultimate section of this chapter will deal with the additive kernel, focussing on remarkable connections with certain random structures (random trees and forests) which have been developed by Pitman.

It should be already clear from the brief presentation above that this text owes much to Jim Pitman. Many of his ideas, results and intuitions form the cornerstones of this study. The reader is strongly encouraged to read his superb Lecture Notes [186] for the St Flour summer school, which covers in particular several topics treated here, sometimes from a different perspective. Needless to say that my warmest thanks go to Jim.

Certain parts presented in this text result directly from collaborations. I would like to express my deep gratitude to Sasha Gnedin, Christina Goldschmidt, Jean-François Le Gall, Servet Martinez, Jim Pitman and Alain Rouault. Working with them in this field has been not only enriching and stimulating, but also very pleasant.

I also address special thanks to Anne-Laure Basdevant, Julien Berestycki, Bénédicte Haas and Grégory Miermont, who have prepared their Ph.D. theses on fragmentation and coalescent processes under my supervision. Some of their results have a key role in this text. From a personal point of view, the strong interest they had in this field, the questions they raised and the clever ideas they developed, have been for me a major source of motivation and support for carrying on this project.

Further, I would like to thank Maria-Emilia Caballero, Pablo Ferrari and Servet Martinez, who offered me the opportunity of giving short courses on fragmentation and coagulation. Earlier drafts of certain portions of this text have been used as unpublished lecture notes corresponding to these courses.

Last but not least, this text has been written while I had a position at the Institut universitaire de France; I would probably not have found time nor energy for undertaking this project without the support of this institution.

# 1
# Self-similar fragmentation chains

Informally, imagine an object that falls apart randomly as time passes. The state of the system at some given time consists in the sequence of the sizes of the pieces, which are often called fragments or particles. Suppose that the evolution is Markovian and obeys the following rules. First, different particles evolve independently of each other, that is the so-called branching property is fulfilled. Second, there is a parameter $\alpha \in \mathbb{R}$, which will be referred to as the index of self-similarity, such that each fragment with size $s$ is stable during an exponential time with parameter proportional to $s^{\alpha}$. In other words, a particle with size $s > 0$ has an exponential lifetime with mean $cs^{-\alpha}$, where $c > 0$ is some constant. At its death, this particle splits and there results a family of fragments, say with sizes $(s_i, i \in \mathbb{N})$, where the sequence of ratios $(s_i/s, i \in \mathbb{N})$ has the same distribution for all particles. The purpose of this chapter is to construct such self-similar fragmentation chains, to shed light on their genealogical structure, and to establish some of their fundamental properties.

## 1.1 Construction of fragmentation chains

In this section, we briefly present some basic elements on Markov chains and branching Markov chains in continuous time which are then used for the construction and the study of fragmentation chains. For convenience, we recall first some standard notation for sets of integers which will be used through this text without further reference.

The sets of positive integers, and respectively of integers, are denoted by

$$\mathbb{N} = \{1, 2, \dots\}\ , \quad \mathbb{Z} = \{\dots, -1, 0, 1, \dots\}\ ,$$

and then

$$\mathbb{Z}_+ = \mathbb{N} \cup \{0\} = \{0, 1, 2, \dots\}$$

designates the set of non-negative integers. When we shall need to consider infinity as an extended integer, we shall use the notation

$$\overline{\mathbb{N}} = \mathbb{N} \cup \{\infty\}, \quad \overline{\mathbb{Z}}_+ = \mathbb{Z}_+ \cup \{\infty\}.$$

### 1.1.1 Preliminaries on Markov chains

Let $(E, \mathrm{d})$ be a Polish space, that is a complete separable metric space, which we also endow with its Borel sigma-field. Consider a collection $(q(x, \cdot), x \in E)$ of finite measures on $E$ which is (weakly) measurable in the variable $x$, in the sense that for every Borel set $B \subseteq E$, the map $x \to q(x, B)$ is measurable. It is well-known that we can use the kernel $(q(x, \cdot), x \in E)$ as the jump rates of some Markov chain in continuous time. Let us briefly recall the main steps and refer, for example, to Norris [174] or Fristedt and Gray [109] for more details.

For every $x \in E$, we write $q(x) := q(x, E)$ for the total mass of the measure $q(x, \cdot)$, and we introduce the normalized probability measure on $E$ given by

$$\bar{q}(x, \cdot) = q(x, \cdot)/q(x)$$

with the convention that $\bar{q}(x, \cdot) = \delta_x$ is the Dirac point mass at $x$ when $q(x) = 0$. So $(\bar{q}(x, \cdot), x \in E)$ is a Markov kernel, that is a (weakly) measurable family of probability measures on $E$. We can think of $(\bar{q}(x, \cdot), x \in E)$ as the transition probabilities of a Markov sequence[1] $Y = (Y(n), n \in \mathbb{Z}_+)$. That is, for every $n \in \mathbb{Z}_+$,

$$\mathbb{P}\left(Y(n+1) \in \cdot \mid Y(0), \dots, Y(n)\right) = \bar{q}(Y(n), \cdot).$$

Next, we shall transform the Markov sequence $Y$ into a Markov process $X = (X(t), t \geq 0)$ in continuous time which visits the same states as the

[1] In the literature, Markov sequences are often called Markov chain in discrete time. However, in order to avoid a possible confusion with Markov chains in continuous time, which are the main object of interest in this section, we shall keep the terminology *chain* for processes in continuous time, and use *sequence* for processes in discrete time. We also mention that in the literature, Markov chains in continuous time generally concern only countable state spaces; some authors prefer to refer to pure-jump Markov processes in the case of a general topological state space. Nonetheless, we shall use here the name *chain* to underline the hold-jump structure (which will be described below), and keep the name *process* for continuous evolutions, that is situations where the process may not remain constant on arbitrarily small time-interval.

sequence $Y$. More precisely, conditionally on the sequence $Y = (y_n, n \in \mathbb{Z}_+)$, we shall replace the unit waiting time at each step $y_n$ by an exponential variable with parameter $q(y_n)$ (thus depending on the state of $Y$ at this step), independently of the other steps. The construction is specified by the following procedure.

Let $\mathbf{e}_0, \mathbf{e}_1, \ldots$ be a sequence of i.i.d. standard exponential variables, which is independent of $Y$. We associate to every sample path of $Y$ the additive functional

$$A(n) := \sum_{i=0}^{n} \mathbf{e}_i / q(Y(i)) , \qquad n \in \mathbb{Z}_+ ,$$

which represents the instant at which $X$ jumps from the state $Y(n)$ to the state $Y(n+1)$. This procedure enables us to define $X(t)$ for any $t \geq 0$ if and only if the series $A(\infty) := \sum_{i=0}^{\infty} \mathbf{e}_i / q(Y(i))$ diverges. In this direction, we recall the following well-known fact.

**Lemma 1.1** *The conditions*

$$A(\infty) := \sum_{i=0}^{\infty} \mathbf{e}_i / q(Y(i)) = \infty \qquad a.s. \tag{1.1}$$

*and*

$$\sum_{i=0}^{\infty} 1/q(Y(i)) = \infty \qquad a.s. \tag{1.2}$$

*are equivalent.*

*Proof* We shall prove a slightly stronger result. Let $(y_i, i \in \mathbb{Z}_+)$ be some sequence of points in $E$, which should be thought of as the sequence of the states visited by $Y$. On the one hand, the identity

$$\mathbb{E}\left(\sum_{i=0}^{\infty} \mathbf{e}_i / q(y_i)\right) = \sum_{i=0}^{\infty} 1/q(y_i)$$

shows that if the series on the right-hand side converges, then $\sum_{i=0}^{\infty} \mathbf{e}_i / q(y_i) < \infty$ a.s. Conversely, taking the Laplace transform, we get

$$\begin{aligned}\mathbb{E}\left(\exp\left(-\sum_{i=0}^{\infty} \mathbf{e}_i / q(y_i)\right)\right) &= \prod_{i=0}^{\infty} \frac{q(y_i)}{1+q(y_i)} \\ &= \exp\left(-\sum_{i=0}^{\infty} \ln\left(1 + 1/q(y_i)\right)\right) .\end{aligned}$$

If the series $\sum_{i=0}^{\infty} \mathbf{e}_i/q(y_i)$ converges with positive probability, then the right-hand side above has to be strictly positive and hence $\sum_{i=0}^{\infty} 1/q(y_i) < \infty$. Note from the first part of the proof that this forces the series $\sum_{i=0}^{\infty} \mathbf{e}_i/q(y_i)$ to converge with probability one, a fact that can also be observed directly from Kolmogorov's 0-1 law. □

Condition (1.2) is plainly fulfilled whenever

$$\sup_{x\in E} q(x) < \infty\,; \tag{1.3}$$

however, in general checking whether (1.2) holds can be tedious. Henceforth, taking (1.1) for granted, we may introduce the time-change

$$\alpha(t) = \min\{n \in \mathbb{Z}_+ : A(n) > t\}\,, \qquad t \geq 0;$$

one says that $\alpha(\cdot)$ is the right-continuous inverse of the additive functional $A(\cdot)$. Then we define a process in continuous time $X = (X(t), t \geq 0)$ by the identity

$$X(t) := Y(\alpha(t))\,, \qquad t \geq 0\,.$$

This construction by random time-substitution can be rephrased in terms of a so-called *hold-jump* description: the states $x \in E$ with $q(x) = 0$ are *absorbing* for $X$, that is

$$\mathbb{P}(X(t) = x \text{ for all } t \geq 0 \mid X(0) = x) = 1\,,$$

and starting from some non-absorbing state $x \in E$ with $q(x) > 0$, the process $X$ stays at $x$ up to the holding time $\mathbf{e}_0/q(x)$ which has an exponential distribution with parameter $q(x)$, and then jumps[2] according to the probability distribution $\bar{q}(x, \cdot)$, independently of the holding time. It is easily seen from the absence of memory of exponential variables that $X$ enjoys the Markov property; one says that $X$ is a Markov chain (in continuous time). Note also that $X$ has right-continuous paths a.s.

The semigroup $(P_t, t \geq 0)$ of $X$ is the family of linear operators on the space of bounded measurable functions $f: E \to \mathbb{R}$ defined by

$$P_t f(x) := \mathbb{E}(f(X(t) \mid X(0) = x)\,, \qquad x \in E;$$

[2] When $q(x, \{x\}) > 0$, the probability that process $X$ stays at the state $x$ after the exponential holding time is positive, so strictly speaking there may be no jump after this first holding time. However, this induces no difficulty whatsoever, and it is convenient not to distinguish this degenerate case.

it satisfies the *Chapman-Kolmogorov equation*

$$P_t \circ P_s = P_{t+s}, \qquad t, s \geq 0.$$

It is easy to check from the hold-jump description that for every bounded measurable function $f : E \to \mathbb{R}$,

$$\begin{aligned}\mathsf{G}f(x) &:= \lim_{t\to 0+} \frac{1}{t}\mathbb{E}(f(X(t)) - f(X(0)) \mid X(0) = x)\\ &= \int_E (f(y) - f(x))\, q(x, dy), \end{aligned} \tag{1.4}$$

which identifies the infinitesimal generator $\mathsf{G}$ of $X$. In particular, combining with the Chapman-Kolmogorov equation yields the classical *backward* equation

$$\frac{dP_t f(x)}{dt} = \mathsf{G}P_t f(x), \qquad t \geq 0. \tag{1.5}$$

Further, when the function $\mathsf{G}f$ is bounded on $E$, we also have the *forward* equation

$$\frac{dP_t f(x)}{dt} = P_t \mathsf{G}f(x), \qquad t \geq 0. \tag{1.6}$$

A well-known alternative characterization of the infinitesimal generator is that for every bounded measurable function $f : E \to \mathbb{R}$ such that $\mathsf{G}f$ is bounded, $\mathsf{G}f$ is the unique bounded measurable function $g : E \to \mathbb{R}$ for which the process

$$f(X(t)) - \int_0^t g(X(s))ds, \qquad t \geq 0$$

is a martingale under $\mathbb{P}(\cdot \mid X(0) = x)$ for every $x \in E$.

Either the construction of $X$ or (1.4), shows that the family $(q(x, \cdot), x \in E)$ can be thought of as the *jump rates* of $X$, and thus entirely characterizes the distribution of the Markov chain $X$. In the same vein, note also that when the space $E$ is discrete and $q(x, \{x\}) = 0$, the jump rates of the chain can be recovered from its one-dimensional distributions by

$$q(x, \{y\}) = \lim_{t\to 0+} \frac{1}{t}\mathbb{P}(X(t) = y \mid X(0) = x), \qquad x \neq y. \tag{1.7}$$

**Example** The so-called *compound Poisson processes* form one of the simplest and best known family of Markov chains in continuous time. Specifically, consider the special case when $E$ is some Euclidean space (or, more generally, some nice topological group) and the jump rates $(q(x, \cdot), x \in E)$ are translation invariant, that is for every $x \in E$, $q(x, \cdot)$ is the image of some given finite measure $\Lambda$ by the translation $y \to x+y$ (in particular $q(0,\cdot)=\Lambda$). The measure $\Lambda$ is known as the *Lévy measure* of the compound Poisson process. Plainly, the transition probabilities $(\bar{q}(x,\cdot), x \in E)$ are also translation invariant, and hence the Markov sequence $Y = (Y(n), n \in \mathbb{Z}_+)$ is a random walk with step distribution $\bar{\Lambda}(\cdot) := \Lambda(\cdot)/\Lambda(E)$. In other words, $Y(n) = Y(0) + \xi_1 + \cdots + \xi_n$, where the increments $\xi_i$ form a sequence of i.i.d. variables with law $\bar{\Lambda}(\cdot)$. Since $c := \Lambda(E) = q(x)$ does not depend on the state $x$, the continuous time Markov chain with jump rates $(q(x,\cdot), x \in E)$ can be obtained as the composition $X = Y \circ N$ where $N = (N_t, t \geq 0)$ is a Poisson process with intensity $c$ which is independent of the random walk $Y$. This construction triggers the name compound Poisson process for such Markov chains. More generally, it is easy to see that a Markov chain in continuous time can be constructed as a Markov sequence time-changed by an independent Poisson process if and only if the total jump rates are bounded, that is if and only if (1.3) holds.

### 1.1.2 Branching Markov chains

In this section, we shall consider an important class of Markov chains with values in a space of measures. First, we call a *finite point measure* on $E$ any measure of the type $m = \sum_{i=1}^{n} \delta_{x_i}$ where $n \in \mathbb{Z}_+$ and $x_i \in E$ ($x_i = x_j$ is allowed, and for $n = 0$, the measure $m = 0$ is trivial). We denote by $E_{\mathrm{p}}$ the space of finite point measures on $E$, endowed with the distance

$$\operatorname{dist}(m, m') := \sup \left| \int_E f(x) m(dx) - \int_E f(x) m'(dx) \right| ,$$

where the supremum is taken over the space of Lipschitz-continuous functions $f : E \to \mathbb{R}$ with $|f(x) - f(y)| \leq d(x,y)$ for every $x, y \in E$. It is easy to check that this distance is equivalent to the Prohorov metric on the space of finite measures on $E$, and that $(E_{\mathrm{p}}, \operatorname{dist})$ is a Polish space; see for example [57].

We shall think of the atoms of point measures as particles; see the picture below. Our goal is to construct a particle system whose dynamics can be described informally as follows. The system is non-interacting, that is different particles have independent evolutions. Each particle *branches*, that is it dies at some rate depending on its location, and at its death it is replaced by some random number of random particles, independently of the lifetime. More precisely, we consider a family $(\nu_x, x \in E)$ of finite measures on $E_{\mathrm{p}}$, which depends measurably on the variable $x$. A particle located at $x$ lives for an

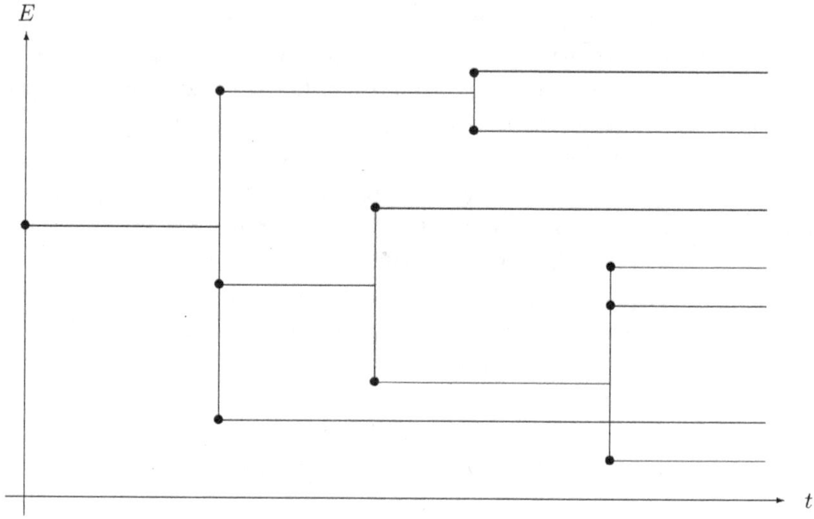


*Sample path of a branching Markov chain started from a single particle*

exponential time with parameter $\nu_x(E_p)$ (so it is immortal when $\nu_x(E_p) = 0$). At its death, the particle is removed and replaced by new particles, say $y_1, \ldots, y_k$, where the finite point measure $\sum_{i=1}^{k} \delta_{y_i}$ is distributed according to the probability measure $\nu_x(\cdot)/\nu_x(E_p)$ on $E_p$.

In order to construct such a particle system as a Markov chain on $E_p$, we introduce some notation related to finite point measures. Given $m \in E_p$, we denote by $\theta_m(\nu)$ the image of $\nu$ by the map $m' \to m + m'$. Then we associate to the family $(\nu_x, x \in E)$, a measurable kernel $(q(m, \cdot), m \in E_p)$ of finite measures on $E_p$ defined as follows. First, $q(0, \cdot) = 0$ and, second, if $m = \sum_{i=1}^{n} \delta_{x_i}$ with $n \geq 1$, we set

$$q(m, \cdot) := \sum_{i=1}^{n} \theta_{m_i}(\nu_{x_i}), \tag{1.8}$$

where $m_i = \sum_{j \neq i} \delta_{x_j}$. The term $\theta_{m_i}(\nu_{x_i})$ in (1.8) corresponds to the rate at which the particle $x_i$ branches in the family $\{x_1, \ldots, x_n\}$. Note that the total mass $q(m) := q(m, E_p)$ is given by

$$q(m) = \sum_{i=1}^{n} \nu_{x_i}(E_p),$$

and as a consequence, an exponential variable with parameter $q(m)$ can be thought of as the minimum of $n$ independent exponential variables with parameters $\nu_{x_1}(E_p), \ldots, \nu_{x_n}(E_p)$. Observe also the important additivity property of this kernel, namely for every point measures $m, m' \in E_p$ and every measurable functional $\Phi : E_p \to \mathbb{R}_+$, there is the identity

$$\int_{E_p} \Phi(m' + \mu) q(m, d\mu) + \int_{E_p} \Phi(m + \mu) q(m', d\mu) = \int_{E_p} \Phi(\mu) q(m + m', d\mu). \tag{1.9}$$

We would like to use the kernel $(q(m, \cdot), m \in E_p)$ as jump rates of a Markov chain. In this direction, let us first describe the evolution of the Markov sequence $Y = (Y(i) : i \in \mathbb{Z}_+)$ with transition probabilities $\bar{q}(m, \cdot) = q(m, \cdot)/q(m)$ for $m \in E_p$. As $q(0) = 0$, the measure 0 is an absorbing state for the sequence. Then, let the sequence start from some finite point measure $m = \sum_{i=1}^{n} \delta_{x_i} \neq 0$. The distribution of the next state of the sequence is obtained by picking an atom $x$ at random among $x_1, \ldots, x_n$, with probability proportional to $\nu_x(E_p)$, and replacing it by the atoms of a random point measure with law $\bar{q}(\delta_x, \cdot)$. Now, we should like to consider a continuous time Markov chain with jump rates $(q(m, \cdot), m \in E_p)$, so that we need conditions ensuring (1.2) (observe that the stronger condition (1.3) cannot hold).

**Lemma 1.2** *In the preceding notation,* (1.2) *is fulfilled whenever there is a finite constant $c > 0$ such that*

$$\int_{E_p} \nu_x(dm) q(m) \le c\nu_x(E_p), \qquad \forall x \in E, \tag{1.10}$$

*where $q(m) = q(m, E_p)$ is the total jump rate from the state $m \in E_p$.*

*Proof* Let the Markov sequence $Y$ start from some finite point measure; we have to check that $\sum_{i=0}^{\infty} 1/q(Y(i)) = \infty$ a.s. For every $n \in \mathbb{N}$, set

$$r(n) := \sum_{i=1}^{k} \nu_{y_i}(E_p),$$

where $y_1, \ldots, y_k$ denote the particles which are created at the $n$-th step of $Y$, and then $\sigma(n) = r(1) + \cdots + r(n)$. Plainly, we have

$$q(Y(n)) \le q(Y(0)) + \sigma(n),$$

as the right-hand side corresponds to the total jump rate from the point measure whose atoms are given by all the particles which are born before the $(n+1)$-th step or existed at time 0 (in other words the particles that have branched are not removed).

Condition (1.10) of the lemma states that the expectation of the total jump rate from a random point measure distributed according to $\bar{q}(x, \cdot)$ is bounded from above by $c$, for any $x \in E$. By conditioning on the particle that branches at the $n$-th step, we see that (1.10) ensures that $\mathbb{E}(r(n)) \leq c$, so that $\mathbb{E}(\sigma(n)) \leq cn$ for all $n$. Thus, by Fatou's lemma, we have

$$\liminf_{n\to\infty} \frac{q(Y(0)) + \sigma(n)}{n} < \infty \qquad \text{a.s.,}$$

or equivalently

$$\limsup_{n\to\infty} \frac{n}{q(Y(0)) + \sigma(n)} > 0 \qquad \text{a.s.}$$

Since the random sequence $\sigma(\cdot)$ is increasing, this implies that

$$\sum_{n=1}^{\infty} 1/(q(Y(0)) + \sigma(n)) = \infty \qquad \text{a.s.}$$

and thus the series $\sum_{n=1}^{\infty} 1/q(Y(n))$ diverges a.s. □

From now on, we shall take the conditions of Lemma 1.2 for granted. The Markov chain in continuous time $X = (X(t), t \geq 0)$ on $E_{\mathrm{p}}$ associated with the family of jump rates $(q(m, \cdot), m \in E_{\mathrm{p}})$ is called a *branching Markov chain*. The measures $\nu_x$ $(x \in E)$ which are used to define the jump rates in (1.8) are referred to as the *branching rates* at locations $x \in E$. Standard properties of independent exponential variables show that its evolution indeed coincides with the dynamics of the non-interacting particle system which we aimed to construct. In particular, this makes the following important statement intuitively obvious.

**Proposition 1.1** (Branching property) *Let $X$ and $X'$ two independent versions of the same branching Markov chain, started respectively from two point measures $m$ and $m'$. Then $X + X'$ is a version of the branching Markov chain started from $m + m'$.*

*Proof* Let $T$ and $T'$ denote the first jump times of $X$ and $X'$, respectively. So $T$ and $T'$ are independent and follow the exponential distribution with parameters $q(m)$ and $q(m')$, respectively. Moreover, $X(T)$ and $X(T')$ are independent with respective laws $\bar{q}(m, \cdot)$ and $\bar{q}(m', \cdot)$, and are jointly independent of $T$ and $T'$. Thus the first jump time of $X + X'$ occurs at time $T \wedge T'$, which has the exponential distribution with parameter $q(m) + q(m') = q(m + m')$. Furthermore the value of $X + X'$ after this jump, namely $X(T \wedge T') + X'(T \wedge T')$, is independent of $T \wedge T'$, and has the law $\bar{q}(m+m',\cdot)$, since for every measurable functional $\Phi : E_{\mathrm{p}} \to \mathbb{R}_+$

$$\begin{aligned}
&\mathbb{E}(\Phi(X(T\wedge T') + X'(T\wedge T'))) \\
&= \mathbb{P}(T < T')\mathbb{E}(\Phi(X(T)+m')) + \mathbb{P}(T' \leq T)\mathbb{E}(\Phi(m+X'(T'))) \\
&= \frac{q(m)}{q(m+m')}\int_{E_{\mathrm{p}}} \Phi(\mu+m')\bar{q}(m,d\mu) \\
&\quad + \frac{q(m')}{q(m+m')}\int_{E_{\mathrm{p}}} \Phi(m+\mu)\bar{q}(m',d\mu) \\
&= \int_{E_{\mathrm{p}}} \Phi(\mu)\bar{q}(m+m',d\mu),
\end{aligned}$$

where we used (1.9) in the third equality. This shows that the distribution of the time and location of the first jump of $X+X'$ is the same as that of the version of the branching Markov chain starting from $m+m'$. An application of the Markov property enables us to extend this identity to the following jumps, which completes the proof. □

**Example** *Branching random walks* (in continuous time) form a special class of branching Markov chains which have been introduced by Uchiyama [208]. Specifically, we now assume that $E$ is some Euclidean space, and we consider a finite measure $\nu$ on the space $E_{\mathrm{p}}$ of finite point measures on $E$. Next, to each site $x \in E$, we associate to $\nu$ its image by the translation of its atoms by $x$. That is, for every finite point measure $m = \sum_{i=1}^{n} \delta_{x_i}$, we denote the shifted measure by $m_x := \sum_{i=1}^{n} \delta_{x+x_i}$, and $\nu_x$ is the image of $\nu$ by this shift. The branching Markov chain with branching rates $(\nu_x, x \in E)$, is called a branching random walk with branching measure $\nu$. In this direction, observe that the condition (1.10) of Lemma 1.2 reduces to

$$\int_{E_{\mathrm{p}}} \nu(dm)m(E) < \infty .$$

It is easy to see that the process of total mass $(\langle X(t), 1\rangle, t \geq 0)$ is a classical branching process. More precisely, each particle lives for an exponential lifetime with parameter $\nu(E_{\mathrm{p}})$ and, at its death, it gives birth to a random number $Z$ of children, where the offspring distribution is specified by

$$\mathbb{P}(Z=n) = \frac{1}{\nu(E_{\mathrm{p}})}\int_{E_{\mathrm{p}}} \nu(dm)\mathbb{1}_{\{m(E)=n\}} .$$

### 1.1.3 Fragmentation chains

Throughout the rest of this chapter, we shall work with the state space of decreasing numerical sequences bounded from above by 1 and with limit 0:

$$\mathcal{S}^{\downarrow} := \{\mathbf{s} = (s_1, s_2, \dots) : 1 \geq s_1 \geq s_2 \geq \cdots \geq 0 \text{ and } \lim s_i = 0\} .$$

We endow $\mathcal{S}^{\downarrow}$ with the uniform distance

$$d(\mathbf{s}, \mathbf{s}') := \max_{i \in \mathbb{N}} |s_i - s'_i| ,$$

which makes $\mathcal{S}^{\downarrow}$ a Polish space. We shall think of a sequence $\mathbf{s} \in \mathcal{S}^{\downarrow}$ as that of the sizes of the fragments resulting from the split of some block with unit size. No term of a sequence $\mathbf{s} \in \mathcal{S}^{\downarrow}$ exceeds 1; the total sum $\sum_{i=1}^{\infty} s_i$ of the series $\mathbf{s} \in \mathcal{S}^{\downarrow}$ may equal 1 (which corresponds to the so-called *conservative* situation), may be less than 1 (*dissipative case*), or even greater than 1 (this situation may occur for instance when the size of an object is the measure of its diameter). It is sometimes convenient to identify the sequence with a Radon point measure on $]0, \infty[$, $\sum_{i:s_i>0} \delta_{s_i}$. Observe that the latter is a finite point measure on $]0, \infty[$ if and only if $s_i = 0$ whenever $i$ is large enough.

We will be interested in a simple family of Markov process with càdlàg paths $X = (X(t), t \geq 0)$ and values in the space $\mathcal{S}^{\downarrow}$, called *fragmentation chains*. Informally, the evolution of $X = (X(t), t \geq 0)$ is given by a non-interacting particle system, in the sense that each particle in the system evolves independently of the others. The dynamics of each particle are the following. A particle lives for an exponential time with parameter depending only on its size and, at the end of its life, it is replaced by a random cloud (possibly infinite) of smaller particles which is independent of the lifetime of the particle. Although this description bears obvious similarities with that for branching Markov chains, it is not completely clear that such a random evolution is well-defined, because we do not require the number of fragments (or particles) at some given time to be finite. As a consequence, the jump rates from a configuration $\mathbf{s} = (s_1, \dots)$ with $s_i > 0$ for every $i \in \mathbb{N}$ may be (and indeed often are) unbounded. In particular, although a fragmentation chain is a Markov process, it is not necessarily a Markov chain in continuous time. However, we keep the terminology *chain* to underline the fact that each particle remains unchanged during some strictly positive time so, in some loose sense, the evolution of the system is discrete.

Let us now explain precisely the construction of this model. To start with, for every $x \in ]0, 1]$, let $\nu_x$ be some finite measure on $\mathcal{S}^{\downarrow}$. We assume that this family depends in a measurable way on the variable $x$. As previously, the total mass $\nu_x(\mathcal{S}^{\downarrow})$ is the parameter of the exponential lifetime of the particle, and the probability law $\nu_x(\cdot)/\nu_x(\mathcal{S}^\downarrow)$ is the law of the random cloud of particles resulting from the dislocation of $x$. We shall now discuss why the evolution of the system is well-defined under the following mild condition: we suppose henceforth first that

$$\nu_x(s_1 > x) = 0\,, \qquad x \geq 0 \tag{1.11}$$

(which means that a.s. the sizes of the fragments cannot be larger than that of the initial particle), and second that for every $\varepsilon > 0$, there exists some finite constant $c_\varepsilon$ such that for every $x > \varepsilon$,

$$\nu_x(\mathcal{S}^\downarrow) < c_\varepsilon \quad \text{and} \quad \int_{\mathcal{S}^\downarrow} \#\{i \in \mathbb{N} : s_i > \varepsilon\}\, \nu_x(d\mathbf{s}) \leq c_\varepsilon \nu_x(\mathcal{S}^\downarrow)\,, \tag{1.12}$$

where # stands for the counting measure on $\mathbb{N}$. Observe that the second requirement in (1.12) is always fulfilled when the measure $\nu_x$ is conservative or dissipative, that is $\sum_{i=1}^\infty s_i \leq x$ for $\nu_x$ almost-every configuration $\mathbf{s}$, since then the bound $\#\{i : s_i > \varepsilon\} \leq x/\varepsilon$ holds $\nu_x(d\mathbf{s})$-a.e.

Call a sequence $\mathbf{s} = (s_1, \dots) \in \mathcal{S}^\downarrow$ *finite* and write $\mathbf{s} \in \mathcal{S}^\downarrow_f$ if $s_j = 0$ for some large enough index $j$; clearly we may (and will) identify a finite sequence $\mathbf{s}$ as a finite point measure on $]0, 1]$, $m = \sum \delta_{s_j}$, where the sum is taken over indices $j$ such that $s_j > 0$. Next, we introduce threshold operators which map $\mathcal{S}^\downarrow$ to the space of finite sequences. Specifically, for every $\varepsilon > 0$, we write $\varphi_\varepsilon :$ $[0, 1] \to [0, 1]$ to be the function such that $\varphi_\varepsilon(x) = x$ if $x > \varepsilon$ and $\varphi_\varepsilon(x) = 0$ otherwise, and, by a slight abuse of notation, we still write $\varphi_\varepsilon : \mathcal{S}^\downarrow \to \mathcal{S}^\downarrow_f$ for its obvious extension to $\mathcal{S}^\downarrow$ (component by component). Plainly, the threshold operators form a projective system, that is $\varphi_\varepsilon \circ \varphi_\eta = \varphi_\eta \circ \varphi_\varepsilon = \varphi_\varepsilon$ for every $0 < \eta \leq \varepsilon$.

Let $\nu_x^\varepsilon$ be the image of $\nu_x$ by the threshold operator $\varphi_\varepsilon$; so we can think of $\nu_x^\varepsilon$ as a finite measure on the space of point measures on $]0, 1]$. In this framework, it is easy to see that whenever (1.12) holds, the conditions of Lemma 1.2 are fulfilled, and thus we can construct a branching Markov chain in continuous time $X^\varepsilon = (X^\varepsilon(t), t \geq 0)$ with branching rates given by the family $(\nu_x^\varepsilon, x \in ]0, 1])$ and started from an arbitrary finite sequence $\mathbf{s}$. This branching Markov chain takes values in the space of finite point measures on $]0, 1]$, but we may also view it as a process with values in $\mathcal{S}^\downarrow_f$ (or even $\mathcal{S}^\downarrow$) by the preceding identification.

We shall now show that one can construct simultaneously the chains $X^\varepsilon$ for different values of the parameter $\varepsilon$ in a consistent way with respect to the threshold operators and, more precisely, that there exists a process $X$ in $\mathcal{S}^\downarrow$ such that its image by the threshold operator $\varphi_\varepsilon$ is a version of $X^\varepsilon$. Clearly, such process is unique in distribution and has the evolution that we wished.

**Lemma 1.3** *For every* $\mathbf{s} \in \mathcal{S}^{\downarrow}$, *there exists a unique (in law) process* $X = (X(t), t \geq 0)$ *with values in* $\mathcal{S}^{\downarrow}$ *such that for every* $\varepsilon > 0$, $\varphi_\varepsilon(X)$ *is a branching Markov chain with the same distribution as* $X^\varepsilon$ *started from* $X^\varepsilon(0) = \varphi_\varepsilon(\mathbf{s})$.

*Proof* Fix $0 < \eta < \varepsilon$, and let $X^\eta$ denote a version of the branching Markov chain with branching rates specified by the family $(\nu_x^\eta, x \in ]0,1])$ and started from the finite configuration $\varphi_\eta(\mathbf{s})$. Recall that our setting requires that the size of a child particle is never larger than that of its parent. As a consequence, for every $s, t \geq 0$, the conditional distribution of $\varphi_\varepsilon(X^\eta(t+s))$ given $X^\eta(t)$ only depends on $\varphi_\varepsilon(X^\eta(t))$. It follows that $\varphi_\varepsilon(X^\eta(\cdot))$ is a Markov process, more precisely a Markov chain in continuous time, and since $\nu_x^\varepsilon$ is the image of $\nu_x^\eta$ by the threshold operator $\varphi_\varepsilon$, the jump rates of $\varphi_\varepsilon(X^\eta(\cdot))$ are the same as those of $X^\varepsilon$ started from $X^\varepsilon(0) = \varphi_\varepsilon(\mathbf{s})$. Thus, the two processes have the same distribution.

This observation enables us to appeal to Kolmogorov's projective theorem, and we obtain the existence of a family of process $(X^\varepsilon(\cdot), \varepsilon > 0)$ such that for every $\varepsilon > \eta > 0$, $\varphi_\varepsilon(X^\eta(\cdot))$ has the same law as $X^\varepsilon(\cdot)$. Plainly, if we are given a family $(m^\varepsilon, \varepsilon > 0)$ of point measures on $]0,1]$ such that for every $\varepsilon > \eta > 0$, $m^\varepsilon$ is the image of $m^\eta$ by the threshold operator $\varphi_\varepsilon$, then there exists a unique sigma-finite measure $m$ on $]0,1]$ such that $m^\varepsilon$ is the image of $m$ by $\varphi_\varepsilon$. Thus the family of processes $(X^\varepsilon(\cdot), \varepsilon > 0)$ can be obtained as the images of the same process $X(\cdot)$ by the threshold operators. □

Let us now turn our attention to self-similarity. Suppose that $\nu$ is some finite measure on $\mathcal{S}^{\downarrow}$ such that

$$\int_{\mathcal{S}^{\downarrow}} \#\{i \in \mathbb{N} : s_i > \varepsilon\} \, \nu(d\mathbf{s}) < \infty \qquad \text{for all } \varepsilon > 0. \tag{1.13}$$

For every $x \in ]0,1]$, write $\nu_x$ for the image of $x^\alpha \nu$ by the dilation $\mathbf{s} \to x\mathbf{s}$. Then, the conditions (1.11) and (1.12) plainly hold. We are now able to introduce the following definition.

**Definition 1.1** (i) *Let* $(\nu_x, 0 < x \leq 1)$ *be a measurable kernel of finite measures on* $\mathcal{S}^{\downarrow}$ *which fulfils (1.11) and (1.12). The process* $X = (X(t), t \geq 0)$ *with values in* $\mathcal{S}^{\downarrow}$ *that has been constructed in Lemma 1.3 is called a* **fragmentation chain** *with dislocation rates* $(\nu_x, 0 < x \leq 1)$.
(ii) *Let* $\nu$ *be some finite measure on* $\mathcal{S}^{\downarrow}$ *such that (1.13) holds, and* $\alpha \in \mathbb{R}$. *For every* $x \in ]0,1]$, *let* $\nu_x$ *denote the image of* $x^\alpha \nu$ *by the dilation* $\mathbf{s} \to x\mathbf{s}$. *The fragmentation chain with dislocation rates* $(\nu_x, 0 < x \leq 1)$ *is called* **self-similar** *with* **index of self-similarity** $\alpha$ *and* **dislocation measure** $\nu$.

Throughout the rest of this chapter, $X$ will denote a self-similar fragmentation chain as defined above. Its law is entirely determined by the index of self-similarity $\alpha$, the dislocation measure $\nu$, and of course the initial configuration $X(0) \in \mathcal{S}^{\downarrow}$. The evolution of the process can be described as follows: a fragment with size $x$ lives for an exponential time with parameter $x^{\alpha}\nu(\mathcal{S}^{\downarrow})$, and then splits and gives rise to a family of smaller fragments distributed as $x\xi$, where $\xi$ has the law $\nu(\cdot)/\nu(\mathcal{S}^{\downarrow})$.

It should be intuitively obvious that the behavior of a self-similar fragmentation depends crucially on its index of self-similarity. Informally, fragments get smaller as time passes; the rate of dislocations thus decreases when the index is positive, whereas it increases when $\alpha < 0$. In particular, we stress that for $\alpha < 0$, this description a priori makes sense only when the size $x$ is non-zero; however, by self-similarity, the children of a particle with size 0 all have size zero. Particles with size zero play no role, and the evolution is thus well-defined in all cases.

As for every $\varepsilon > 0$, the infinitesimal generator of the Markov chain $\varphi_{\varepsilon}(X)$, which is obtained from $X$ by discarding the fragments with size less than or equal to $\varepsilon$, is given in terms of its jump rates, we immediately derive from (1.4) explicit expressions for the infinitesimal generator $\mathsf{G}$ of a self-similar fragmentation chain with index $\alpha$ and dislocation measure $\nu$. Typically, consider a bounded measurable functional $\mathbf{f} : \mathcal{S}^{\downarrow} \to \mathbb{R}$ which only depends on fragments with size at least $\varepsilon$, in the sense that $\mathbf{f} = \mathbf{f} \circ \varphi_{\varepsilon}$. For every configurations $\mathbf{x}, \mathbf{s} \in \mathcal{S}^{\downarrow}$ and every integer $i$, introduce the notation $\mathrm{Frag}_i(\mathbf{x}, \mathbf{s})$ to designate the sequence obtained from $\mathbf{x}$ by removing its $i$-th term $x_i$, replacing it by the terms of the configuration $x_i\mathbf{s}$, and reordering all the terms in decreasing order. Then it is easily checked that in this situation, the infinitesimal generator $\mathsf{G}$ is given by

$$\mathsf{G}\mathbf{f}(\mathbf{x}) = \sum_{i=1}^{\infty} \mathbb{1}_{\{x_i>0\}} x_i^{\alpha} \int_{\mathcal{S}^{\downarrow}} (\mathbf{f}(\mathrm{Frag}_i(\mathbf{x}, \mathbf{s})) - \mathbf{f}(\mathbf{x}))\nu(d\mathbf{s}) . \qquad (1.14)$$

This expression has an interesting application to the so-called *fragmentation equation* which appears in various models in physics (see for example [52] and references therein).

**Corollary 1.1** *Assume that the dislocation measure $\nu$ is conservative or dissipative, that is $\nu(\mathbf{s} \in \mathcal{S}^{\downarrow} : \sum_{i=1}^{\infty} s_i > 1) = 0$. For every $t \geq 0$ and $0 < x \leq 1$, define a Radon measure $\mu_t^x$ on $]0, 1]$ by*

$$\langle \mu_t^x, f \rangle = \mathbb{E}_x \left( \sum_{i=1}^{\infty} f(X_i(t)) \mathbb{1}_{\{X_i(t)>0\}} \right) ,$$

*where $f : ]0,1] \to \mathbb{R}_+$ denotes a generic measurable function with compact support and $\langle \mu_t^x, f\rangle$ the integral of $f$ with respect to $\mu_t^x$. Then the family $(\mu_t^x)_{t\geq 0}$ solves the system of partial differential equations*

$$\frac{\partial}{\partial t}\langle \mu_t^x, f\rangle = \int_{]0,1]} \mu_t^x(dy) y^\alpha \int_{\mathcal{S}^\downarrow} \nu(d\mathbf{s}) \left( -f(y) + \sum_{i=1}^\infty f(ys_i) \right),$$

*with initial condition $\mu_0^x = \delta_x$.*

More generally, by linearity of the fragmentation, we obtain a solution $(\mu_t^m)_{t\geq 0}$ with initial condition $m$, where $m$ denotes an arbitrary Radon measure on $]0,1]$, in the form

$$\mu_t^m = \int_{]0,1]} \mu_t^x m(dx).$$

We refer to Haas [119] for a much deeper study of the applications of fragmentation processes to the fragmentation equation.

*Proof* Let $\varepsilon > 0$ such that $f = 0$ on $]0,\varepsilon]$. The dislocation measure $\nu$ being conservative or dissipative, the process of the total mass, $t \to \sum_{i=1}^\infty X_i(t)$ is non-increasing a.s.; in particular the number of fragments of size at least $\varepsilon$ in $X(t)$ is bounded from above by $x/\varepsilon$, $\mathbb{P}_x$-a.s. Define an additive functional $\mathbf{f} : \mathcal{S}^\downarrow \to \mathbb{R}$ by

$$\mathbf{f}(\mathbf{s}) = \sum_{i=1}^\infty f(s_i) \mathbb{1}_{\{s_i>0\}}, \qquad \mathbf{s} = (s_1, s_2, \ldots).$$

Plainly the functional $\mathbf{f}$ only depends on fragments with size at least $\varepsilon$, but is not bounded on $\mathcal{S}^\downarrow$. However, the observations above show that $\mathbf{f}(X(t)) \leq \varepsilon^{-1} x \max f$ for all $t \geq 0$, $\mathbb{P}_x$-a.s., which enables us to work with the bounded functional $\tilde{\mathbf{f}} := \mathbf{f} \wedge (\varepsilon^{-1} x \max f)$. Specializing (1.14) yields that for every configuration $\mathbf{x} = (x_1, \ldots) \in \mathcal{S}^\downarrow$ which has at most $\lfloor x/\varepsilon \rfloor$ fragments of size greater than $\varepsilon$, we have

$$\mathsf{G}\tilde{\mathbf{f}}(\mathbf{x}) = \sum_{i=1}^\infty \mathbb{1}_{\{x_i>0\}} x_i^\alpha \int_{\mathcal{S}^\downarrow} (\mathbf{f}(x_i \mathbf{s}) - f(x_i)) \nu(d\mathbf{s}). \tag{1.15}$$

It is readily checked that our assumptions ensure that $\mathsf{G}\tilde{\mathbf{f}}$ is a bounded functional. Combining (1.15) with the forward equation (1.6) immediately yields the statement. □

Similarly, consider a measurable function $g : [0, 1] \to ]0, 1]$ with $g = 1$ on some neighborhood of 0, and define a multiplicative functional $\mathbf{g} : \mathcal{S}^{\downarrow} \to ]0, \infty[$ by

$$\mathbf{g}(\mathbf{s}) = \prod_{i=1}^{\infty} g(s_i), \qquad \mathbf{s} = (s_1, s_2, \ldots).$$

Plainly, $0 \leq \mathbf{g}(\mathbf{s}) \leq 1$ for every $\mathbf{s} \in \mathcal{S}^{\downarrow}$, for every $\mathbf{x} = (x_1, \ldots) \in \mathcal{S}^{\downarrow}$, (1.14) yields

$$\mathsf{G}\mathbf{g}(\mathbf{x}) = \sum_{i=1}^{\infty} \mathbb{1}_{\{x_i>0\}} x_i^{\alpha} \frac{\mathbf{g}(\mathbf{x})}{g(x_i)} \int_{\mathcal{S}^{\downarrow}} (\mathbf{g}(x_i \mathbf{s}) - g(x_i)) \nu(d\mathbf{s}).$$

Later in the text, it will be convenient to denote for every $x \in [0, 1]$ by $\mathbb{P}_x$ the law of $X$ started from the configuration $(x, 0, \ldots)$. In other words, $\mathbb{P}_x$ is the distribution of the fragmentation chain when at the initial time, there is just one particle with size $x$. We shall further simply write $\mathbb{P} = \mathbb{P}_1$ when the size is $x = 1$. Quite often we shall work under the law $\mathbb{P}$; essentially this does not induce any loss of generality, thanks to the following basic properties.

**Proposition 1.2** (i) *Any fragmentation chain $X$ has the* **branching property**, *namely for every sequence $\mathbf{s} = (s_1, \ldots) \in \mathcal{S}^{\downarrow}$ and every $t > 0$, the distribution of $X(t)$ given that $X(0) = \mathbf{s}$ is the same as that of the decreasing rearrangement of the terms of independent random sequences $X^{(1)}(t), X^{(2)}(t), \ldots$, where for each $i \in \mathbb{N}$, $X^{(i)}(t)$ is distributed as $X(t)$ under $\mathbb{P}_{s_i}$.*
(ii) *Any self-similar fragmentation chain $X$ has the* **scaling property**, *namely for every $x \in [0, 1]$, the distribution of the rescaled process $(xX(x^{\alpha}t), t \geq 0)$ under $\mathbb{P}_1$ is $\mathbb{P}_x$.*

We now conclude this section by discussing a few special cases and examples.

First, in the special case $\alpha = 0$, where the total dislocation rate of a fragment does not depend on its size, we say that the fragmentation chain is *homogeneous*. In this case, there is a natural connection with branching random walks (see Section 1.1.2). Specifically, let $\mu$ be some finite measure on the space of finite point measures on $]0, \infty[$, and let $Z$ denote a branching random walk in continuous time with branching measure $\mu$. Consider the process $X$ valued in $\mathcal{S}^{\downarrow}$ obtained by shifting the atoms of $Z$ by the exponential map $z \to \mathrm{e}^{-z}$, that is $X(t) = (\mathrm{e}^{-z_1}, \ldots, \mathrm{e}^{-z_k}, 0, \ldots)$, where $z_1, \ldots, z_k$ denote the atoms of the branching random walk at time $t$, ranked in increasing order. It should be plain that $X$ is a homogeneous fragmentation chain with dislocation measure $\nu$, where $\nu$ is the image of the branching measure $\mu$ by the exponential map $z \to e^{-z}$. Conversely, consider a homogeneous fragmentation chain $X$ with dislocation measure $\nu$ charging only finite sequences. Then the process

$$Z^{(t)}(dx) := \sum \delta_{-\ln X_i(t)}(dx),$$

where the sum is taken over the fragments with strictly positive size, is a branching random walk with branching measure the image of $\nu$ by the map $x \to -\ln x$. This elementary connection has a number of interesting consequences as it essentially reduces the study of the class of homogeneous fragmentations associated to a dislocation measure charging only finite configurations, to that of branching random walks on $]0, \infty[$, for which many results are known; see in particular the forthcoming Section 5 in this chapter.

Let us now present two simple examples of self-similar fragmentation chains with index $\alpha \neq 0$.

**Example** (Poissonian rain) Consider a Poisson point process with values in the unit interval and characteristic measure given by the Lebesgue measure on $]0, 1[$. In other words, we have a sequence $U_1, \ldots$ of i.i.d. uniform variables on $[0, 1]$; the times when they appear are the jump times of some independent Poisson process $(N_t, t \geq 0)$. Now, think of these Poissonian points as drops of rain, and consider for every time $t \geq 0$ the subset of $]0, 1[$ which is dry, that is to say $\vartheta(t) := ]0, 1[ \setminus \{U_i : i \leq N_t\}$. So $\vartheta(t)$ is a random open set which consists of $N_t + 1$ disjoint interval components. If we write $X_1(t) \geq \cdots \geq X_{N_t+1}(t)$ for the sequence of their lengths ranked in decreasing order and $X_i(t) = 0$ for $i > N_t + 1$, then it is easy to check that the process $X = (X(t), t \geq 0)$ is a (conservative) self-similar fragmentation chain with index $\alpha = 1$, with dislocation measure $\nu$ given by the distribution of $(1 - V, V, 0, \ldots)$ with $V$ uniformly distributed on $[0, 1/2]$.

The construction of the Poissonian rain can be extended to higher dimensions. For instance, in dimension 2, we start from a triangle $\mathbb{T}$ with unit area and consider a Poisson point process in $\mathbb{T}$ with Lebesgue intensity. That is we have a sequence $U_1, \ldots$ of i.i.d. uniform variables on $\mathbb{T}$, which we call atoms; the times when they appear are the jump times of some independent Poisson process with unit rate. Each time an atom of the point process arises, we split $\mathbb{T}$ into a sequence of smaller triangles and obtain a triangulation as follows. Specifically, at the instant when some atom $U$ occurs, $U$ belongs to some triangle, say $(A, B, C)$ of the current triangulation, and we split $(A, B, C)$ into $(U, B, C)$, $(A, U, C)$ and $(A, B, U)$. So at time $t$, we obtain a triangulation $\tau(t) = (\mathbb{T}_{1,n}, \ldots, \mathbb{T}_{2n+1,n})$ of $\mathbb{T}$, where $n = N(t)$ is the number of atoms that have occurred before time $t$. Denote by $X(t) = (X_1(t), X_2(t), \ldots)$ the mass-partition given by the ordered sequence of the areas of the triangles in the triangulation $\tau(t)$ and completed by an infinite sequence of 0s. Then it is easy to check that $X = (X(t), t \geq 0)$ is a self-similar fragmentation with index $\alpha = 1$, and that its dislocation measure is the law of the decreasing rearrangement of a variable which is uniformly distributed on the simplex $\{\mathbf{x} = (x_1, x_2, x_3) : x_i \geq 0 \text{ and } x_1 + x_2 + x_3 = 1\}$.

**Example** Let us briefly present a generalization of the Poissonian rain model above which has been considered by Baryshnikov and Gnedizn concerning a packing problem related to communication networks; see [19] for much more on this example. As previously, let $U_1, \ldots$ be a sequence of i.i.d. uniform variables, and $\ell_1, \ell_2, \ldots$ a sequence of i.i.d. random variables, which is independent of the $U_i$ s. We throw successively the random intervals $[U_i, U_i + \ell_i]$ on $]0, 1[$ at the jump times $T_1 < T_2 < \cdots$ of an independent rate 1 Poisson process $N = (N_t, t \geq 0)$. We construct a process of nested open subsets $\vartheta := (\vartheta(t), t \geq 0)$ as follows. The process starts from $\vartheta(0) := ]0, 1[$ and remains constant except at times $T_1, T_2, \ldots$ where it may jump. Specifically, if $[U_i, U_i + \ell_i]$ is entirely contained in $\vartheta(T_i-)$, then $\vartheta(T_i) = \vartheta(T_i-) \setminus [U_i, U_i + \ell_i]$. Otherwise $\vartheta(T_i) = \vartheta(T_i-)$. Elementary properties of Poisson point processes easily imply that the process of the ranked lengths of the interval components of $\vartheta(\cdot)$ is a dissipative fragmentation chain. For instance, when the $\ell_i$ are uniformly distributed, it is easily checked that the fragmentation is self-similar with index $\alpha = 2$.

## 1.2 Genealogical structure

Throughout the rest of this chapter, we shall consider a self-similar fragmentation chain $X = (X(t), t \geq 0)$ with index $\alpha \in \mathbb{R}$ and dislocation measure $\nu$ which fulfills (1.13). In order to avoid uninteresting discussions of degenerate cases, we shall always implicitly assume that

$$\nu(s_i \in ]0, 1[) > 0 \qquad \text{for some } i \in \mathbb{N}.$$

In this section, we point at a representation of fragmentation chains as random infinite marked trees. This representation can be viewed as a different parametrization of the process, in which the natural time is replaced by the generation of the different particles.

### 1.2.1 The tree of generations

We start by introducing some notation. We consider the infinite tree (see the figure below)

$$\mathcal{U} := \bigcup_{n=0}^{\infty} \mathbb{N}^n ,$$

with the convention $\mathbb{N}^0 = \{\emptyset\}$. We will refer to $\mathcal{U}$ as the *genealogical tree*; its elements are called *nodes* (or sometimes also *individuals*) and the distinguished node $\emptyset$ is the *root*. Nodes will be used to label the particles produced by a fragmentation chain.

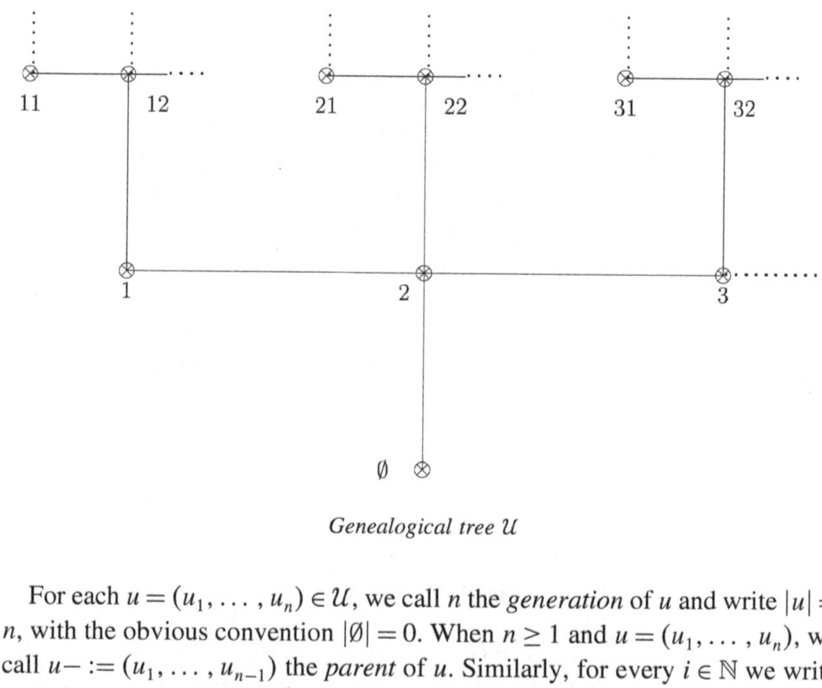


*Genealogical tree* $\mathcal{U}$

For each $u = (u_1, \ldots, u_n) \in \mathcal{U}$, we call $n$ the *generation* of $u$ and write $|u| = n$, with the obvious convention $|\emptyset| = 0$. When $n \geq 1$ and $u = (u_1, \ldots, u_n)$, we call $u- := (u_1, \ldots, u_{n-1})$ the *parent* of $u$. Similarly, for every $i \in \mathbb{N}$ we write $ui = (u_1, \ldots, u_n, i) \in \mathbb{N}^{n+1}$ for the $i$-th *child* of $u$. Finally, any map from $\mathcal{U}$ to some (measurable) set is called a *mark* .

Now, consider a self-similar fragmentation chain $X = (X(t), t \geq 0)$ with index $\alpha \in \mathbb{R}$ and dislocation measure $\nu$. Suppose for simplicity that the process starts from a single fragment with size $x > 0$, that is we work under the law $\mathbb{P}_x$. We associate to each path of the process a mark on the infinite tree $\mathcal{U}$; roughly the mark at a node $u$ is the triple $(\xi_u, a_u, \zeta_u)$ where $\xi_u$ is the size, $a_u$ the birth-time and $\zeta_u$ the lifetime of the particle with label $u$. More precisely, the initial particle $x$ corresponds to the ancestor $\emptyset$ of the tree $\mathcal{U}$, and the mark at $\emptyset$ is the triple $(x, 0, \zeta_\emptyset)$ where $\zeta_\emptyset$ is the lifetime of the initial particle (in particular, $\zeta_\emptyset$ has the exponential law with parameter $x^\alpha \nu(\mathcal{S}^\downarrow)$). The nodes of the tree at the first generation are used as the labels of the particles arising at the first split. Again, the mark associated to each of the nodes $i \in \mathbb{N}^1$ at the first generation, is the triple $(\xi_i, a_i, \zeta_i)$, where $\xi_i$ is the size of the $i$-th child of the ancestor, $a_i = a_\emptyset + \zeta_\emptyset$ (the birth-time of a child particle coincides with the death-time of the parent), and $\zeta_i$ stands for the lifetime of the $i$-th child. And we iterate the same construction with each particle at each generation.

Clearly, the description of the dynamics of fragmentation implies that its genealogical representation also enjoys the branching property. Specifically, the distribution of the random mark can be described recursively as follows.

**Proposition 1.3** *There exist two independent families of i.i.d. variables indexed by the nodes of the genealogical tree,* $\left((\tilde{\xi}_{ui})_{i\in\mathbb{N}}, u \in \mathcal{U}\right)$ *and* $((\mathbf{e}_{ui})_{i\in\mathbb{N}}, u \in \mathcal{U})$*, where each* $(\tilde{\xi}_{ui})_{i\in\mathbb{N}}$ *is distributed according to the law* $\nu(\cdot)/\nu(\mathcal{S}^\downarrow)$*, and each* $(\mathbf{e}_{ui})_{i\in\mathbb{N}}$ *is a sequence of i.i.d. exponential variables with parameter* $\nu(\mathcal{S}^\downarrow)$*, and such that the following holds:*
*Given the marks* $((\xi_v, a_v, \zeta_v), |v| \le n)$ *of the first* $n$ *generations, the marks at generation* $n+1$ *are given by*

$$\xi_{ui} = \tilde{\xi}_{ui}\xi_u \,, \quad a_{ui} = a_u + \zeta_u \,, \quad \zeta_{ui} = \xi_{ui}^{-\alpha}\mathbf{e}_{ui} \,,$$

*where* $u = (u_1, \ldots, u_n)$ *and* $ui = (u_1, \ldots, u_n, i)$ *is the* $i$*-th child of* $u$*.*

The genealogical coding of a self-similar fragmentation chain yields an elementary transformation of the latter, which only affects the lifetime of particles, and enables us to change the index of self-similarity.

**Corollary 1.2** *Let* $(\xi, a, \zeta)$ *be the random mark on the genealogical tree induced by a self-similar fragmentation chain with index* $\alpha$ *and dislocation measure* $\nu$*. Fix* $\beta \in \mathbb{R}$ *and consider the random mark* $(\xi^{(\beta)}, a^{(\beta)}, \zeta^{(\beta)})$ *defined by*

$$\xi_u^{(\beta)} = \xi_u \,, \quad \zeta_u^{(\beta)} = \xi_u^{-\beta}\zeta_u \,, \quad a_u^{(\beta)} = \sum_{i=0}^{n-1} \zeta^{(\beta)}_{(u_1,\ldots,u_i)} \,,$$

*where* $u = (u_1, \ldots, u_n)$ *denotes a generic node. Then* $(\xi^{(\beta)}, a^{(\beta)}, \zeta^{(\beta)})$ *is distributed as the genealogical coding of a self-similar fragmentation chain with index* $\alpha + \beta$ *and dislocation measure* $\nu$*.*

*Proof* This is obvious from Proposition 1.3 as conditionally on the mass $\xi_u = \xi_u^{(\beta)}$ of the particle labeled by the node $u \in \mathcal{U}$, the lifetime $\zeta_u^{(\beta)}$ has an exponential distribution with parameter $\xi_u^{\alpha+\beta}\nu(\mathcal{S}^{\downarrow})$. □

Now to every node $u$ of the genealogical tree, we can associate the interval $I_u := [a_u, a_u + \zeta_u[$ during which the particle labeled by $u$ is alive. Putting the pieces together, we may express the fragmentation chain at time $t$ as the ranked sequence of the particles which are alive at time $t$:

**Proposition 1.4** *With probability one, for every $t \geq 0$ there is the identity between point measures on $]0, \infty[$:*

$$\sum_{i=1}^{\infty} \mathbb{1}_{\{X_i(t)>0\}} \delta_{X_i(t)} = \sum_{u \in \mathcal{U}} \mathbb{1}_{\{t \in I_u\}} \delta_{\xi_u} .$$

*Proof* We have to check that all the fragments with positive size which are present at time $t$ have a finite generation, that is result from finitely many dislocations of the initial particle. In this direction, let us fix some arbitrarily small $\varepsilon > 0$, and consider the threshold operator $\varphi_\varepsilon$ which consists of removing all the fragments with size less than or equal to $\varepsilon$. Recall from Section 1.1.3 that $\varphi_\varepsilon(X)$ is a Markov chain, in particular the number of jumps accomplished by this chain before time $t$ is finite a.s., and this number obviously is an upper bound for the generation of all fragments with size greater than $\varepsilon$. □

### 1.2.2 Malthusian hypotheses and the intrinsic martingale

The purpose of this section is to introduce the so-called *intrinsic martingale* which is naturally induced by the tree representation of fragmentations. This martingale will play a crucial role when we investigate the asymptotic behavior of self-similar fragmentation chains.

We start by introducing the notation

$$\underline{p} := \inf \left\{ p > 0 : \int_{\mathcal{S}^{\downarrow}} \sum_{i=1}^{\infty} s_i^p \nu(ds) < \infty \right\} \tag{1.16}$$

(with the convention $\inf \emptyset = \infty$) and then for every $p > \underline{p}$

$$\kappa(p) := \int_{\mathcal{S}^{\downarrow}} \left( 1 - \sum_{i=1}^{\infty} s_i^p \right) \nu(ds) . \tag{1.17}$$

Note that $\kappa$ is always a continuous strictly increasing function on $]\underline{p}, \infty[$; $\kappa(\underline{p}+)$ may be finite or equal to $-\infty$. We stress that $\kappa$ and $\underline{p}$ depend on the dislocation measure $\nu$ but not on the parameter of self-similarity $\alpha$. Let us discuss a couple of examples:

First, recall that for the so-called Poissonian rain introduced in Section 1.1.3, the dislocation measure is the law of $(1-V, V, 0, \ldots)$ where $V$ is a uniform random variable on $[0, 1/2]$. One readily finds that $\underline{p} = 0$ and $\kappa(p) = (p-1)/(p+1)$.

Second, consider the so-called uniform stick-breaking scheme (see the forthcoming Corollary 2.3 for much more on this). That is, cut the unit interval at a uniformly distributed random variable, keep the left portion, cut the right one at an independent uniformly distributed random variable, keep the left portion, and so on. The sequence of the lengths of the resulting intervals (ordered from the left to the right) is thus $U_1, (1-U_1)U_2, (1-U_1)(1-U_2)U_3, \ldots$, where $U_1, U_2, \ldots$ are i.i.d. uniform variables; in particular the $p$-th moment of the $k$-th length is thus $(1+p)^{-k}$. When the distribution of the sequence of the lengths ranked in decreasing order is used as the dislocation measure $\nu$ of some fragmentation chain, we get that $\underline{p} = 0$ and $\kappa(p) = 1 - 1/p$ for $p > 0$.

We now make the fundamental:

**Malthusian Hypotheses.** *There exists a (unique) solution $p^* \geq \underline{p}$ to the equation*

$$\kappa(p^*) = 0\,,$$

*which is called the* **Malthusian exponent**. *Furthermore the integral*

$$\int_{\mathcal{S}^\downarrow} \left( \sum_{i=1}^{\infty} s_i^{p^*} \right)^p \nu(d\mathbf{s})\,,$$

*is finite for some $p > 1$.*

Throughout the rest of this chapter, the Malthusian hypotheses will always be taken for granted.

In order to claim one of the basic results in this setting, it will be convenient to call by the name *extinction* the event that for some large enough $n \in \mathbb{N}$, all nodes $u$ at the $n$-th generation have zero size, and by the name *non-extinction* the complementary event. Clearly, extinction may occur if and only if $\nu(s_1 = 0) > 0$ (i.e. a particle may disappear entirely after a dislocation). Moreover it follows from Proposition 1.3 that the number of fragments with positive size at the $n$-th generation

$$\#(n) := \#\left\{u \in \mathcal{U} : |u| = n \text{ and } \xi_u > 0\right\}\,, \qquad n \in \mathbb{Z}_+$$

is a Galton-Watson process with reproduction law given by the distribution of $\max\{i : s_i > 0\}$ under $\nu(\cdot)/\nu(\mathcal{S}^{\downarrow})$. Combining the inequality

$$\sum_{i=1}^{\infty} s_i^{p^*} \leq \#\{i \in \mathbb{N} : s_i > 0\}, \qquad \text{whenever } 0 \leq \cdots \leq s_2 \leq s_1 \leq 1$$

and the Malthusian hypotheses, we see that this Galton-Watson process is super-critical. By a classical result, the probability of extinction is the unique solution $p \in [0, 1[$ to the equation $g(p) = p$, where $g$ is the moment generation function of the reproduction law. To summarize, the probability of non-extinction is always strictly positive, and equals one if and only if $\nu(s_1 = 0) = 0$.

For the sake of simplicity, we henceforth work under $\mathbb{P} = \mathbb{P}_1$, that is we assume that at the initial time there is a single fragment with size 1. Recall that this does not induce any significant loss of generality. We may now state.

**Theorem 1.1** *The process*

$$\mathcal{M}_n := \sum_{|u|=n} \xi_u^{p^*}, \qquad n \in \mathbb{Z}_+$$

*is a martingale which is bounded in $L^p$ for some $p > 1$, and in particular is uniformly integrable. Moreover, the terminal value $\mathcal{M}_\infty$ is strictly positive a.s. conditionally on non-extinction.*

Later in the text, we will refer to $(\mathcal{M}_n, n \in \mathbb{Z}_+)$ as the *intrinsic martingale*. Observe that in the important case when dislocations are conservative, in the sense that $\sum_{i=1}^{\infty} s_i = 1$, $\nu(d\mathbf{s})$-a.e., then the Malthusian hypotheses are automatically fulfilled with $p^* = 1$; further $\mathcal{M}_n = 1$ for all $n \in \mathbb{Z}_+$, and the statement in Theorem 1.1 is trivial.

*Proof* Denote by $\mathcal{G}_n$ the sigma field generated by $(\xi_u, |u| \leq n)$, so $(\mathcal{G}_n)$ is a filtration. It should be plain from the description of the dynamics of the random marks in Proposition 1.3 that for every $q > \underline{p}$,

$$\mathbb{E}\left(\sum_{|u|=n+1} \xi_u^q \mid \mathcal{G}_n\right) = \mathbb{E}\left(\sum_{|v|=n} \xi_v^q \sum_{i=1}^{\infty} \tilde{\xi}_{vi}^q \mid \mathcal{G}_n\right) = c(q) \sum_{|v|=n} \xi_v^q$$

where

$$c(q) = \int_{\mathcal{S}^{\downarrow}} \sum_{i=1}^{\infty} s_i^q \nu(d\mathbf{s})/\nu(\mathcal{S}^{\downarrow}) = 1 - \kappa(q)/\nu(\mathcal{S}^{\downarrow}).$$

In particular for the Malthusian exponent $q = p^*$, one has $c(p^*) = 1$ and the martingale property is proven.

In order to establish the boundedness of the martingale in $L^p(\mathbb{P})$, we shall use the following well-known consequence of the Burkholder-Davis-Gundy inequalities (see for instance [152]). For every $p \in ]1, 2]$, there exists a universal constant $c_p > 0$ such that for every martingale $(M_n, n \in \mathbb{Z}_+)$ with $M_0 = 0$, there is the inequality

$$\mathbb{E}\left(\sup_n |M_n|^p\right) \leq c_p \mathbb{E}\left(\sum_{n=0}^{\infty} |M_{n+1} - M_n|^p\right).$$

In particular, if $(\beta_i, i \in \mathbb{N})$ is a sequence of independent centered variables, then

$$\mathbb{E}\left(\left|\sum_{i=1}^{\infty} \beta_i\right|^p\right) \leq c_p \sum_{i=1}^{\infty} \mathbb{E}\left(|\beta_i|^p\right), \tag{1.18}$$

in the sense that whenever the right-hand side is finite, the series in the left-hand side converges a.s. and inequality (1.18) holds.

So all that we need is to check that the sum of the jumps of the intrinsic martingale raised to some power $p \in ]1, 2]$ has a finite mean, that is

$$\mathbb{E}\left(\sum_{n=0}^{\infty} |\mathcal{M}_{n+1} - \mathcal{M}_n|^p\right) < \infty. \tag{1.19}$$

It this direction, we use Proposition 1.3 and express the $n$-th jump in the form

$$\mathcal{M}_{n+1} - \mathcal{M}_n = \sum_{|u|=n} \xi_u^{p^*} \left(\left(\sum_{j=1}^{\infty} \tilde{\xi}_{u,j}^{p^*}\right) - 1\right)$$

where $(\tilde{\xi}_{u,\cdot}, |u| = n)$ is a family of i.i.d. variables with the law $\nu(\cdot)/\nu(\mathcal{S}^{\downarrow})$, which is independent of $\mathcal{G}_n$. We raise this quantity to the power $p > 1$ and then take the conditional expectation given $\mathcal{G}_n$. By definition of the Malthusian exponent $p^*$, the variables $\left(\sum_{j=1}^{\infty} \tilde{\xi}_{u,j}^{p^*}\right) - 1$ are centered, so inequality (1.18) yields

$$\mathbb{E}\left(|\mathcal{M}_{n+1} - \mathcal{M}_n|^p \mid \mathcal{G}_n\right) \leq c_p b \sum_{|u|=n} \xi_u^{p^* p}$$

where

$$b := \int_{\mathcal{S}^{\downarrow}} \left|1 - \sum_{i=1}^{\infty} s_i^{p^*}\right|^p \nu(ds)/\nu(\mathcal{S}^{\downarrow})$$

is some finite constant, thanks to the second condition of the Malthusian hypotheses.

Now we know from the first part of the proof that

$$\mathbb{E}\left(\sum_{|u|=n} \xi_u^{p^* p}\right) = c(p^* p)^n = \left(1 - \kappa(pp^*)/\nu(\mathcal{S}^\downarrow)\right)^n .$$

Since $p^* p > p^*$ (because $p^* > \underline{p} \geq 0$) and $\kappa$ is strictly increasing, $\kappa(p^* p) > 0$ and thus $c(p^* p) < 1$. This yields (1.19) and completes the proof of the first part of the statement.

Finally, let us now check that $\mathcal{M}_\infty > 0$ a.s. conditionally on non-extinction. Write $\varrho = \mathbb{P}(\mathcal{M}_\infty = 0)$; the fact that $\mathbb{E}(\mathcal{M}_\infty) = 1$ ensures that $\varrho < 1$. On the other hand, an application of the branching property yields

$$\mathbb{E}\left(\varrho^{\#(n)}\right) = \varrho ,$$

where $\#(n)$ is the number of fragments with positive size at the $n$-th generation. Clearly $(\#(n), n \in \mathbb{Z}_+)$ is a Galton-Watson process and it follows that $\varrho$ is its probability of extinction. Since $\mathcal{M}_\infty = 0$ on extinction, the two events coincide a.s. □

The terminal value $\mathcal{M}_\infty$ of the intrinsic martingale will appear in many limit theorems for the fragmentation. In general its distribution is not known explicitly. However, it is straightforward from the branching property that there is the identity in law

$$\mathcal{M}_\infty \overset{(d)}{=} \sum_{j=1}^{\infty} \xi_j^{p^*} \mathcal{M}_\infty^{(j)} \tag{1.20}$$

where $\xi = (\xi_j, j \in \mathbb{N})$ has the law $\nu(\cdot)/\nu(\mathcal{S}^\downarrow)$, and $\mathcal{M}_\infty^{(j)}$ are independent copies of $\mathcal{M}_\infty$, also independent of $\xi$. It is known that under fairly general conditions, such an equation characterizes the law $\mathcal{M}_\infty$ uniquely, see for example [194, 154]. We also refer to Liu [153] for information of the tail behavior of the solution.

The intrinsic martingale $\mathcal{M}_n$ is indexed by the generations; it will also be convenient to consider its analog in continuous time, that is

$$\mathcal{M}(t) := \sum_{i=1}^{\infty} X_i^{p^*}(t) = \sum_{u \in \mathcal{U}} \mathbb{1}_{\{t \in I_u\}} \xi_u^{p^*} , \qquad t \geq 0 ,$$

where in the right-hand side, $I_u$ denotes the life-interval of the particle indexed by the node $u$. It is straightforward to check that when the index of self-similarity is positive, $(\mathcal{M}(t), t \geq 0)$ is again a martingale in the natural filtration $(\mathcal{F}_t)_{t \geq 0}$ of the fragmentation $(X(t), t \geq 0)$; and more precisely, we have the following.

**Proposition 1.5** *Assume that the index of self-similarity $\alpha$ is non-negative. Then*

$$\mathcal{M}(t) = \mathbb{E}(\mathcal{M}_\infty \mid \mathcal{F}_t) ,$$

*where $\mathcal{M}_\infty$ is the terminal value of the intrinsic martingale $(\mathcal{M}_n, n \in \mathbb{N})$, and $(\mathcal{F}_t)_{t\geq 0}$ the natural filtration of $(X(t), t \geq 0)$. In particular $\mathcal{M}(t)$ converges in $L^p(\mathbb{P})$ to $\mathcal{M}_\infty$ for some $p > 1$.*

*Proof* We know that $\mathcal{M}_n$ converges in $L^p(\mathbb{P})$ to $\mathcal{M}_\infty$ as $n$ tends to $\infty$, so

$$\mathbb{E}(\mathcal{M}_\infty \mid \mathcal{F}_t) = \lim_{n\to\infty} \mathbb{E}(\mathcal{M}_n \mid \mathcal{F}_t) .$$

On the other hand, it is easy to deduce from the Markov property applied at time $t$ that

$$\mathbb{E}(\mathcal{M}_n \mid \mathcal{F}_t) = \sum_{i=1}^{\infty} X_i^{p^*}(t) \mathbb{1}_{\{G(X_i(t))\leq n\}} + \sum_{|u|=n} \xi_u^{p^*} \mathbb{1}_{\{a_u+\zeta_u<t\}} ,$$

where $G(x)$ stands for the generation of the particle $x$ (i.e. $G(\xi_u) = |u|$), and $a_u + \zeta_u$ is the instant when the particle corresponding to the node $u$ splits. We can express the latter quantity in the form

$$a_u + \zeta_u = x_0\mathbf{e}_0 + x_1^{-\alpha}\mathbf{e}_1 + \cdots + x_{|u|}^{-\alpha}\mathbf{e}_{|u|}$$

where $\mathbf{e}_0, \ldots$ is a sequence of independent exponential variables with parameter $\nu(\mathcal{S}^{\downarrow})$, which is also independent of $\xi_u$, and $x_i$ stands for the size of particle labeled by the ancestor of $u$ at the $i$-th generation. When the index of self-similarity is non-negative, $x_i^{-\alpha} \geq 1$ and hence for each fixed node $u \in \mathcal{U}$, $a_u + \zeta_u$ is bounded from below by the sum of $|u|+1$ independent exponential variables with parameter $\nu(\mathcal{S}^{\downarrow})$ which are independent of $\xi_u$. It follows that

$$\lim_{n\to\infty} \mathbb{E}\left( \sum_{|u|=n} \xi_u^{p^*} \mathbb{1}_{\{a_u+\zeta_u<t\}} \right) = 0 ,$$

and we conclude that $\mathbb{E}(\mathcal{M}_\infty \mid \mathcal{F}_t) = \mathcal{M}(t)$. □

We stress that the statement *fails* when $\alpha < 0$; more precisely we shall see in Section 1.3 that then $\mathcal{M}(t) = 0$ whenever $t$ is sufficiently large.

### 1.2.3 A randomly tagged branch

Throughout this section, we shall implicitly suppose for simplicity that the fragmentation starts from a single fragment with unit size, that is we shall work under $\mathbb{P} = \mathbb{P}_1$. Our purpose is to introduce a fundamental tool which enables explicit calculations for the first moment of certain functionals of the fragmentation. Informally, the idea is to follow the evolution of a fragment picked at random; one of the issues being of course to specify the meaning of 'picking a fragment at random'.

In this direction, let us say that a *branch* of the genealogical tree $\mathcal{U}$ is an infinite sequence of positive integers $b = (i_1, \ldots)$, which we can think of as the line of ancestors of some leaf of the tree, in the sense that for each $n$, we associate to $b$ its ancestor $b_n := (i_1, \ldots, i_n)$ at the generation $n$. Following an original idea developed by Lyons, Pemantle and Peres [157], we shall enrich the probabilistic structure by distinguishing at random a branch, called the *tagged branch*.

Specifically, we consider a pair $(M, \beta)$ where $M : \mathcal{U} \to [0, 1] \times \mathbb{R}_+ \times \mathbb{R}_+$ is a random mark on the genealogical tree and $\beta$ is a random branch of $\mathcal{U}$, whose joint distribution denoted by $\mathbb{P}^*$ is specified as follows. Let $\mathcal{H}_n$ stand for the space of bounded functionals $\Phi$ which depend on the mark $M$ and the branch $\beta$ only up to the $n$-th generation, that is such that $\Phi(M, \beta) = \Phi(M', \beta')$ if $\beta_n = \beta'_n$ and $M(u) = M'(u)$ whenever $|u| \le n$. For such functionals, it will be convenient to use the slightly abusing notation $\Phi(M, \beta) = \Phi(M, \beta_n)$. It is immediately seen from the dynamics of the random mark $M$ described in Proposition 1.3 and the definition of the Malthusian exponent that

$$\mathbb{E}\left(\sum_{|u|=n} \Phi(M, u)\xi_u^{p*}\right) = \mathbb{E}\left(\sum_{|u|=n} \Phi(M, u)\xi_u^{p*} \sum_{i=1}^{\infty} \tilde{\xi}_{ui}^{p^*}\right)$$
$$= \mathbb{E}\left(\sum_{|v|=n+1} \Phi(M, v)\xi_v^{p*}\right),$$

for every $\Phi \in \mathcal{H}_n$. By Kolmogorov's consistency theorem, this allows us to define unambiguously a probability measure $\mathbb{P}^*$ viewed as the joint distribution of a random mark $M$ and a random branch $\beta$ by

$$\mathbb{E}^*\left(\Phi(M, \beta)\right) = \mathbb{E}\left(\sum_{|u|=n} \Phi(M, u)\xi_u^{p*}\right), \qquad \Phi \in \mathcal{H}_n. \tag{1.21}$$

Note that since the intrinsic martingale $(\mathcal{M}_n, n \in \mathbb{Z}_+)$ is uniformly integrable (cf. Theorem 1.1), the first marginal of $\mathbb{P}^*$ is absolutely continuous with respect to the law of the random mark $M$ under $\mathbb{P}$, with density $\mathcal{M}_\infty$.

The sizes of particles on the tagged branch will play an important role in the analysis of fragmentation chains. More precisely, recall that $a_{\beta_n}$ and $\zeta_{\beta_n}$ denote respectively the birth-time and lifetime of the particle labeled by tagged node $\beta_n$ (i.e. $\beta_n$ is the node of the tagged branch at the $n$-th generation). We write $\chi_n = \xi_{\beta_n}$ for the size of the particle corresponding to the node $\beta_n$, and $\chi(t)$ for the size of the tagged particle alive at time $t$, that is to say

$$\chi(t) = \chi_n \quad \text{if } a_{\beta_n} \leq t < a_{\beta_n} + \zeta_{\beta_n},$$

and

$$\chi(t) = 0 \quad \text{if } t \geq a_{\beta_\infty} := \lim_{n\to\infty} a_{\beta_n}.$$

In this direction, we make the easy observation that $a_{\beta_\infty} = \infty$ $\mathbb{P}^*$-a.s. when the index of self-similarity $\alpha$ is non-negative (whereas $a_{\beta_\infty} < \infty$ $\mathbb{P}^*$-a.s. when $\alpha < 0$ as we shall see later on).

The following lemma shows that the first moment of additive functionals of the fragmentation are easily expressed in terms of the tagged particle.

**Lemma 1.4** *Let $f : \mathbb{R}_+ \to \mathbb{R}_+$ be a measurable function. Then we have for every $n \in \mathbb{N}$*

$$\mathbb{E}^*(f(\chi_n)) = \mathbb{E}\left(\sum_{|u|=n} \xi_u^{p^*} f(\xi_u)\right).$$

*Moreover, if $f(0) = 0$, then for every $t \geq 0$*

$$\mathbb{E}^*(f(\chi(t))) = \mathbb{E}\left(\sum_{i=1}^{\infty} X_i^{p^*}(t) f(X_i(t))\right).$$

*Proof* The identity (1.21) yields the first formula. The second also derives readily from (1.21) by conditioning on the generation $\gamma(t)$ of the tagged particle at time $t$. Indeed, we have

$$\begin{aligned}
\mathbb{E}^*(f(\chi(t))) &= \mathbb{E}^*\left(\sum_{n=0}^{\infty} f(\chi_n) \mathbb{1}_{\{\gamma(t)=n\}}\right) \\
&= \sum_{n=0}^{\infty} \mathbb{E}^*\left(f(\chi_n) \mathbb{1}_{\{a_{\beta_n} \leq t < a_{\beta_n} + \zeta_{\beta_n}\}}\right) \\
&= \sum_{n=0}^{\infty} \mathbb{E}\left(\sum_{|u|=n} f(\xi_u) \xi_u^{p^*} \mathbb{1}_{\{a_u \leq t < a_u + \zeta_u\}}\right) \\
&= \mathbb{E}\left(\sum_{i=1}^{\infty} X_i^{p^*}(t) f(X_i(t))\right),
\end{aligned}$$

where the last identity stems from Proposition 1.4. □

Lemma 1.4 will be useful for computing first moments for fragmentations in combination with the following proposition.

**Proposition 1.6** *Under* $\mathbb{P}^*$,

$$S_n := -\ln \chi_n \,, \qquad n \in \mathbb{Z}_+$$

*is a random walk on* $\mathbb{R}_+$ *with step distribution*

$$\mathbb{P}\left(\ln \chi_n - \ln \chi_{n+1} \in dy\right) = \tilde{\nu}(dy)/\nu(\mathcal{S}^{\downarrow}) \,,$$

*where the finite measure* $\tilde{\nu}$ *is defined by*

$$\int_{]0,\infty[} f(y)\tilde{\nu}(dy) = \int_{\mathcal{S}^{\downarrow}} \left( \sum_{i=1}^{\infty} x_i^{p^*} f(-\ln x_i) \right) \nu(d\mathbf{x}) \,.$$

*Equivalently, the Laplace transform of the step distribution is given by*

$$\mathbb{E}^*\left(\mathrm{e}^{-pS_1}\right) = \mathbb{E}^*\left(\chi_1^p\right) = 1 - \kappa(p+p^*)/\nu(\mathcal{S}^{\downarrow}) \,, \qquad p \geq 0 \,.$$

*Moreover, conditionally on* $(\chi_n, n \in \mathbb{Z}_+)$ *the sequence of the lifetimes* $(\zeta_{\beta_0}, \zeta_{\beta_1}, \ldots)$ *along the tagged branch is a sequence of independent exponential variables with respective parameters* $\chi_0^{\alpha}\nu(\mathcal{S}^{\downarrow}), \chi_1^{\alpha}\nu(\mathcal{S}^{\downarrow}), \ldots$

*Proof* Consider a functional $\Phi \in \mathcal{H}_n$. We see from (1.21) that for every $q \geq 0$

$$\begin{aligned} \mathbb{E}^*\left(\exp(-q(S_{n+1} - S_n))\Phi(M, \beta)\right) &= \mathbb{E}^*\left(\chi_{n+1}^q \chi_n^{-q} \Phi(M, \beta)\right) \\ &= \mathbb{E}\left( \sum_{|u|=n} \sum_{i=1}^{\infty} \tilde{\xi}_{ui}^q (\xi_u \tilde{\xi}_{ui})^{p^*} \Phi(M, u) \right) , \end{aligned}$$

where each $(\tilde{\xi}_{ui})_{i \in \mathbb{N}}$ is a random variable in $\mathcal{S}^{\downarrow}$ with law $\nu(\cdot)/\nu(\mathcal{S}^{\downarrow})$, which is independent of $\Phi(M, u)$. This shows that under $\mathbb{P}^*$, $S_n$ is a random walk with step distribution given by

$$\mathbb{E}^*(f(S_1)) = \int_{\mathcal{S}^{\downarrow}} \left( \sum_{i=1}^{\infty} f(-\ln s_i) s_i^{p^*} \right) \frac{\nu(d\mathbf{s})}{\nu(\mathcal{S}^{\downarrow})} ,$$

and thus establishes the first claim. The second is obvious. □

In particular, in the so-called homogeneous case when the index of self-similarity is $\alpha = 0$, Proposition 1.6 shows that the size of the tagged particle $(\chi(t), t \geq 0)$ can be expressed in the form $\chi(t) = \exp(-\eta_t)$, where

$$\eta_t = S \circ N_t \,, \qquad t \geq 0 \,,$$

with $N$ a Poisson process with parameter $\nu(\mathcal{S}^{\downarrow})$ which is independent of the random walk $S$. In other words, the tagged particle is the exponential of the opposite of the $\eta = S \circ N$.

In the case $\alpha \neq 0$, the process $(\chi(t), t \geq 0)$ is Markovian and enjoys an obvious scaling property. More precisely, an immediate check can be made that it can be expressed in the form

$$\chi(t) = \exp(-\eta_{\tau(t)}), \qquad t \geq 0,$$

where $\eta$ is the compound Poisson defined above and $\tau$ the time-change given as the inverse of the functional

$$t \to \int_0^t \exp(\alpha \eta_s) ds .$$

Such transformations of compound Poisson processes, and more generally of Lévy processes, were first considered by Lamperti [148], as the fundamental representation of self-similar Markov processes on $]0, \infty[$. We stress that the time-change $\tau(t)$ is finite for all $t \geq 0$ when $\alpha \geq 0$, whereas when $\alpha < 0$,

$$\tau(t) < \infty \Longleftrightarrow t < I := \int_0^\infty \exp(\alpha \eta_s) ds ,$$

and then

$$\chi(t) = 0 \qquad \text{whenever } t \geq I .$$

More precisely, $I < \infty$ a.s. when $\alpha < 0$, and this random variable has the same distribution as $a_{\beta_\infty}$, the limit of the birth-time $a_{\beta_n}$ of the tagged particle at the $n$ th generation as $n \to \infty$.

These observations enable us in particular to compute the moments of power sums of self-similar fragmentations.

**Proposition 1.7** *We have for every $p \geq p^*$ and $t \geq 0$:*

(i) *in the homogeneous case $\alpha = 0$,*

$$\mathbb{E}\left(\sum_{i=1}^{\infty} X_i^p(t)\right) = \mathbb{E}^*(\chi(t)^{p-p^*}) = \exp(-t\kappa(p)) ,$$

(ii) *in the case $\alpha > 0$ when the index of self-similarity is positive,*

$$\mathbb{E}\left(\sum_{i=1}^{\infty} X_i^p(t)\right) = \mathbb{E}^*(\chi(t)^{p-p^*}) = \sum_{n=0}^{\infty} \frac{(-t)^n}{n!} \Gamma(n, p) ,$$

*where* $\Gamma(0,p)=1$ *and for* $n\geq 1$

$$\Gamma(n,p)=\prod_{k=0}^{n-1}\kappa(p+\alpha k)\,.$$

*Proof* (i) Thanks to Proposition 1.6, we have whenever $p+p^*>\underline{p}$ that

$$\begin{aligned}\mathbb{E}^*(\chi(t)^p) &= \mathbb{E}^*(\exp(-pS_{N_t}))\\ &= \exp(-t\nu(\mathcal{S}^{\downarrow}))\sum_{k=0}^{\infty}\frac{(t\nu(\mathcal{S}^{\downarrow}))^k}{k!}\left(\int \mathrm{e}^{-py}\tilde{\nu}(dy)/\nu(\mathcal{S}^{\downarrow})\right)^k\\ &= \exp(-t\kappa(p+p^*))\,.\end{aligned}$$

In other words, the so-called Laplace exponent of the compound Poisson process $S\circ N=\eta$ is $\kappa(p^*+\cdot)$, and this characterizes its law. The stated formula now derives from Lemma 1.4.

(ii) We start by observing from Lemma 1.4 that for every sufficiently large $p$

$$\begin{aligned}\int_0^\infty \mathbb{E}\left(\sum_{i=1}^{\infty}X_i^{p+p^*}(t)\right)dt &= \int_0^\infty \mathbb{E}^*\left(\chi(t)^p\right)dt\\ &= \mathbb{E}^*\left(\int_0^\infty \exp(-p\eta_{\tau(t)})dt\right)\\ &= \mathbb{E}^*\left(\int_0^\infty \exp(-(p-\alpha)\eta_t)dt\right)\\ &= \int_0^\infty \exp(-t\kappa(p^*+p-\alpha))dt\\ &= 1/\kappa(p^*+p-\alpha)\,,\end{aligned}$$

where the fourth equality stems from (i). Next, we apply the self-similarity and branching properties of the fragmentation to see that for every $t\geq 0$

$$\begin{aligned}\mathbb{E}\left(\int_t^\infty ds\sum_{i=1}^{\infty}X_i^p(s)\right) &= \mathbb{E}\left(\sum_{i=1}^{\infty}X_i^{p-\alpha}(t)\right)\mathbb{E}\left(\int_0^\infty ds\sum_{i=1}^{\infty}X_i^p(s)\right)\\ &= \mathbb{E}\left(\sum_{i=1}^{\infty}X_i^{p-\alpha}(t)\right)/\kappa(p-\alpha)\,,\end{aligned}$$

where the second equality follows from the identity above. Taking the derivative in the variable $t$, we arrive at

$$\frac{d}{dt}\mathbb{E}\left(\sum_{i=1}^{\infty} X_i^{p-\alpha}(t)\right) = -\kappa(p-\alpha)\mathbb{E}\left(\sum_{i=1}^{\infty} X_i^{p}(t)\right).$$

This equation implies that whenever $p > p^*$, the function $t \to \mathbb{E}\left(\sum_{i=1}^{\infty} X_i^p(t)\right)$ is completely monotone, and thus, by Bernstein's theorem, can be expressed as the Laplace transform of some measure $\mu$ on $\mathbb{R}_+$:

$$\mathbb{E}\left(\sum_{i=1}^{\infty} X_i^{p}(t)\right) = \int_{\mathbb{R}_+} \mathrm{e}^{-tx}\mu(dx).$$

More precisely, as the moments of $\mu$ can be recovered from the derivatives of its Laplace transform, we get

$$\int_{\mathbb{R}_+} x^k \mu(dx) = \kappa(p)\ldots\kappa(p+(k-1)\alpha).$$

By the series expansion of $\mathrm{e}^{-tx}$, we finally arrive at the expression

$$\mathbb{E}\left(\sum_{i=1}^{\infty} X_i^{p}(t)\right) = \sum_{k=0}^{\infty}\frac{(-t)^k}{k!}\int_{\mathbb{R}_+} x^k \mu(dx)$$

which yields the desired formula. □

As a check, it may be interesting to recover the formulas in Proposition 1.7 using the expression (1.15) for the infinitesimal generator of the fragmentation chain applied to additive functionals and the backward equation.

## 1.3 Extinction and formation of dust for $\alpha < 0$

In this section, we shall again implicitly assume that at the initial time, the fragmentation chain starts from a single particle with unit size. The purpose of this section is to point to the following phenomenon. Under a natural and fairly general assumption, any self-similar fragmentation with a negative index of self-similarity almost surely reaches the absorbing state $(0,0,\ldots)$ at a finite time (see Proposition 1.8 below). In the situation where the dislocation measure is conservative, that is when each time a particle splits, the sum of the sizes of the children particles is the same as the size of their parent, it is convenient to think of the size of a particle as a mass. So, when the index of self-similarity is negative, even if the dislocation measure is conservative, the initial mass is eventually reduced to dust (i.e. fragments of infinitesimal mass), and in the final section, we will investigate the phenomenon of formation of dust as time passes.

### 1.3.1 Extinction

Intuitively, when the index of self-similarity is negative, fragments with small sizes are subject to high splitting rates, and this makes them vanish entirely quickly. To make a rigorous statement, we shall suppose throughout this section that

$$\int_{\mathcal{S}^{\downarrow}} \#\{i \in \mathbb{N} : s_i = 1\}\, \nu(d\mathbf{s})/\nu(\mathcal{S}^{\downarrow}) < 1 . \tag{1.22}$$

To explain the role of this condition, observe that the process $\#\{i \in \mathbb{N} : X_i(t) = 1\}$ that counts the number of fragments with unit size as time passes, is a branching process in continuous time, and (1.22) amounts to assuming that this branching process is sub-critical so that it becomes extinct a.s. Plainly, in the super-critical situation where

$$\int_{\mathcal{S}^{\downarrow}} \#\{i \in \mathbb{N} : s_i = 1\}\, \nu(d\mathbf{s})/\nu(\mathcal{S}^{\downarrow}) > 1 ,$$

this branching process has a positive probability of surviving forever, which clearly would impede the extinction of the fragmentation.

**Proposition 1.8** *Let (1.22) be fulfilled. Then the following assertions hold with probability one:*

(i) *For $\alpha < 0$, $X(t) = (0, \ldots)$ for all sufficiently large $t$.*
(ii) *When $\kappa(-\alpha) > 0$, for almost every $t > 0$*

$$\#\{j \in \mathbb{N} : X_j(t) > 0\} < \infty .$$

We stress that in general, no matter what the value of $\alpha$ is, there may exist random instants $t$ at which

$$\#\{j \in \mathbb{N} : X_j(t) > 0\} = \infty .$$

For instance when the dislocation measure fulfills

$$\nu\left(x_j = 0 \text{ for some } j \in \mathbb{N}\right) = 0 ,$$

then with probability one, each dislocation in the fragmentation produces infinitely many terms. This does not induce any contradiction with Proposition 1.8 (ii) when $\kappa(-\alpha) > 0$, because informally, as the index of self-similarity is negative, we know that fragments with small size vanish quickly.

*Proof* Recall that $\{\xi_u, |u| = k\}$ denotes the set of particles at the $k$-th generation. We get from Lemma 1.4 and Proposition 1.6 that

$$\mathbb{E}\left(\sum_{|u|=k}^{\infty} |\xi_u|^p\right) = \gamma(p)^k, \tag{1.23}$$

where

$$\gamma(p) = \int_{\mathcal{S}^{\downarrow}} \sum_{i=1}^{\infty} s_i^p \nu(d\mathbf{s})/\nu(\mathcal{S}^{\downarrow}) = \frac{\nu(\mathcal{S}^{\downarrow}) - \kappa(p)}{\nu(\mathcal{S}^{\downarrow})}.$$

By monotone convergence, (1.22) yields

$$\lim_{p\to\infty} \kappa(p) = \int_{\mathcal{S}^{\downarrow}} (1 - \#\{i \in \mathbb{N} : s_i = 1\})\, \nu(d\mathbf{s}) > 0.$$

This enables us to choose $p$ sufficiently large such that $\kappa(p) > 0$, that is such that $\gamma(p) < 1$. In particular, the series

$$\sum_{k=1}^{\infty} \sum_{|u|=k}^{\infty} |\xi_u|^p$$

has a finite mean and thus converges a.s., and a fortiori

$$\lim_{k\to\infty} \max_{|u|=k} |\xi_u| = 0. \tag{1.24}$$

Now suppose for simplicity that $\nu(\mathcal{S}^{\downarrow}) = 1$, pick $a > 0$ arbitrary and consider the event that for some generation $k$, there exists at least one particle $\xi_u$, $|u| = k$, with lifetime $\zeta_u > a/k^2$. Because each particle with size $x$ has a lifetime which is exponentially distributed with parameter $x^\alpha$, the probability of this event can be bounded from above by

$$\sum_{k=1}^{\infty} \mathbb{E}\left(\sum_{|u|=k} \exp\left(-ak^{-2}|\xi_u|^\alpha\right)\right) \le c_p a^{-p} \sum_{k=1}^{\infty} k^{2p} \mathbb{E}\left(\sum_{|u|=k} |\xi_u|^{-\alpha p}\right),$$

where $c_p$ is some constant which depends only on $p$. Now use (1.23) and take $p$ sufficiently large, so that the right-hand side can be bounded from above by $c'_p a^{-p}$, where $c'_p$ is some constant which depends only on $p$ and $\nu$.

We see that provided that $a$ is chosen large enough, the probability that for all $k$ there are no particles of the $k$-th generation alive at time

$$t := \zeta_\varnothing + a \sum_{k=1}^{\infty} k^{-2},$$

can be made as close to 1 as we wish. Recalling (1.24), this completes the proof of (i).

Next, suppose $\kappa(-\alpha) > 0$. In this situation, we have by (1.23) that

$$\mathbb{E}\left(\sum_{u\in\mathcal{U}} \zeta_u\right) = \mathbb{E}\left(\sum_{k=0}^{\infty}\sum_{|u|=k} \xi_u^{-\alpha}\right) < \infty .$$

So, if $I_u$ denotes the time-interval during which the particle with label $u$ is alive (so the length of $I_u$ is the lifetime $\zeta_u$ of this particle), we have

$$\mathbb{E}\left(\int_0^{\infty} dt \sum_{u\in\mathcal{U}} \mathbb{1}_{\{t\in I_u\}}\right) = \mathbb{E}\left(\sum_{u\in\mathcal{U}} \zeta_u\right) < \infty ,$$

which implies that for almost every $t \geq 0$, there are only finitely many particles alive at time $t$. □

It would be interesting to have information on the distribution of the extinction time

$$T := \inf\{t \geq 0 : X(t) = (0, 0, \ldots)\} ,$$

however, it does not seem possible to express this law in a closed form. Nonetheless, we point out that an application of the branching property at the first dislocation (cf. Proposition 1.3) yields the identity in distribution

$$T \stackrel{(d)}{=} \mathbf{e} + \max_{j\in\mathbb{N}} \xi_j^{-\alpha} T_j , \tag{1.25}$$

where $\mathbf{e}$ is a standard exponential variable, $(\xi_j, j \in \mathbb{N})$ is distributed according to $\nu(\cdot)/\nu(\mathcal{S}^{\downarrow})$, $(T_j, j \in \mathbb{N})$ is a sequence of independent copies of $T$, and $\mathbf{e}$, $(\xi_j, j \in \mathbb{N})$ and $(T_j, j \in \mathbb{N})$ are independent. We refer to [7] for a survey of this type of equation in distribution.

### 1.3.2 Formation of dust

In this section, we focus on the case when the dislocation measure is conservative, that is when

$$\nu\left(\sum_{i=1}^{\infty} s_i \neq 1\right) = 0 . \tag{1.26}$$

It is easy to deduce by iteration that for every $n \in \mathbb{N}$, the total mass of particles at the $n$-th generation is conserved, that is

$$\sum_{|u|=n} \xi_u = 1 , \qquad \text{a.s.}$$

Turning our interest to the total mass of particles at time $t$, we introduce the quantity

$$\mathrm{D}(t) := 1 - \sum_{n=1}^{\infty} X_i(t),$$

which can be viewed as the total mass of *dust*, that is of infinitesimal particles at time $t$. One could be tempted to believe that the assumption (1.26) would yield $\mathrm{D} \equiv 0$; indeed the argument in the proof of Proposition 1.5 shows that this holds when the index of self-similarity of the fragmentation is non-negative. However, Proposition 1.8 shows that for negative indices of self-similarity, $\mathrm{D}$ reaches 1 at a finite time a.s.

Furthermore, one readily sees that the process $\mathrm{D}$ increases and has right-continuous paths a.s. Indeed, let us fix $\varepsilon > 0$ and write for every $t \geq 0$

$$\mathrm{D}_\varepsilon(t) := 1 - \sum_{i=1}^{\infty} X_i(t) \mathbb{1}_{\{X_i(t) > \varepsilon\}}.$$

So $\mathrm{D}_\varepsilon(t)$ is a functional of the Markov chain $\varphi_\varepsilon(X(t))$, where $\varphi_\varepsilon : \mathcal{S}^\downarrow \to \mathcal{S}^\downarrow$ is the threshold operator which consists in removing fragments with size less than or equal to $\varepsilon$ (cf. Section 1.1.3). It is immediately seen that the process $\mathrm{D}_\varepsilon(\cdot)$ is increasing with right-continuous paths, and, by monotone convergence, that it decreases to $\mathrm{D}(\cdot)$ when $\varepsilon \downarrow 0$. As a consequence, the process $\mathrm{D}(\cdot)$ increases. Moreover, for every $t \geq 0$ we have

$$\mathrm{D}(t+) := \lim_{s \downarrow t} \mathrm{D}(s) = \inf_{s>t} \inf_{\varepsilon>0} \mathrm{D}_\varepsilon(s),$$

and we can rewrite the right-hand side as

$$\inf_{\varepsilon>0} \inf_{s>t} \mathrm{D}_\varepsilon(s) = \lim_{\varepsilon \to 0} \mathrm{D}_\varepsilon(t) = \mathrm{D}(t),$$

which shows that $\mathrm{D}(\cdot)$ is right-continuous. The following proposition gathers a couple of simple observations on this phenomenon of formation of dust.

**Proposition 1.9** *Suppose (1.26) holds and $\alpha < 0$. The following assertions hold with probability one:*

(i) $\mathrm{D}$ *is a continuous increasing process which reaches* 1 *in finite time.*
(ii) *If* $\#(t) := \#\{i \in \mathbb{N} : X_i(t) > 0\}$ *denotes the number of fragments with positive mass at time $t$, then*

$$\int_0^\infty \mathbb{1}_{\{\#(t) < \infty\}} d\mathrm{D}(t) = 0.$$

This statement again reflects the fact that, informally, dislocations occur faster and faster as time passes. Observe that it implies that almost-surely, there exist uncountable many times at which there are infinitely many fragments with positive size, which may be rather surprising (for instance in the case when dislocations are binary, that is produce exactly two fragments; see also Proposition 1.8).

*Proof* (i) We have already seen that $\mathrm{D}$ has right-continuous increasing paths that reach 1 a.s.; let us check the absence of jumps. In this direction, it is convenient to write $\partial\mathcal{U} = \mathbb{N}^{\mathbb{N}}$ for the boundary of the genealogical tree, that is the set of infinite sequences of integers. An element $\ell \in \partial\mathcal{U}$ can be thought of as a leaf of the tree, and can also be identified with the branch connecting $\ell$ to the root $\emptyset$. For every $\ell, \ell' \in \partial\mathcal{U}$, we can define the distance $\mathrm{d}(\ell, \ell') = 2^{-g(\ell,\ell')}$, where $g(\ell, \ell')$ is the generation of the last common ancestor of $\ell$ and $\ell'$ (i.e. the sequences $\ell$ and $\ell'$ coincide up to the $g(\ell, \ell')$-th term and then differ).

We can use the random mark on $\mathcal{U}$ to define the length of the branch connecting a leaf $\ell$ to the root as

$$\lambda(\ell) := \sum_{n=0}^{\infty} \zeta_{\ell_n},$$

where $\ell_n$ is the node of the branch at the $n$-th generation (i.e. the sequence of the $n$ first terms of $\ell$), and $\zeta_{u_n}$ the lifetime of the fragment labeled by $u_n$. We stress that with probability one, $\lambda(\ell) < \infty$ for all leaves $\ell \in \partial\mathcal{U}$, see the proof of Proposition 1.8(i). On the other hand, the fact that dislocations preserve masses also enables us to endow $\partial\mathcal{U}$ with a natural (random) probability measure $\pi$: given a leaf $\ell \in \partial\mathcal{U}$ and an integer $n \in \mathbb{N}$, the ball $B(\ell, 2^{-n})$ centered at $\ell$ with radius $2^{-n}$ consists in the set of leaves $\ell'$ that have the same ancestor $\ell_n \in \mathbb{N}^n$ as $\ell$ at generation $n$, and we set $\pi(B(\ell, 2^{-n})) = \xi_{\ell_n}$, where $\xi_{\ell_n}$ is the size of the fragment with label $\ell_n$.

Recall from Proposition 1.4 that

$$\sum_{n=1}^{\infty} X_i(t) = \sum_{u\in\mathcal{U}} \xi_u \mathbb{1}_{\{t\in I_u\}}$$

where $I_u$ stands for the life-interval of the fragment labeled by $u$, and observe that in this setting the right-hand side can be expressed as $\pi(\{\lambda(\ell) > t\})$. Thus there is the identity

$$\mathrm{D}(t) = \pi(\{\lambda(\ell) \le t\}), \tag{1.27}$$

and all that is needed now is to check that $\mathbb{P}$-a.s.

$$\pi \otimes \pi (\{\ell, \ell' \in \partial\mathcal{U} : \lambda(\ell') = \lambda(\ell)\}) = 0 . \tag{1.28}$$

Given the mark on the genealogical tree induced by the fragmentation, let us pick a leaf $L$ at random by the same procedure as for the randomly tagged branch in Section 1.2.3. The ancestor $L_0$ of $L$ at generation 0 is $\emptyset$, and for every $n \in \mathbb{N}$, given the ancestor $L_{n-1}$ of $L$ at generation $n-1$, the ancestor $L_n$ at generation $n$ is chosen at random among the children of $L_{n-1}$ with probability proportional to their sizes (i.e. by size-biased sampling). It should be plain from this construction that $L$ has law $\pi$. Next pick a second leaf $L'$ at random, independently of the first, and according to the same procedure, so that given the marks, $L$ and $L'$ are two independent variables distributed according to $\pi$. On the one hand, it follows from (1.24) that the random measure $\pi$ has no atoms, that is it does not charge $\{\ell\}$ for any $\ell \in \partial\mathcal{U}$, so $\pi(L = L') = 0$. Now by conditioning on the last common ancestor of $L$ and $L'$, we see that the lengths $\lambda(L)$ and $\lambda(L')$ are different $\pi$-a.s., which establishes (1.28).

(ii) As above, we work with a leaf $L$ picked at random according to $\pi$, and write $L_n$ for the ancestor of $L$ at generation $n$. For each $n \in \mathbb{N}$, consider the event $\Lambda_n$ that among the nodes at the $n$-th generation with parent $L_{n-1}$ (i.e. the brothers of $L_n$), at least one of them has a lifetime which is larger than the distance from $L_n$ to $L$. Observe that, by construction, the fragment corresponding to such a node is alive at time $\lambda(L)$, and that its size cannot be zero, as nodes with zero size have lifetime $\zeta = 0$. From the description of the dynamics of the random marks in Proposition 1.3, it is readily seen that the probability of $\Lambda_n$ is bounded away from 0 as $n \to \infty$, and then, by Kolmogorov's 0-1 law, that $\limsup \Lambda_n$ has probability 1. This shows that at time $\lambda(L)$, there are infinitely many fragments with positive size, and by (1.27), this completes the proof. □

## 1.4 Some strong laws for $\alpha \geq 0$

In this section, we consider the asymptotic behavior of self-similar fragmentation chains, and shall establish some strong limit theorems for (a weighted version of) the empirical distribution of the fragments. Again, for the sake of simplicity, we assume implicitly that at the initial time the process starts from a single fragment with unit size; the general case follows easily. The approach relies crucially on an extension of the classical law of large numbers that we now present.

### 1.4.1 A variation of the law of large numbers

To start with, let us specify the setting. For each $t \geq 0$, let $\lambda(t) = (\lambda_i(t), i \in \mathbb{N})$ be a sequence of non-negative random variables such that for some fixed $p > 1$

$$\sup_{t\geq 0} \mathbb{E}\left(\left(\sum_{i=1}^{\infty} \lambda_i(t)\right)^p\right) < \infty \quad \text{and} \quad \lim_{t\to\infty} \mathbb{E}\left(\sum_{i=1}^{\infty} \lambda_i^p(t)\right) = 0.$$

Let also $(Y_i(t), i \in \mathbb{N})$ be a sequence of random variables which are independent conditionally on $\lambda(t)$. Finally, assume there is a sequence $(\bar{Y}_i, i \in \mathbb{N})$ of independent and identically distributed variables in $L^p(\mathbb{P})$, which is independent of $\lambda(t)$ for each fixed $t$, and such that $|Y_i(t)| \leq \bar{Y}_i$ for all $i \in \mathbb{N}$ and $t \geq 0$. We can now state:

**Lemma 1.5** *Under the preceding assumptions,*

$$\lim_{t\to\infty} \sum_{i=1}^{\infty} \lambda_i(t)(Y_i(t) - \mathbb{E}(Y_i(t) \mid \lambda(t))) = 0 \qquad \text{in } L^p(\mathbb{P}).$$

Before establishing this result, let us make the connection with the classical law of large numbers. Let $(Y_i, i \in \mathbb{N})$ be a sequence of i.i.d. variables with a finite $p$-th moment. We set $Y_i(t) = Y_i$ for all $t$, and $\lambda_i(t) = 1/t$ if $i \leq t$ and $\lambda_i(t) = 0$ otherwise. Then the assumptions above are clearly fulfilled, and an application of Lemma 1.5 gives

$$\lim_{t\to\infty} \frac{1}{t} \sum_{i=1}^{[t]} (Y_i - \mathbb{E}(Y_i)) = 0 \qquad \text{in } L^p(\mathbb{P}),$$

or equivalently $\lim_{n\to\infty} n^{-1} \sum_{i=1}^{n} Y_i = \mathbb{E}(Y_1)$ in $L^p(\mathbb{P})$.

*Proof* Let $a > 0$ be an arbitrarily large real number. Introduce for every $i \in \mathbb{N}$ and $t \geq 0$ the truncated variables $Y_i(t, a) = \mathbb{1}_{\{|Y_i(t)| < a\}} Y_i(t)$. To start with, there is the upper bound

$$\begin{aligned} \left|\sum_{i=1}^{\infty} \lambda_i(t)(Y_i(t) - \mathbb{E}(Y_i(t) \mid \lambda(t)))\right| &\leq \left|\sum_{i=1}^{\infty} \lambda_i(t)(Y_i(t) - Y_i(t, a))\right| \\ &+ \left|\sum_{i=1}^{\infty} \lambda_i(t)(Y_i(t, a) - \mathbb{E}(Y_i(t, a) \mid \lambda(t)))\right| \\ &+ \left|\sum_{i=1}^{\infty} \lambda_i(t)\mathbb{E}(Y_i(t, a) - Y_i(t) \mid \lambda(t))\right| . \end{aligned}$$

Consider the first series in the right-hand side. The bounds $|Y_i(t) - Y_i(t, a)| \leq \mathbb{1}_{\{\bar{Y}_i > a\}} \bar{Y}_i$ and the independence of $(\bar{Y}_i, i \in \mathbb{N})$ and $\lambda(t)$ yield

$$\mathbb{E}\left(\left|\sum_{i=1}^{\infty} \lambda_i(t)(Y_i(t) - Y_i(t, a))\right|^p\right) \leq \mathbb{E}(\mathbb{1}_{\{\bar{Y}_1 > a\}} \bar{Y}_1^p)\mathbb{E}\left(\left(\sum_{i=1}^{\infty} \lambda_i(t)\right)^p\right),$$

and the latter quantity converges to 0 as $a \to \infty$, uniformly for $t \geq 0$. The same argument also shows that

$$\lim_{a\to\infty} \sup_{t\geq 0} \mathbb{E}\left(\left|\sum_{i=1}^{\infty} \lambda_i(t)\mathbb{E}\left(Y_i(t) - Y_i(t, a) \mid \lambda(t)\right)\right|^p\right) = 0.$$

Finally, conditionally on $\lambda(t)$, the variables $Y_i(t, a) - \mathbb{E}(Y_i(t, a) \mid \lambda(t))$ are centered, independent, and bounded in absolute value by $a$. Thus, conditionally on $\lambda(t)$,

$$\sum_{i=1}^{n} \lambda_i(t)(Y_i(t, a) - \mathbb{E}(Y_i(t, a)) \mid \lambda(t)), \qquad n \in \mathbb{N}$$

is a martingale bounded in $L^p$ and there exists a universal constant $c_p$ such that

$$\mathbb{E}\left(\left|\sum_{i=1}^{\infty} \lambda_i(t)(Y_i(t, a) - \mathbb{E}(Y_i(t, a) \mid \lambda(t))\right|^p \mid \lambda(t)\right) \leq c_p a^p \sum_{i=1}^{\infty} \lambda_i^p(t).$$

Our assumptions ensure that the latter quantity converges to 0 as $t \to \infty$ in $L^1(\mathbb{P})$, so putting the pieces together this completes the proof of the statement. □

Let us now explain how we shall apply Lemma 1.5 in the rest of this section. We shall be interested in limit theorems involving functionals of the fragmentation of the type

$$A(t) := \sum_{i=1}^{\infty} X_i^{p^*}(t) g(X_i(t), t),$$

where $p^*$ is the Malthusian exponent and $g$ a certain measurable function. The first step of the analysis consists in considering this functional at time $t + s$ and applying the branching property of the fragmentation at time $t$. This yields an expression of the form

$$A(t+s) = \sum_{k=1}^{\infty} \lambda_k(t) Y_k(t, s)$$

where $\lambda_k(t) = X_k^{p^*}(t)$ and $Y_k(t, s)$ is a quantity depending of the fragmentation started at time $t$ from a single particle with size $X_k(t)$. More precisely, writing $\tilde{X}$ for the latter, we have

$$Y_k(t, s) = \sum_{j=1}^{\infty} \frac{\tilde{X}_j^{p^*}(s)}{X_k^{p^*}(t)} g(\tilde{X}_j(s), t+s).$$

We then use Lemma 1.5 to reduce the study of the asymptotic behavior of the latter quantity as both $t, s \to \infty$ to that of

$$\sum_{k=1}^{\infty} \lambda_k(t) \mathbb{E}(Y_k(t, s) \mid X_k(t)).$$

Essentially, this reduction amounts to getting estimates for the first moment of additive functionals of the fragmentation, which are then obtained by limit theorems for the tagged particle (cf. Section 1.2.3). Roughly speaking, one shows that $\mathbb{E}(Y_k(t, s) \mid X_k(t)) \sim c$ as $s, t \to \infty$ for some constant $c$ depending on $g$, so, using Proposition 1.5, we can conclude that

$$A(t+s) \sim c \sum_{k=1}^{\infty} \lambda_k(t) = c \sum_{k=1}^{\infty} X_k^{p^*}(t) \sim c\mathcal{M}_\infty.$$

Of course, one has to check carefully the estimates above, which is somewhat technical.

### 1.4.2 The homogeneous case ($\alpha = 0$)

We suppose throughout this section that $X = (X(t), t \geq 0)$ is a homogeneous fragmentation (i.e. the index of self-similarity is $\alpha = 0$) for which the Malthusian hypotheses of Theorem 1.1 hold. We are interested in the asymptotic behavior of the empirical distribution of the fragments as time tends to infinity. The following pathwise limit theorem extends a result due to Kolmogorov [141] in the conservative case, which seems to have appeared in the very first probabilistic work on fragmentation. See also Asmussen and Kaplan [14] for a closely related result.

Introduce the first and second right-derivatives of $\kappa$ at the Malthusian exponent $p^*$, which are given in terms of the dislocation measure by

$$\kappa'(p^*) = -\int_{\mathcal{S}^\downarrow} \left( \sum_{i=1}^{\infty} x_i^{p^*} \ln x_i \right) \nu(d\mathbf{x}), \quad \kappa''(p^*) = -\int_{\mathcal{S}^\downarrow} \left( \sum_{i=1}^{\infty} x_i^{p^*} (\ln x_i)^2 \right) \nu(d\mathbf{x}).$$

Note that these quantities are finite since $p^* > \underline{p}$. Recall that $\mathcal{M}_\infty$ denotes the terminal value of the intrinsic martingale in Theorem 1.1.

**Theorem 1.2** *Let $f: \mathbb{R} \to \mathbb{R}$ be a continuous bounded function.*

(i) *We have*

$$\lim_{t\to\infty} \sum_{i=1}^{\infty} X_i^{p^*}(t) f(t^{-1} \ln X_i(t)) = \mathcal{M}_\infty f(-\kappa'(p^*)),$$

*in $L^p(\mathbb{P})$ for some $p > 1$.*

(ii) *Denote by $\mathcal{N}(0, -\kappa''(p^*))$ a centered Gaussian distribution with variance $-\kappa''(p^*)$. Then*

$$\lim_{t\to\infty} \sum_{i=1}^{\infty} X_i^{p^*}(t) f(t^{-1/2}(\ln X_i(t) + \kappa'(p^*)t)) = \mathcal{M}_\infty \mathbb{E}(f(\mathcal{N}(0, -\kappa''(p^*)))),$$

*in $L^p(\mathbb{P})$ for some $p > 1$.*

Loosely speaking, the first part of the statement shows that most fragments decay exponentially fast with rate $\kappa'(p^*)$ as time tends to infinity, and the second part provides a sharper estimate of the second order.

*Proof* We shall apply Lemma 1.5 in the following situation. Let $f: \mathbb{R}_+ \to [0, 1]$ be a continuous function; we are interested in

$$\sum_{k=1}^{\infty} X_k^{p^*}(t+t^2) f\left((t+t^2)^{-1} \ln X_k(t+t^2)\right).$$

By an application of the Markov property at time $t$ and self-similarity, we can re-express this variable in the form

$$\sum_{i=1}^{\infty} \lambda_i(t) Y_i(t)$$

where $\lambda_i(t) = X_i^{p^*}(t)$ and

$$Y_i(t) = \sum_{j=1}^{\infty} X_{ij}^{p^*}(t^2) f\left((t+t^2)^{-1} \ln(X_i(t) X_{ij}(t^2))\right),$$

with $X_{1\cdot}, X_{2\cdot}, \ldots$ a sequence of i.i.d. copies of $X$ which is independent of $X(t) = (X_1(t), \ldots)$.

It follows from Proposition 1.5 that the sequence $(\lambda_i(t), i \in \mathbb{N})$ fulfills the requirement of Lemma 1.5. Let us now consider the sequence $(Y_i(t), i \in \mathbb{N})$ conditionally on $X(t)$. Plainly, it is given by independent variables, each of which is bounded from above by

$$\bar{Y}_i := \sup_{s\geq 0} \sum_{j=1}^{\infty} X_{ij}^{p^*}(s).$$

On the one hand, the $\bar{Y}_i$ are clearly i.i.d. On the other hand, because $\sum_{j=1}^{\infty} X_{ij}^{p^*}(s)$ is a martingale which is bounded in $L^p(\mathbb{P})$ for some $p > 1$, it follows from Doob's inequality that its overall supremum belongs to $L^p(\mathbb{P})$.

Thus we may apply Lemma 1.5, which reduces the study to that of the asymptotic behavior of

$$\sum_{i=1}^{\infty} \lambda_i(t)\mathbb{E}(Y_i(t) \mid X(t))$$

as $t$ tends to $\infty$. In this direction, we thus compute the conditional expectation of $Y_i(t)$ given $X(t)$; we easily get on the event $\{X_i(t) = x\}$ that

$$\mathbb{E}(Y_i(t) \mid X(t)) = \mathbb{E}\left(\sum_{j=1}^{\infty} X_j^{p^*}(t^2) f((t+t^2)^{-1}(\ln X_j(t^2) + \ln x))\right).$$

Now recall the notion of tagged particle $(\chi(t), t \geq 0)$ in Section 1.2.3. In particular we know from Lemma 1.4 that there is the identity

$$\mathbb{E}\left(\sum_{j=1}^{\infty} X_j^{p^*}(t^2) f((t+t^2)^{-1}(\ln X_j(t^2) + \ln x))\right)$$
$$= \mathbb{E}^*(f((t+t^2)^{-1}(\ln \chi(t^2) + \ln x)).$$

Moreover, recall from Proposition 1.6 that the process of the logarithm of size of the tagged particle is a compound Poisson process, $-\ln \chi(t) = S(N_t)$, where $S$ is a random walk with step distribution $\tilde{\nu}(\cdot)/\nu(\mathcal{S}^{\downarrow})$ and $N$ an independent Poisson process with parameter $\nu(\mathcal{S}^{\downarrow})$. In particular, $S(1)$ has finite mean $\kappa'(p^*)/\nu(\mathcal{S}^{\downarrow})$, and it follows from the law of large numbers that

$$\lim_{t\to\infty} \mathbb{E}^*(f((t+t^2)^{-1}(\ln \chi(t^2) + \ln x)) = f(-\kappa'(p^*)),$$

where the limit is uniform in $x$ such that, say, $-\ln x \leq t^{3/2}$. On the other hand, using again Lemma 1.4, we have

$$\mathbb{E}\left(\sum_{i=1}^{\infty} X_i^{p^*}(t) \mathbb{1}_{\{-\ln X_i(t)) > t^{3/2}\}}\right) = \mathbb{P}^*(-\ln \chi(t) > t^{3/2}),$$

and the latter quantity tends to 0 as $t \to \infty$.

Recall from Proposition 1.5 that $\sum_{i=1}^{\infty} \lambda_i(t)$ converges to $\mathcal{M}_\infty$ in $L^p(\mathbb{P})$. Putting the pieces together, we get that as $t \to \infty$

$$\sum_{i=1}^{\infty} \lambda_i(t)\mathbb{E}(Y_i(t) \mid X(t)) \sim f(-\kappa'(p^*)) \sum_{i=1}^{\infty} \lambda_i(t) \sim \mathcal{M}_\infty f(-\kappa'(p^*)).$$

(ii) The proof is similar; the arguments above are easily adapted to reduce the proof to asymptotics for the first moment

$$\mathbb{E}^*(f(t^{-1/2}(\ln \chi(t) + t\kappa'(p^*))),$$

where $f$ is a continuous bounded function. We may then use the central limit theorem for the compound random walk $-\ln \chi(t) = S \circ N_t = \eta_t$ to show that the preceding quantity converges to $\mathbb{E}(f(\mathcal{N}(0, -\kappa''(p^*))))$ when $t \to \infty$. Details are left to the reader. □

### 1.4.3 The case $\alpha > 0$

We now suppose throughout this section that $X = (X(t), t \geq 0)$ is a self-similar fragmentation with index $\alpha > 0$, with a dislocation measure $\nu$. Again, we shall assume that the Malthusian hypotheses of Theorem 1.1 hold and we will be interested in the asymptotic behavior of the process as time tends to infinity. Because for positive indices of self-similarity, small fragments split more slowly than large fragments, we may expect some homogenization phenomenon.

We now state the main result of this section, which has been obtained first by Filippov [103] in the conservative case. Its extension to non-conservative self-similar fragmentations has been established recently by Bertoin and Gnedin [41] (see also the forthcoming Section 1.6 for further references related to this result). Roughly, it shows that most fragments decay like $t^{-1/\alpha}$ as time $t$ tends to infinity, which contrasts with the exponential decay in the homogeneous case (see Theorem 1.2).

**Theorem 1.3** *Suppose that $\alpha > 0$ and that the step distribution of the random walk $S_n = -\ln \chi_n$ is not arithmetic. Then for every bounded continuous function $f : \mathbb{R}_+ \to \mathbb{R}$*

$$\lim_{t\to\infty} \sum_{i=1}^{\infty} X_i^{p^*}(t) f(t^{1/\alpha} X_i(t)) = \mathcal{M}_\infty \int_0^\infty f(y)\rho(dy), \qquad \text{in } L^1(\mathbb{P}),$$

*where $\mathcal{M}_\infty$ is the terminal value of the intrinsic martingale and $\rho$ is a deterministic probability measure. More precisely, $\rho$ is determined by the moments*

$$\int_{]0,\infty[} y^{\alpha k} \rho(dy) = \frac{(k-1)!}{\alpha\kappa'(p^*)\,\kappa(p^*+\alpha)\cdots\kappa(p^*+(k-1)\alpha)} \qquad \text{for } k \in \mathbb{N},$$

*(with the usual convention that the right-hand side above equals $1/(\alpha\kappa'(p^*))$ for $k = 1$).*

Before proving this result, let us consider a couple of examples. First, recall the Poissonian rain model of Section 1.1.3, for which $\alpha = 1$ and $\kappa(p) = (p-1)/(p+1)$. The dislocation measure is conservative, so $p^* = 1$ and the intrinsic martingale $\mathcal{M}$ is constant. We find

$$\int_{]0,\infty[} y^k \rho(dy) = \frac{(k-1)!}{\frac{1}{2} \times \frac{1}{3} \times \frac{2}{4} \times \cdots \times \frac{k-1}{k+1}} = (k+1)!\,,$$

and we conclude that $\rho$ has density $\rho(dy)/dy = ye^{-y}$, a result which can also be checked directly by more elementary calculations for this specific case. More generally, for self-similar fragmentation chains with index $\alpha > 0$ and $\kappa(p) = (p-1)/(p+1)$, one gets that $\rho$ is the distribution of $Y^{1/\alpha}$ where $Y$ has the gamma$(2/\alpha)$ law; see [63].

Second, suppose that the dislocation measure $\nu$ is the distribution induced by the uniform stick-breaking scheme as described after equation (1.17). So $\kappa(p) = 1 - 1/p$, and the Malthusian exponent is $p^* = 1$. Moreover, the dislocation measure is conservative, and the intrinsic martingale is thus trivial, $\mathcal{M}_n \equiv 1$. Suppose further that the index of self-similarity is $\alpha = 1$. One then gets

$$\frac{(k-1)!}{\alpha\kappa'(p^*)\,\kappa(p^*+\alpha)\cdots\kappa(p^*+(k-1)\alpha)} = k!\,, \qquad k \in \mathbb{N},$$

so the probability measure $\rho$ appearing in Theorem 1.3 is simply the standard exponential distribution.

Just as in the homogeneous case, Lemma 1.5 reduces the proof to the analysis of the so-called tagged particle; we shall only provide details on the latter aspect. Recall from Proposition 1.6 that the process $(\chi(t), t \geq 0)$ is a continuous time Markov chain, which enjoys an obvious scaling property. The classical renewal theory yields an important limit theorem for $\chi(t)$ as $t$ tends to infinity, which is due to Brennan and Durrett [63].

**Proposition 1.10** *Suppose that the step distribution of the random walk $S_n = -\ln \chi_n$ is not arithmetic. Then under $\mathbb{P}^*$, $t^{1/\alpha}\chi(t)$ converges in distribution as $t \to \infty$ towards some variable $Y_\alpha$ which can be expressed in the form*

$$Y_\alpha = \left( \sum_{n=0}^{\infty} \exp(-\alpha R_n) \mathbf{e}_n \right)^{1/\alpha},$$

*where $(R_0, R_1, \ldots)$ is a random walk with the same step distribution as $S$ and with initial distribution*

$$\frac{\nu(\mathcal{S}^{\downarrow})}{\kappa'(p^*)} \mathbb{P}^*(S_1 > y)dy\,,$$

*and* $\mathbf{e}_0, \mathbf{e}_1, \ldots$ *a sequence of i.i.d. exponential variables with parameter* $\nu(\mathcal{S}^{\downarrow})$, *which is independent of the random walk* $(R_0, R_1, \ldots)$.

This result can be proven by taking limits as $t \to \infty$ in the explicit moment calculations of Proposition 1.7(ii), using complex analysis and contour integral, see [41]. We develop below a more probabilistic approach.

*Proof* There is no loss of generality in assuming that the fragmentation starts from a single fragment with unit size. Write $T(y) := \min\{n \in \mathbb{N} : S_n > -\ln y\}$ for every $y \in ]0, 1]$. We thus have

$$\mathbb{P}^*(\chi(t) < y) = \mathbb{P}^*\left(\sum_{n=0}^{T(y)} \exp(\alpha S_n)\mathbf{e}'_n \leq t\right),$$

where $\mathbf{e}'_0, \mathbf{e}'_1, \ldots$ is a sequence of i.i.d. exponential variables with parameter $\nu(\mathcal{S}^{\downarrow})$, which is independent of the random walk (so that $\zeta_n = \exp(\alpha S_n)\mathbf{e}'_n$ is the lifetime of the tagged particle at the $n$-th generation). Reversing the indices, we may re-express this quantity in the form

$$\mathbb{P}^*(\chi(t) < y) = \mathbb{P}^*\left(\sum_{n=0}^{T(y)} \exp(-\alpha R_n(y))\mathbf{e}_n \leq y^{\alpha} t\right),$$

where $R_n(y) := -\ln y - S_{T(y)-n}$ and $\mathbf{e}_0, \mathbf{e}_1, \ldots$ is a new sequence of i.i.d. exponential variables with parameter $\nu(\mathcal{S}^{\downarrow})$, which is again independent of $(R_n(y), n \in \mathbb{N})$. Rescaling thus gives

$$\mathbb{P}^*(t^{1/\alpha}\chi(t) < y) = \mathbb{P}^*\left(\sum_{n=0}^{T(t^{-1/\alpha}y)} \exp(-\alpha R_n(t^{-1/\alpha}y))\mathbf{e}_n \leq y^{\alpha}\right).$$

When $t$ tends to infinity, so does $T(t^{-1/\alpha}y)$, and, by renewal theory (see for example Chapter 10 in [118]), the sequence $(R_0(t^{-1/\alpha}y), R_1(t^{-1/\alpha}y), \ldots)$ converges in law to $(R_0, R_1, \ldots)$ where the latter is a random walk with the same step distribution as $S$, and the initial variable $R_0$ has the limiting law of the so-called age in a renewal process, that is to say

$$\mathbb{P}^*(R_0 \in dy) = \frac{\mathbb{P}^*(S_1 > y)}{\mathbb{E}(S_1)} dy.$$

This easily yields the statement. □

We now complete the proof of Theorem 1.3 by pointing out that the distribution of the limiting variable $Y_\alpha$ can be characterized by its moments as follows. Recall that the function $\kappa$ is defined in (1.17).

**Proposition 1.11** *For every integer $k \geq 1$, we have*

$$\mathbb{E}^*\left(Y_\alpha^{k\alpha}\right) = \frac{(k-1)!}{\alpha\kappa'(p^*)\kappa(p^*+\alpha)\cdots\kappa(p^*+(k-1)\alpha)},$$

*and this determines uniquely the law of $Y_\alpha$.*

*Proof* It is convenient to combine Propositions 1.6 and 1.10 and re-express

$$Y_\alpha^\alpha = \exp(-\alpha R_0)\int_0^\infty \mathrm{e}^{-\alpha \Upsilon_t} dt, \tag{1.29}$$

where $\Upsilon = (\Upsilon_t, t \geq 0)$ is distributed as the increasing compound Poisson process $S \circ N$, where $N$ is a Poisson process with rate $\nu(\mathcal{S}^\downarrow)$ which is independent of $R_0$ and the random walk $S$ (recall the example at the end of Section 1.1.1). In particular, the Laplace transform of $\Upsilon_t$ is given by

$$\begin{aligned}\mathbb{E}^*\left(\mathrm{e}^{-q\Upsilon_t}\right) &= \exp\left(-t\nu(\mathcal{S}^\downarrow)\mathbb{E}(1-\mathrm{e}^{-qS_1})\right) \\ &= \exp\left(-t\int_{\mathcal{S}^\downarrow}\nu(d\mathbf{x})\sum_{i=1}^\infty\left(x_i^{p^*}-x_i^{p^*+q}\right)\right),\end{aligned}$$

and finally

$$\mathbb{E}^*\left(\mathrm{e}^{-q\Upsilon_t}\right) = \exp(-t\kappa(q+p^*)). \tag{1.30}$$

On the one hand, it is immediate that

$$\mathbb{E}^*\left(\exp(-qR_0)\right) = \frac{\kappa(q+p^*)}{q\kappa'(p^*)}, \qquad q > 0. \tag{1.31}$$

On the other hand we shall check that

$$\mathbb{E}^*\left(\left(\int_0^\infty \exp(-\alpha\Upsilon_s)\,ds\right)^k\right) = \frac{k!}{\kappa(p^*+\alpha)\cdots\kappa(p^*+\alpha k)}, \qquad k = 1, 2, \ldots \tag{1.32}$$

For this purpose, set

$$I_t = \int_t^\infty \exp(-\alpha\Upsilon_s)ds$$

for every $t \geq 0$. On the one hand, for every positive real number $r > 0$, we have the identity

$$I_0^r - I_t^r = r\int_0^t \exp(-\alpha\Upsilon_s)I_s^{r-1}ds. \tag{1.33}$$

On the other hand, we may express $I_s$ in the form $I_s = \exp(-\alpha\Upsilon_s)I_0'$, where

$$I_0' = \int_0^\infty \exp(-\alpha\Upsilon_u')du \quad \text{and} \quad \Upsilon_u' = \Upsilon_{s+u} - \Upsilon_s. \tag{1.34}$$

From the independence and stationarity of the increments of the Lévy process, we see that $I'_0$ has the same law as $I_0 = I$ and is independent of $\Upsilon_s$. Plugging this into (1.33) and taking expectations, we get using (1.30) that

$$\mathbb{E}^*(I^r)(1-\exp(-t\kappa(p^*+\alpha r))) = r\int_0^t \exp(-s\kappa(p^*+\alpha r))\mathbb{E}^*(I^{r-1})\,ds$$
$$= \frac{r}{\kappa(p^*+\alpha r)}\left(1-\mathrm{e}^{-t\kappa(p^*+\alpha r)}\right)\mathbb{E}^*\left(I^{r-1}\right).$$

Finally

$$\mathbb{E}^*(I^r) = \frac{r}{\kappa(p^*+\alpha r)}\mathbb{E}^*\left(I^{r-1}\right),$$

and since $\mathbb{E}^*(I^0) = 1$, we get the formula (1.32) by iteration, taking $r = k \in \mathbb{N}$. Combining (1.29), (1.31) and (1.32) completes the proof of the first statement.

Finally, as $\lim_{q\to\infty}\kappa(q) = \nu(\mathcal{S}^{\downarrow})$, we see that $Y^{\alpha}_{\alpha}$ possesses exponential moments of any order less than $\nu(\mathcal{S}^{\downarrow})$, and therefore is determined by its entire moments. □

### 1.4.4 Another strong law via renewal theory

Finally, we turn our interest to a question which is motivated by the mining industry. Specifically, we are concerned with the stage during which mineral blocks are crushed in mills to produce a thin powder. Mineral grains are screened during the process, so that when they become smaller than the diameter of the mesh of a thin grid, they are removed from the mill.

We use a fragmentation chain to model the crushing process, and we are interested in the distribution of the small mineral grains that go across the grid. In other words, we would like to get information about the distribution of the state of a fragmentation chain when we stop each particle at the instant when it becomes smaller than some small parameter $\varepsilon > 0$. Clearly, this does not depend on the index of self-similarity $\alpha$, but only on the dislocation measure $\nu$. More precisely, we use the genealogical tree representation and consider

$$\varphi_\varepsilon(da) := \sum_{u\in\mathcal{U},u\neq\emptyset} \mathbb{1}_{\{\xi_{u-}\geq\varepsilon,\xi_u<\varepsilon\}}\xi_u^{p^*}\delta_{\xi_u/\varepsilon}(da),$$

where $u-$ stands for the parent of $u$. So $\varphi_\varepsilon$ is a random finite measure on $]0,1[$, which can be viewed as a weighted version of the empirical measure of the particles taken at the instant when they become smaller than $\varepsilon$ and rescaled. Henceforth, we shall suppose that the fragmentation is not arithmetic as in Proposition 1.10. Recall also that $\mathcal{M}_\infty$ denotes the terminal value of the intrinsic martingale.

**Proposition 1.12** *As $\varepsilon \to 0$, $\varphi_\varepsilon$ converges in probability to $\mathcal{M}_\infty \varphi$, where $\varphi$ is a deterministic probability measure on $[0, 1]$ given by*

$$\varphi(da) = \left( \int_{\mathcal{S}^\downarrow} \sum_{i=1}^{\infty} \mathbb{1}_{\{s_i < a\}} s_i^{p^*} \nu(ds) \right) \frac{da}{a\kappa'(p^*)} .$$

*Proof* We start by considering the quantities

$$\sum_{u \in \mathcal{U}, u \neq \emptyset} \mathbb{1}_{\{\xi_{u-} \geq \varepsilon, \xi_u < \varepsilon\}} \xi_u^{p^*}, \qquad \varepsilon > 0.$$

Write $\mathcal{G}_\varepsilon$ for the sigma-field generated by the variables $(\mathbb{1}_{\{\xi_{u-} \geq \varepsilon\}} \xi_u, u- \in \mathcal{U})$, so $(\mathcal{G}_\varepsilon)_{\varepsilon > 0}$ is a reversed filtration. An easy application of the branching property of the marked tree, similar to that in Proposition 1.5, implies that

$$\sum_{u \in \mathcal{U}, u \neq \emptyset} \mathbb{1}_{\{\xi_{u-} \geq \varepsilon, \xi_u < \varepsilon\}} \xi_u^{p^*} = \mathbb{E}\left(\mathcal{M}_\infty \mid \mathcal{G}_\varepsilon\right),$$

and it follows that

$$\lim_{\varepsilon \to 0} \sum_{u \in \mathcal{U}, u \neq \emptyset} \mathbb{1}_{\{\xi_{u-} \geq \varepsilon, \xi_u < \varepsilon\}} \xi_u^{p^*} = \mathcal{M}_\infty \qquad \text{in } L^p(\mathbb{P})$$

for some $p > 1$.

Next, we turn our attention to first moment estimates. Let $f: [0, 1] \to \mathbb{R}$ be a continuous function with support in $]0, 1[$, and consider

$$\langle \varphi_\varepsilon, f \rangle = \sum_{u \in \mathcal{U}, u \neq \emptyset} \mathbb{1}_{\{\xi_{u-} \geq \varepsilon, \xi_u < \varepsilon\}} \xi_u^{p^*} f(\xi_u / \varepsilon).$$

Fix $\eta > 0$ and work under $\mathbb{P}_\eta$, that is when at the initial time there is a unique fragment with size $\eta$. We compute the expectation of this variable by conditioning on the mark of the parent $u- = v$ of $u$ and applying the branching property. We get

$$\begin{aligned} \mathbb{E}_\eta(\langle \varphi_\varepsilon, f \rangle) &= \mathbb{E}_\eta \left( \sum_{v \in \mathcal{U}} \mathbb{1}_{\{\xi_v \geq \varepsilon\}} \xi_v^{p^*} \sum_{i=1}^{\infty} \mathbb{1}_{\{\xi_v \tilde{\xi}_i < \varepsilon\}} \tilde{\xi}_i^{p^*} f(\xi_v \tilde{\xi}_i / \varepsilon) \right) \\ &= \mathbb{E}_\eta \left( \sum_{v \in \mathcal{U}} \mathbb{1}_{\{\ln \xi_v \geq \ln \varepsilon\}} \xi_v^{p^*} \sum_{i=1}^{\infty} \mathbb{1}_{\{\tilde{\xi}_i < \varepsilon / \xi_v\}} \tilde{\xi}_i^{p^*} f(\xi_v \tilde{\xi}_i / \varepsilon) \right), \end{aligned}$$

where $(\tilde{\xi}_i)_{i \in \mathbb{N}}$ has the law $\nu(\cdot)/\nu(\mathcal{S}^\downarrow)$ and is independent of $\xi_v$. Integrating with respect to the latter gives

$$\mathbb{E}_\eta(\langle \varphi_\varepsilon, f \rangle) = \mathbb{E}_\eta \left( \sum_{n=0}^{\infty} \sum_{|v|=n} \mathbb{1}_{\{\ln \xi_v \geq \ln \varepsilon\}} \xi_v^{p^*} h(\ln \xi_v - \ln \varepsilon) \right),$$

where for $a \geq 0$

$$h(a) = \mathbb{E}\left(\sum_{i=1}^{\infty} \mathbb{1}_{\{\tilde{\xi}_i < e^{-a}\}} \tilde{\xi}_i^{p^*} f\left(\tilde{\xi}_i e^a\right)\right).$$

Now we can evaluate this expression using the tagged branch and the random walk $S_n = -\ln \chi_n$, thanks to Lemma 1.4:

$$\mathbb{E}_\eta\left(\langle \varphi_\varepsilon, f \rangle\right) = \sum_{n=0}^{\infty} \mathbb{E}^*\left(\mathbb{1}_{\{S_n \leq \ln \eta - \ln \varepsilon\}} h(\ln \eta - S_n - \ln \epsilon)\right).$$

Our assumptions enable us to apply the renewal theorem to the renewal process $S_n$, and we get the estimate

$$\lim_{\varepsilon \to 0} \mathbb{E}_\eta\left(\langle \varphi_\varepsilon, f \rangle\right) = \frac{1}{\mathbb{E}^*(S_1)} \int_0^\infty h(a) da,$$

where the convergence is uniform in $\eta$ as long as $\varepsilon/\eta \to 0$.

Then using the extension of the law of large numbers stated in Lemma 1.5 in a similar way as in the proof of Theorem 1.2 (again technical details are left to the reader), we can check that

$$\lim_{\varepsilon \to 0} \langle \varphi_\varepsilon, f \rangle = \frac{\mathcal{M}_\infty}{\mathbb{E}^*(S_1)} \int_0^\infty h(a) da \qquad \text{in } L^1(\mathbb{P}).$$

Finally, we already know that $\mathbb{E}^*(S_1) = \kappa'(p^*)/\nu(\mathcal{S}^\downarrow)$, and on the other hand, an easy computation gives

$$\int_0^\infty h(a) da = \int_0^1 f(b) b^{-1} \mathbb{E}\left(\sum_{i=1}^{\infty} \mathbb{1}_{\{\tilde{\xi}_i < b\}} \tilde{\xi}_i^{p^*}\right) db,$$

which completes the proof. □

## 1.5 Additive martingales (homogeneous case $\alpha = 0$)

In this section, we consider a homogeneous fragmentation chain (i.e. self-similar with index $\alpha = 0$), say $X = (X(t), t \geq 0)$ with dislocation measure $\nu$. The starting point of our study lies in the observation that one can associate to $X$ a whole family of natural martingales, which includes the intrinsic martingale as a special case. We shall specify which of these martingales are uniformly integrable, and derive further information on the asymptotic behavior of homogeneous fragmentation chains. In this direction, recall from Section 1.1.3 that when $\nu$ only charges finite sequences, there is a simple connection with branching random walks in continuous time; and the properties that we shall establish here essentially rephrase parts of the folklore of the theory of branching random walks in the setting of homogeneous fragmentation chains.

Recall also that the Malthusian hypotheses are enforced. It will be convenient to introduce the following terminology: we say that the fragmentation becomes extinct if $T := \inf\{t \geq 0 : X(t) = (0, \dots)\}$ is finite, and that it survives forever otherwise. Plainly, the fragmentation survives forever a.s. whenever $\nu(s_1 = 0) = 0$, and the Malthusian hypotheses imply that the probability of extinction is always strictly less than 1; see Proposition 1.5.

### 1.5.1 Convergence of additive martingales

A crucial fact for the study of homogeneous fragmentation is that there is a simple formula for the moments of the process. Indeed, we know from Proposition 1.7(i) and self-similarity that for every $p > \underline{p}$ and $t \geq 0$,

$$\mathbb{E}_x\left(\sum_{i=1}^{\infty} X_i^p(t)\right) = x^p \exp(-t\kappa(p)) .$$

It follows immediately from the branching property and scaling that:

**Corollary 1.3** *For every $p > \underline{p}$, the process*

$$M(p, t) := \exp(t\kappa(p)) \sum_{i=1}^{\infty} X_i^p(t)$$

*is a non-negative martingale which converges a.s.*

In order to investigate the asymptotic behavior of homogeneous fragmentation chains, it is crucial to know if the limit of the martingale above is strictly positive or zero. A first step in the analysis is the following elementary lemma.

**Lemma 1.6** *The function $p \to \kappa(p)/p$ reaches its maximum at a unique location $\bar{p} > p^*$, which is the unique solution to the equation*

$$p\kappa'(p) = \kappa(p) .$$

*More precisely, the function $p \to \kappa(p)/p$ increases on $]\underline{p}, \bar{p}[$ and decreases on $]\bar{p}, \infty[$, and the value of its maximum is $\kappa'(\bar{p}) = \kappa(\bar{p})/\bar{p}$.*

*Proof* We first point out that the function $\kappa$ is concave and increasing. It follows that

$$\text{the function } p \to p\kappa'(p) - \kappa(p) \text{ decreases on } ]\underline{p}, \infty[. \tag{1.35}$$

Indeed, this function has derivative $p\kappa''(p)$, which is negative since $\kappa$ is concave. Recall that $\kappa(p^*) = 0$ by the definition of the Malthusian exponent $p^*$. On the other hand, it is obvious that $\lim_{q\to\infty} \kappa(q)/q = 0$, hence the function $p \to \kappa(p)/p$ has the same limit at $p^*$ and at $\infty$, so it reaches its overall maximum at a unique point $\bar{p} > p^*$. In particular, we deduce from (1.35) that the derivative of $p \to \kappa(p)/p$ is positive on $]\underline{p}, \bar{p}[$ and negative on $]\bar{p}, \infty[$. Finally, the derivative must be zero at $\bar{p}$, which implies that the overall maximum is given by $\kappa'(\bar{p}) = \kappa(\bar{p})/\bar{p}$. □

We may now state the main result of this section.

**Theorem 1.4** *Assume that there exists some constants $a, b > 0$ such that*

$$\nu\left(\sum_{i=1}^{\infty} s_i^a > b\right) = 0. \tag{1.36}$$

*Then for every $p \in ]\underline{p}, \bar{p}[$, the martingale $M(p, \cdot)$ is bounded in $L^q(\mathbb{P})$ for some $q > 1$. Moreover its terminal value is strictly positive conditionally on non-extinction.*

*Proof* The proof uses the same route as that of Theorem 1.1. Recall that all that we need is to check that the sum of the jumps of the martingale raised to some power $q > 1$ has a finite mean, that is

$$\mathbb{E}\left(\sum_{t>0} |M(p,t) - M(p,t-)|^q\right) < \infty. \tag{1.37}$$

It is convenient to re-express the left-hand side in terms of the generation of the fragments. Specifically, denote by $\xi_1^{(k)}, \ldots$ the fragments of the $k$-th generation, and by $T_i^{(k)}$ the instant when $\xi_i^{(k)}$ splits. The jump of $M(p,t)$ at time $T_i^{(k)}$ is

$$\exp(\kappa(p)T_i^{(k)})|\xi_i^{(k)}|^p \left(\sum_{j=1}^{\infty} \tilde{\xi}_j^p - 1\right)$$

where $\tilde{\xi}$ is independent of $T_i^{(k)}$ and $\xi_i^{(k)}$ and has the law $\nu(\cdot)/\nu(\mathcal{S}^{\downarrow})$. The conditional expectation of this quantity raised to the power $q$, given the splitting time $T_i^{(k)}$ and $\xi_i^{(k)}$ is

$$c \exp(q\kappa(p)T_i^{(k)})|\xi_i^{(k)}|^{pq},$$

where

$$c := \int_{\mathcal{S}^{\downarrow}} \left|1 - \sum_{i=1}^{\infty} s_i^p\right|^q \nu(d\mathbf{s})/\nu(\mathcal{S}^{\downarrow}).$$

We point out that $c < \infty$; indeed (1.36) and Jensen's inequality yield

$$\begin{aligned}\left|\sum_{i=1}^{\infty} s_i^p\right|^q &= b^q \left|\sum_{i=1}^{\infty} \frac{s_i^a}{b} s_i^{p-a}\right|^q \\ &\leq b^{q-1} \sum_{i=1}^{\infty} s_i^{q(p-a)+a},\end{aligned}$$

which yields the claim since $q(p-a)+a > \underline{p}$ provided that $q$ is chosen sufficiently close to 1.

On the one hand, we know that $T_i^{(k)}$ is the sum of $k+1$ independent exponential variables with parameter $\nu(\mathcal{S}^{\downarrow})$; in other words it has the gamma distribution with parameters $k+1$ and $\nu(\mathcal{S}^{\downarrow})$. In particular,

$$\mathbb{E}\left(\exp(q\kappa(p)T_i^{(k)})\right) = \left(\frac{\nu(\mathcal{S}^{\downarrow})}{\nu(\mathcal{S}^{\downarrow}) - q\kappa(p)}\right)^{k+1}.$$

On the other hand, we have already seen in (1.23) that

$$\mathbb{E}\left(\sum_{i=1}^{\infty} |\xi_i^{(k)}|^{pq}\right) = \left(\frac{\nu(\mathcal{S}^{\downarrow}) - \kappa(pq)}{\nu(\mathcal{S}^{\downarrow})}\right)^k.$$

Because $p < \bar{p}$, thanks to Lemma 1.6 we may choose $q > 1$ small enough so that $q\kappa(p) < \kappa(pq)$, and then the series

$$\sum_{k=0}^{\infty} \mathbb{E}\left(\sum_{i=1}^{\infty} \exp(q\kappa(p)T_i^{(k)})|\xi_i^{(k)}|^{pq}\right)$$

converges, which completes the proof of (1.37).

Finally, checking that the terminal value $M(p,\infty)$ is strictly positive conditionally on non-extinction follows the same argument as in the proof of Theorem 1.1. □

### 1.5.2 Some applications

In this section, we develop some applications of the preceding theorem to the asymptotic behavior of homogeneous fragmentation. First, we consider the largest fragment.

**Corollary 1.4** *The assumptions are the same as in Theorem 1.4. Then, conditionally on non-extinction, it holds with probability one that*

$$\lim_{t\to\infty} \frac{1}{t} \ln X_1(t) = -\kappa'(\bar{p}) = -\frac{\kappa(\bar{p})}{\bar{p}}.$$

In other words, the largest fragment $X_1(t)$ decays exponentially fast as $t \to \infty$, with rate $\kappa'(\bar{p})$. Observe that this is smaller than $\kappa'(p^*)$ (because $p^* < \bar{p}$ and $\kappa'$ is decreasing), which is the exponential rate of decay of a typical fragment; see Theorem 1.2(i).

*Proof* For every $p > \underline{p}$, we have

$$\exp(t\kappa(p))X_1^p(t) \le \exp(t\kappa(p)) \sum_{i=1}^{\infty} X_i^p(t)$$

and the right-hand side remains bounded as $t$ tends to infinity. Hence

$$\limsup_{t\to\infty} \frac{1}{t} \ln X_1(t) \le -\frac{\kappa(p)}{p},$$

and optimizing over $p$ yields

$$\limsup_{t\to\infty} \frac{1}{t} \ln X_1(t) \le -\frac{\kappa(\bar{p})}{\bar{p}}.$$

On the other hand, for every $p \in ]\underline{p}, \bar{p}[$ and $\varepsilon > 0$ sufficiently small, we have the lower bound

$$\exp(t\kappa(p)) \sum_{i=1}^{\infty} X_i^p(t) \le X_1^{\varepsilon}(t) \exp(t\kappa(p)) \sum_{i=1}^{\infty} X_i^{p-\varepsilon}(t).$$

We know that both limits

$$\lim_{t\to\infty} \exp(t\kappa(p)) \sum_{i=1}^{\infty} X_i^p(t) \quad \text{and} \quad \lim_{t\to\infty} \exp(t\kappa(p-\varepsilon)) \sum_{i=1}^{\infty} X_i^{p-\varepsilon}(t)$$

are finite and strictly positive a.s. conditionally on non-extinction, and we deduce that

$$\liminf_{t\to\infty} \frac{1}{t} \ln X_1(t) \ge -\frac{\kappa(p) - \kappa(p-\varepsilon)}{\varepsilon}.$$

We take the limit of the right-hand side as $\varepsilon \to 0+$ and then as $p$ tends to $\bar{p}$ to conclude that

$$\liminf_{t\to\infty} \frac{1}{t} \ln X_1(t) \ge -\kappa'(\bar{p}).$$

Now, this quantity coincides with $-\kappa(\bar{p})/\bar{p}$, as we know from Lemma 1.6. □

We point out that the argument of the proof above also shows that the martingale $M(p,t)$ converges to 0 a.s. (and a fortiori is not uniformly integrable) for $p > \bar{p}$. In fact, it can even be shown that the same remains true for $p = \bar{p}$.

Finally we conclude this section by an application to the asymptotic behavior of homogeneous fragmentations which is easily deduced from the martingales that we considered and classical large deviations techniques. Again, we shall focus for the sake of simplicity on the case when the dislocation measure $\nu$ is conservative. Further, it will be convenient here to represent the random sequence $X(t) = (X_1(t), \dots)$ by the empirical distribution,

$$\rho_t(dy) := \sum_{i=1}^{\infty} \delta_{\frac{1}{t}\ln X_i(t)}(dy). \tag{1.38}$$

Define the convex decreasing function $\Lambda$ on $]\underline{p}, \infty[$ by

$$\Lambda(p) = \begin{cases} -\kappa(p) & \text{if } \underline{p} < p < \bar{p}, \\ -p\kappa'(\bar{p}) & \text{if } p \geq \bar{p}. \end{cases}$$

**Corollary 1.5** *The assumptions are the same as in Theorem 1.4. It holds a.s. conditionally on non-extinction that*

$$\lim_{t\to\infty} \frac{1}{t} \ln \int_{\mathbb{R}} e^{tpy} \rho_t(dy) = \Lambda(p)$$

*for every $p > \underline{p}$.*

*Proof* Let us first prove the statement for a fixed $p > \underline{p}$. Observe that

$$\int_{\mathbb{R}} e^{tpy} \rho_t(dy) = \sum_{i=1}^{\infty} X_i^p(t).$$

The case $\underline{p} < p < \bar{p}$ follows from Theorem 1.4, so suppose that $\bar{p} \leq p$. We use the bounds

$$X_1^p(t) \leq \sum_{i=1}^{\infty} X_i^p(t) \leq X_1^{p-\bar{p}}(t) \sum_{i=1}^{\infty} X_i^{\bar{p}}(t).$$

Recall first from Corollary 1.4 that $\ln X_1(t) \sim -t\kappa'(\bar{p})$ as $t \to \infty$, and then from Lemma 1.6 and Corollary 1.3 that $e^{t\bar{p}\kappa'(\bar{p})} \sum_{i=1}^{\infty} X_i^{\bar{p}}(t)$ is a martingale which converges a.s. It follows immediately that

$$\lim_{t\to\infty} \frac{1}{t} \ln \sum_{i=1}^{\infty} X_i^p(t) = \lim_{t\to\infty} \frac{1}{t} \ln \int_{\mathbb{R}} e^{tpy} \rho_t(dy) = -p\kappa'(\bar{p}) = \Lambda(p) \qquad \text{a.s.}$$

The limit above holds a.s. simultaneously for every rational number $p > \underline{p}$, and by an immediate monotonicity argument, the proof is complete. □

Pathwise large deviation estimates for the family of random measures $(\rho_t, t \geq 0)$ follow from Corollary 1.5. Introducing the Fenchel-Legendre transform of $\Lambda$,

$$\Lambda^*(a) = \sup_{p > \underline{p}} (ap - \Lambda(p)) \,,$$

the classical duality for the Fenchel-Legendre transform (see for example [75]) yields the identity

$$\Lambda^*(a) = \kappa(p) - p\kappa'(p)\,, \qquad \text{for every } p \in ]\underline{p}, \bar{p}[ \text{ and } a = -\kappa'(p).$$

Note also that $\Lambda^*(a) = \infty$ for every $a > -\kappa'(\bar{p})$ and that $\Lambda^*$ is left-continuous at $-\kappa'(\bar{p})$.

**Corollary 1.6** *The assumptions are the same as in Theorem 1.4. The following holds a.s. conditionally on non-extinction:*

(i) *For any closed set $F \subseteq ]\underline{p}, \infty[$,*

$$\limsup_{t \to \infty} \frac{1}{t} \ln \rho_t(F) \leq -\inf \{\Lambda^*(a), a \in F\} \,.$$

(ii) *For any open set $G \subseteq \mathbb{R}$,*

$$\liminf_{t \to \infty} \frac{1}{t} \ln \rho_t(G) \geq -\inf \left\{\Lambda^*(a), a > -\kappa'(\underline{p}+) \text{ and } a \in G\right\} .$$

(iii) *If moreover $\kappa'(\underline{p}+) = \infty$, then $(\rho_t, t \geq 0)$ satisfy the Large Deviation Principle with the good convex rate function $\Lambda^*$ (see for instance [75] for the terminology).*

*Proof* We aim to apply the Gärtner-Ellis theorem (see Section 2.3 in Dembo and Zeitouni [75]). The fundamental condition on the behavior of the Laplace transform of $\rho_t$ (see Assumption 2.3.2 in [75]), is the conclusion of Corollary 1.5. Note that the assumption $\underline{p} < 1$ ensures that 0 belongs to the interior of the domain of $\Lambda$. According to Lemma 2.3.9 in [75], every $x \in ]-\kappa'(\underline{p}+), -\kappa'(\bar{p})[$ is a so-called exposed point of the Fenchel-Legendre transform $\Lambda^*$, and since $\Lambda^*(a) = \infty$ for every $a > -\kappa'(\bar{p})$, statements (i) and (ii) of Corollary 1.6

merely rephrase Theorem 2.3.6 in [75]. The last statement follows from the first two and Lemma 2.3.9 in [75]. □

One can interpret Corollary 1.6 as a multi-scale limit theorem for numbers of fragments. Specifically, it is immediately seen that when the hypothesis of Corollary 1.6 is fulfilled, then we have for every $a > -\kappa'(\underline{p}+)$ that

$$\lim_{\varepsilon\to 0+}\lim_{t\to\infty}\frac{1}{t}\ln\#\left\{i\in\mathbb{N}:\mathrm{e}^{(a-\varepsilon)t}\le X_i(t)\le\mathrm{e}^{(a+\varepsilon)t}\right\} = -\Lambda^*(a).$$

Observe that this quantity equals $-\infty$ for $a > -\kappa'(\bar{p})$, in agreement with Corollary 1.3. On the other hand, taking $a = -\kappa'(p)$ for some $p \in ]\underline{p}, \bar{p}[$, we see that, roughly speaking, the number of fragments of size approximately $\mathrm{e}^{-\kappa'(p)t}$ at time $t$ is approximately $\exp(t(p\kappa'(p) - \kappa(p)))$ when $t$ is large. We refer to Berestycki [25] for further developments in this vein, related to the so-called multifractal spectrum of homogeneous fragmentations (see also Krell [145] for a related work).

## 1.6 Comments

It seems that theoretical works on random fragmentation chains have been motivated initially by the study of the crushing of blocks of mineral in the mining industry.[3] In particular, the first significant probabilistic contribution in this field was due to Kolmogorov [141] himself in 1941, who provided an explanation for the statistical observation that the logarithms of the sizes of mineral grains are often normally distributed. In this direction, Kolmogorov introduced a version of homogeneous fragmentation chains in discrete time as follows. At the initial time consider a mass, say $m > 0$. At time 1, this mass is broken randomly, which produces smaller masses, say $m\xi_1 \ge m\xi_2, \ge \cdots \ge 0$ where $\xi = (\xi_1, \dots)$ has a fixed distribution with $\sum_{i=1}^{\infty}\xi_i = 1$ a.s. (in other words, the dislocation is conservative). The next steps consist of independent iterations, that is each mass that results from the previous step is broken independently of each other and according to the same law. Kolmogorov established in this framework the discrete time analog of Theorem 1.2 (note that then the intrinsic martingale $\mathcal{M}$ is trivial since the dislocations are conservative). Nowadays, such a result should be viewed as a special case of the central limit theorem for branching random walks, see for example [14] and [54].

[3] In this direction it may be interesting to mention that a significant proportion of the energy consumption in the world is used for particle size reduction in the mining industry.

Further important developments in this vein were made in 1961 by one of Kolmogorov's students, Filippov [103], who considered self-similar fragmentation chains as described in the present chapter. In particular, Filippov proved the version of Theorem 1.3 in the conservative case. This result was then re-discovered independently by Brennan and Durrett [63] for the binary and conservative dislocations; more precisely these authors obtained a strong limit theorem in this setting, with almost-sure convergence instead of convergence in probability. Some special cases of these mathematical results have also appeared in the literature in physics, see for example [23, 144] and references therein. Baryshnikov and Gnedin [19] studied instances of dissipative self-similar fragmentation chains and obtained the convergence of the mean measures associated with the empirical distribution of fragments. The present statement of Theorem 1.3 for general non-conservative dislocation measures was established quite recently in [41].

As it has been briefly mentioned in the Introduction, fragmentation phenomena occur frequently in physics and have thus motivated a variety of works in this field (see for example [6, 23, 52, 143, 144] and the references therein). Self-similar fragmentation chains, as introduced in this chapter, also arise in different areas such as analysis of algorithms (for example quick-search [71], recursive trees [78], ...), degradation of polymer chains (see [22], [63], ...), packing problems in communication networks [19], ... In these applications, the interests concern both deterministic and statistical aspects of fragmentation. The former are often considered via so-called *fragmentation equations* that are meant to describe the evolution of the density of particles in media where particles break at certain rates; a simple example of such a fragmentation equation is given in Corollary 1.1. More general systems where the self-similarity assumption is dropped and further physical phenomena such as coagulation of particles or spatial motions can be incorporated, have been studied intensively in the literature, see for instance [96, 97, 150] and references therein. We stress that stochastic models of fragmentations provide an efficient tool for establishing the existence and studying properties of pure fragmentation equations; see in particular [105] and [119].

The genealogical structure of self-similar fragmentation as described in Proposition 1.3 has arisen first in the context of so-called multiplicative cascades and random fractal constructions; see the pioneering works of Mandelbrot [160], Kahane and Peyrière [132], Mauldin and Williams [164], and also Barral [18] and Liu [153] for further references. Of course, it plays also an important part in branching processes, see for example Athreya and Ney [15], Jagers [128], and Haccou, Jagers and Vatutin [123]. The so-called

Malthusian hypotheses,[4] the intrinsic martingale, and techniques based on randomly tagged branches, are classical cornerstones of these theories.

The phenomenon of formation of dust in self-similar fragmentation chains with negative indices of self-similarity was first observed by Filippov [103]. In the setting of (deterministic) fragmentation equations (cf. Corollary 1.1) with conservative dislocation measures, it corresponds to a phenomenon of loss of mass or shattering; see [11] and [17]. The sharper results about a.s. extinction such as stated in Proposition 1.8 appear in [38]; see also [105], [119] and [130]. We refer to the recent work of Haas [120] for several deep results about the regularity of the formation of dust in self-similar fragmentations with negative indices and conservative dislocation measures, and to Wagner [211] for the study of this phenomenon for more general processes.

Lemma 1.5, which provides a key technical step to the limit theorems in Section 1.4, is due to Nerman [171], who used it to develop a renewal theory for supercritical general branching processes. The results of Nerman play a crucial role in the asymptotic analysis of the fragmentation energy (a problem motivated by the mining industry) in [47], from which Proposition 1.12 is an excerpt.

Additive martingales are one of the most powerful tools for the study of branching random walks, both in discrete time (see the well-known article by Biggins [53]) and in continuous time (cf. in particular [55, 146, 208]). The approach to convergence which is used in Section 1.5.1 relies on a standard application of stochastic calculus for purely discontinuous martingales in continuous time. Alternatively, one can also establish Theorem 1.4 by adapting the conceptual proof of Lyons *et al.* [157] (see also Lyons [156]) which is based on a clever argument of change of probability measures. We further point out that Biggins [55] has obtained a stronger limit theorem for additive martingales associated to branching random walks, in which convergence holds uniformly in the parameter $p$, and this reinforcement yields precise large deviation estimates for the empirical measure of a branching random walk. Such results can be shifted to homogeneous fragmentation chains by time-discretization techniques, see [51]. In a somewhat different direction, one can obtain precise large deviation estimates for the probability of presence of abnormally large fragments as time goes to infinity. See [50, 51], which extend earlier results in this vein by Rouault [195] for branching random walks in the sub-critical region.

[4] We stress that hypotheses that we make in Section 1.2.2 are slightly stronger than the usual $L \log L$ conditions, but they are easier to handle for the applications we have in mind.

Finally, fragmentation chains are clearly Markov processes of transitive type, in the sense that they never return to a state they visited before. Nonetheless, combining fragmentation either with random coagulation, or with immigration of particles, may produce recurrent processes. We refer to the book by Whittle [213], and to the recent works of Ben-Naim and Krapivsky [23], Berestycki [26], Diaconis *et al.* [77], Durrett *et al.* [88], Erlihson and Granovsky [95] and Haas [121], for some studies of such models.

# 2
# Random partitions

Fragmentation chains which have been discussed in the preceding chapter, only form a special and rather simple sub-class of fragmentation processes which enjoy the self-similar and branching properties. The construction and the study of the general family are harder, since we can no longer rely on a discrete genealogy. We shall now prepare material to circumvent this fundamental difficulty, at least in the conservative or dissipative case. In this direction, we shall first introduce several notions of partitions (for masses, for intervals, and for the set of natural integers) and develop their connections. Mass-partitions induced by Poisson random measures, and in particular the so-called Poisson-Dirichlet partitions, will receive special attention. Finally, Kingman's theory for exchangeable random partitions of $\mathbb{N}$ will be presented in the ultimate section.

## 2.1 Mass-partitions

In this section, we introduce some elementary material on the simple notion of partition of a unit mass.

### 2.1.1 Partitions of a unit mass

A partition of some set $E$ is a collection of disjoint subsets whose union is $E$. When one focusses on finite sets and their cardinals, this yields the notion of *partition of integers* in combinatorics: A partition of $n \in \mathbb{N} := \{1, 2, \dots\}$ refers to a finite family of positive integers, say $\{p_1, \dots, p_k\}$, with $p_1 + \cdots + p_k = n$. We stress that this family is unordered; for instance $\{2, 3\}$ and $\{3, 2\}$ represent the same partition of 5. In practice, it is convenient to enumerate the elements of a partition, and since one does not want to induce any special order on these elements, one makes the convention to rank them systematically in decreasing order.

Plainly, we can rewrite a partition $\{p_1, \dots, p_k\}$ of $n$ in the form $1 = s_1 + \cdots + s_k$ with $s_j = p_j/n$ for $j = 1, \dots, k$. When $n$ tends to infinity, this yields the following notion:

**Definition 2.1** *A* **mass-partition** *is an infinite numerical sequence*

$$\mathbf{s} = (s_1, s_2, \dots)$$

*such that*

$$s_1 \geq s_2 \geq \cdots \geq 0 \quad \textit{and} \quad \sum_{i=1}^{\infty} s_i \leq 1.$$

*The space of mass-partitions is denoted by $\mathcal{P}_{\mathrm{m}}$.*

In other words, mass-partitions form the subset in $\mathcal{S}^{\downarrow}$ of conservative or dissipative configurations. We may think of a generic element $\mathbf{s} \in \mathcal{P}_{\mathrm{m}}$ as the sequence, ranked in decreasing order, of the masses of clusters in a universe with unit total mass. We stress that the total mass of the clusters, $\sum_{i=1}^{\infty} s_i$, can be strictly less than 1. In this direction, it is convenient to define

$$s_0 := 1 - \sum_{i=1}^{\infty} s_i,$$

a quantity which can be thought of as the total mass of *dust* (i.e. infinitesimal particles) in the universe. A mass-partition $\mathbf{s}$ is called *proper* if there is no dust,[1] that is if $s_0 = 0$, and *improper* otherwise.

It is easy to check that the space of mass-partitions enjoys nice topological properties.

**Proposition 2.1** *The space $\mathcal{P}_{\mathrm{m}}$, endowed with the uniform distance*

$$d(\mathbf{s}, \mathbf{s}') = \max\{|s_i - s'_i|, i \in \mathbb{N}\}, \qquad \mathbf{s}, \mathbf{s}' \in \mathcal{P}_{\mathrm{m}},$$

*is compact, and the induced topology coincides with that of pointwise convergence.*

*Proof* Because terms of a mass-partition are ranked in decreasing order and their sum does not exceed 1, we have

$$|s_i - s'_i| \leq 1/i \qquad \text{for all } i \in \mathbb{N} \text{ and } \mathbf{s}, \mathbf{s}' \in \mathcal{P}_{\mathrm{m}}.$$

[1] The terminology is perhaps better understood in French: *une partition de masse est dite propre si elle n'a pas de poussière.*

This implies that pointwise convergence in $\mathcal{P}_m$ is equivalent to uniform convergence.

Consider now some sequence $(\mathbf{s}^{(n)}, n \in \mathbb{N})$ in $\mathcal{P}_m$. By the diagonal procedure, we may extract some subsequence, which we still denote by $(\mathbf{s}^{(n)}, n \in \mathbb{N})$ for convenience, such that $\lim_{n\to\infty} s_i^{(n)} := s_i^{(\infty)}$ exists for each $i \in \mathbb{N}$. Plainly, one has $s_1^{(\infty)} \geq s_2^{(\infty)} \geq \cdots \geq 0$ and, by Fatou's lemma, $\sum_{i=1}^{\infty} s_i^{(\infty)} \leq 1$. By the theorem of Bolzano-Weierstrass, this establishes the compactness. □

The fact that $\mathcal{P}_m$ is a compact metric space will be especially useful later in the text when we deal with the construction of $\mathcal{P}_m$-valued random variables or processes. We now turn our attention to some alternative representations of mass-partitions, which we shall also use for different purposes.

### 2.1.2 Interval-partitions

It is well-known that every open subset of the unit interval $I = ]0, 1[$ can be decomposed into a unique (at most) countable collection of disjoint open intervals. In our framework, we may thus view such open sets as interval-partitions; points in the complementary closed set can be thought of as dust, that is isolated infinitesimal particles.

**Definition 2.2** *The collection of the interval-components of an arbitrary open set* $\vartheta \subseteq ]0, 1[$ *is called an* **interval-partition**. *By a slight abuse of notation, we shall often identify this interval-partition with the open set* $\vartheta$. *The space of interval-partitions is denoted by* $\mathcal{P}_I$.

*The lengths of the interval components of an interval-partition* $\vartheta$ *are called* **spacings**. *We denote by* $|\vartheta|^{\downarrow}$ *the sequence of spacings, ranked in decreasing order and completed by an infinite sequence of* 0 *when* $\vartheta$ *only has finitely many interval components, so that* $|\vartheta|^{\downarrow}$ *is a mass-partition.*

Given an arbitrary mass-partition $\mathbf{s} \in \mathcal{P}_m$, it is not difficult to construct an interval-partition $\vartheta \in \mathcal{P}_I$ with $|\vartheta|^{\downarrow} = \mathbf{s}$; such $\vartheta$ will then be called an *interval-representation* of $\mathbf{s}$. Clearly, there are in general many different interval-representations of the same mass-partition. Observe also that a mass-partition is proper if and only if some (and then all) of its interval-representations has full Lebesgue measure.

There is a natural distance on $\mathcal{P}_I$ that we now introduce. Each interval-partition $\vartheta$ is determined by the function

$$\chi_\vartheta(x) = \max\{|y - x|, y \in \vartheta^c\}, \qquad x \in [0, 1],$$

where $\vartheta^c = [0, 1]\backslash\vartheta$ stands for the complementary set of $\vartheta$ in $[0, 1]$ (so $\vartheta^c$ is a closed set which always contains the boundary points 0 and 1).

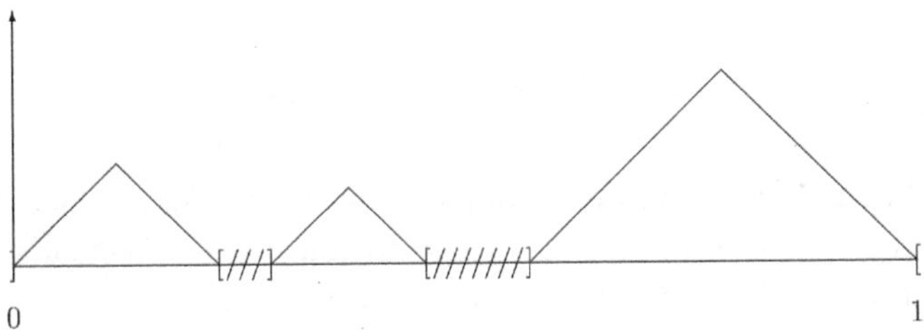


*Interval-partition $\vartheta$ with 3 interval components and graph of $\chi_\vartheta$*

We then define the distance

$$d(\vartheta, \vartheta') = \max\{|\chi_\vartheta(x) - \chi_{\vartheta'}(x)|, x \in [0, 1]\}.$$

We mention that $d(\vartheta, \vartheta')$ can also be viewed as the Hausdorff distance between the closed sets $\vartheta^c$ and $\vartheta'^c$; see for example [163].

**Proposition 2.2** *The space $(\mathcal{P}_{\mathrm{I}}, d)$ is compact, and the map $\vartheta \to |\vartheta|^\downarrow$ is continuous from $\mathcal{P}_{\mathrm{I}}$ to $\mathcal{P}_{\mathrm{m}}$.*

*Proof* We have to check sequential compactness, that is for every sequence $(\vartheta_n, n \in \mathbb{N})$ in $\mathcal{P}_{\mathrm{I}}$, we can extract a subsequence which converges in $\mathcal{P}_{\mathrm{I}}$. For each $n \in \mathbb{N}$, let $]a_{1,n}, b_{1,n}[, ]a_{2,n}, b_{2,n}[, \ldots$ denote the sequence of the interval components of $\vartheta_n$ ordered by decreasing lengths (if $\vartheta_n$ has only $k$ interval components, then we agree that $a_{j,n} = b_{j,n} = 0$ for $j > k$, and if two or more interval components have the same positive length, then we order them from the left to the right). By a diagonal extraction procedure, we suppose that for each $i$, the sequences $(a_{i,n}, n \in \mathbb{N})$ and $(b_{i,n}, n \in \mathbb{N})$ converge, say to $a_i$ and $b_i$, where $0 \le a_i \le b_i \le 1$. Plainly the intervals $]a_i, b_i[$, $i \in \mathbb{N}$, must be pairwise disjoint, so we may consider the interval-partition

$$\vartheta := \bigcup_{i \in \mathbb{N}} ]a_i, b_i[ .$$

Now, by the triangle inequality, we have that for every integers $n, k$

$$d(\vartheta_n, \vartheta) \le \max_{i=1,\ldots,k} (|a_{i,n} - a_i| + |b_{i,n} - b_i|) + \max_{j>k}(|b_{j,n} - a_{j,n}| + |b_j - a_j|).$$

As the $j$-th largest spacing of an interval-partition is at most $1/j$, we see that the second maximum in the right-hand side is bounded by $2/k$. Furthermore, for each fixed $k$, the first maximum converges to 0 as $n \to \infty$, so $\lim_{n\to\infty} d(\vartheta_n, \vartheta) = 0$, and the compactness is proven.

Next, observe that for every interval-partition $\vartheta \in \mathcal{P}_{\mathrm{I}}$ and every integer $k$, the sum of the lengths of the $k$ largest interval components of $\vartheta$ can be expressed as

$$\sum_{i=1}^{k} |\vartheta|_i^{\downarrow} = \max \sum_{j=1}^{2k} |\chi_{\vartheta}(x_j) - \chi_{\vartheta}(x_{j-1})| \,,$$

where in the right-hand side, the maximum is taken over the set of subdivisions $0 = x_0 < x_1 < \cdots < x_{2k} = 1$ of $[0, 1]$. We deduce that if $(\vartheta_n, n \in \mathbb{N})$ is a sequence of interval-partitions which converges to $\vartheta \in \mathcal{P}_{\mathrm{I}}$, then

$$\lim_{n\to\infty} \sum_{i=1}^{k} |\vartheta_n|_i^{\downarrow} = \sum_{i=1}^{k} |\vartheta|_i^{\downarrow} \,,$$

so $|\vartheta_n|^{\downarrow}$ converges pointwise to $|\vartheta|^{\downarrow}$. □

The Poissonian rain (cf. Section 1.1.3) provides an example of a representation of a conservative fragmentation chain as a fragmentation of the unit interval. More generally, interval-partitions are well-suited for describing fragmentation chains with values in the space of mass-partitions. Typically, consider a self-similar fragmentation chain $X = (X(t), t \geq 0)$ with index $\alpha \in \mathbb{R}$, and suppose that its dislocation measure $\nu$ is conservative or dissipative, that is has support in the space $\mathcal{P}_{\mathrm{m}}$ of mass-partitions. As usual, we also assume that the chain starts from the configuration $(1, 0, \ldots)$. We may think of $\nu$ as the image of some measure $\nu_{\mathrm{I}}$ on the space $\mathcal{P}_{\mathrm{I}}$ of interval-partitions by the function $\vartheta \to |\vartheta|^{\downarrow}$ that maps an interval-partition to the ranked sequence of its spacing (see Proposition 2.2).

Then it is straightforward to construct a *self-similar interval-fragmentation* $\Theta = (\Theta(t), t \geq 0)$ with dislocation measure $\nu_{\mathrm{I}}$ and index $\alpha$. This means that $\Theta$ is a Markovian family of nested interval-partitions (i.e. $\Theta(t) \subseteq \Theta(s)$ for $s \leq t$), started from $\Theta(0) = ]0, 1[$, and that its dynamics can be described as follows. The intervals components of $\Theta(\cdot)$ are viewed as particles which evolve independently of each other. A particle $]a, b[\neq \emptyset$ lives for an exponential lifetime with parameter $(b-a)^{\alpha}\nu_{\mathrm{I}}(\mathcal{P}_{\mathrm{I}}) = (b-a)^{\alpha}\nu(\mathcal{P}_{\mathrm{m}})$, and at its death it is replaced by a random interval-partition $\vartheta$ whose image by the normalizing function $x \to (x-a)/(b-a)$ which maps $]a, b[$ to $]0, 1[$, has the distribution $\nu_{\mathrm{I}}(\cdot)/\nu_{\mathrm{I}}(\mathcal{P}_{\mathrm{I}})$. The existence and uniqueness (in law) of such a Markov chain can be checked by the same argument as in Section 1.1.3. Furthermore, it is immediately seen that the process $|\Theta(\cdot)|^{\downarrow}$ of the ranked lengths of the interval components of $\Theta(\cdot)$ is a version of $X$, that is a self-similar fragmentation chain with index $\alpha$ and dislocation measure $\nu$.

### 2.1.3 Size-biased sampling and reordering

The order relation $\leq$ on the unit interval induces a natural order (from the left to the right) for the components of an interval-partition, and thus we can think of an interval representation $\vartheta \in \mathcal{P}_{\mathrm{I}}$ of a mass-partition $\mathbf{s} \in \mathcal{P}_{\mathrm{m}}$ as reordering the terms of $\mathbf{s}$. We shall now present a useful procedure for reordering randomly the terms of mass-partitions.

A *proper* mass-partition $\mathbf{s} \in \mathcal{P}_{\mathrm{m}}$ can be viewed as a discrete probability measure; more precisely we may associate to $\mathbf{s}$ an integer-valued random variable $1^*$ with distribution

$$\mathbb{P}(1^* = i) = s_i\,, \qquad i \in \mathbb{N}\,.$$

One then calls any random variable $s_1^*$ distributed as $s_{1^*}$ a *size-biased sample* from $\mathbf{s}$. The law of a size-biased sample $s_1^*$ is referred to as the *structural distribution* of $\mathbf{s}$. In other words, the structural distribution of $\mathbf{s}$ is given by $\sum_{i=1}^{\infty} s_i \delta_{s_i}$, where $\delta_s$ stands for the Dirac point mass at $s$, and plainly one can recover $\mathbf{s}$ from its structural distribution. When $S$ is a *random* mass-partition which is proper a.s., we call a variable whose conditional distribution given $S = \mathbf{s}$ is that of a size-biased sample of $\mathbf{s}$ a size-biased sample of $S$. We stress that in general, one cannot recover the law of a random mass-partition from that of its size-biased sample.

It will also be convenient to define size-biased sampling for possibly *improper* mass-partitions. In this more general situation, we consider a random variable $1^*$ with values in $\overline{\mathbb{N}} := \mathbb{N} \cup \{\infty\}$ with law given by

$$\mathbb{P}(1^* = i) = s_i \quad \text{for} \quad i \in \mathbb{N}\,, \quad \text{and} \quad \mathbb{P}(1^* = \infty) = s_0 = 1 - \sum_{i=1}^{\infty} s_i\,,$$

and we agree that $s_\infty = 0$. Just as above, we call a random variable $s_1^*$ distributed as $s_{1^*}$ a size-biased sample of $\mathbf{s}$.

By sampling recursively with a size-bias and without replacement the terms of some proper mass-partition $\mathbf{s}$, we obtain a so-called size-biased reordering of $\mathbf{s}$. Before giving a formal definition, it is convenient to introduce the space $\mathcal{S}_{[0,1]}$ of numerical sequences $\mathbf{x} = (x_1, \ldots)$ with values in $[0, 1]$. This space is equipped with the distance

$$\delta(\mathbf{x}, \mathbf{x}') := \sum_{i=1}^{\infty} 2^{-i} |x_i - x_i'|\,, \qquad \mathbf{x} = (x_1, \ldots), \mathbf{x}' = (x_1', \ldots)\,.$$

This makes $\mathcal{S}_{[0,1]}$ compact, and the induced topology coincides with that of pointwise convergence.

For the sake of simplicity, we again focus on *proper* mass-partitions $\mathbf{s} = (s_1, \dots)$ and give a formal definition.[2]

**Definition 2.3** *Let* $\mathbf{s} \in \mathcal{P}_m$ *be a proper mass-partition. A* **size-biased permutation** *(based on* $\mathbf{s}$*) is a random map* $\sigma : \mathbb{N} \to \mathbb{N}$ *whose finite-dimensional distributions are given as follows:*

- *For every* $n \in \mathbb{N}$ *such that* $s_n > 0$ *and every n-tuple* $k_1, \dots, k_n$ *of distinct integers,*

$$\mathbb{P}(\sigma(1) = k_1, \dots, \sigma(n) = k_n) = \prod_{i=1}^{n} \frac{s_{k_i}}{1 - (s_{k_1} + \cdots + s_{k_{i-1}})},$$

  *where by convention, the term in the product above corresponding to* $i = 1$ *is equal to* $s_{k_1}$.
- *When* $\ell := \inf\{n \in \mathbb{N} : s_n = 0\} < \infty$, *we agree that for every* $n \geq \ell$ *and every n-tuple* $k_1, \dots, k_n$ *of distinct integers with* $k_i = i$ *for every* $i \geq \ell$,

$$\mathbb{P}(\sigma(1) = k_1, \dots, \sigma(n) = k_n) = \prod_{i=1}^{\ell} \frac{s_{k_i}}{1 - (s_{k_1} + \cdots + s_{k_{i-1}})}.$$

*The random sequence* $\mathbf{s}^* = (s_{\sigma(1)}, s_{\sigma(2)}, \dots)$ *is then called a* **size-biased reordering** *of* $\mathbf{s}$.

More generally, when $S$ is a random mass-partition which is proper a.s., we call a random variable $S^*$ with values in $\mathcal{S}_{[0,1]}$ whose conditional distribution given $S = \mathbf{s}$ is that of a size-biased reordering of $\mathbf{s}$ a size-biased reordering of $S$. Note in particular that $S_1^*$ is a size-biased sample from $S$ (so our notation is coherent), and moreover that with probability one, the unordered families $\{S_i, i \in \mathbb{N}\}$ and $\{S_i^*, i \in \mathbb{N}\}$ coincide. In particular, we may recover $S$ from $S^*$ by ranking the terms of the latter in decreasing order.

Here is a simple procedure for constructing size-biased reordering. Fix a proper mass-partition $\mathbf{s} \in \mathcal{P}_m$, and let $\vartheta_{\mathbf{s}}$ be some interval representation of $\mathbf{s}$. So $\vartheta_{\mathbf{s}} \subseteq ]0, 1[$ is an open set with Lebesgue measure 1. Let $U_1, \dots$ be a sequence of i.i.d. uniform variables on $[0, 1]$. Next define $I_1^*$ as the interval component of $\vartheta_{\mathbf{s}}$ which contains $U_1$, and then recursively $I_n^*$ for $n \geq 2$ as the interval component of $\vartheta_{\mathbf{s}}$ which contains $U_{n^*}$, where

$$n^* := \inf\left\{k \in \mathbb{N} : U_k \in \vartheta_{\mathbf{s}} \setminus \bigcup_{j=1}^{n-1} I_j^*\right\}.$$

[2] There would be at least two natural definitions for size-biased reordering for possibly improper mass-partitions, which only coincide in the proper case. Rather than choosing one arbitrarily, we focus here on the proper case.

In words, $I_n^*$ is the $n$-th interval component of $\vartheta_s$ which is visited by the sequence $U_1, \ldots$ As usual, we write $|I_n^*|$ for its length. When $\ell := \inf\{j \in \mathbb{N} : s_j > 0\} < \infty$, $\vartheta_s$ has exactly $\ell - 1$ non-empty interval-components, and we agree that $I_n^* = \emptyset$ and $|I_n^*| = 0$ for $n \geq \ell$.

**Lemma 2.1** *The sequence* $(|I_1^*|, \ldots)$ *is a size-biased reordering of* $\mathbf{s}$.

*Proof* As the variable $U_1$ is uniformly distributed on $\vartheta_s$, the length $|I_{1*}|$ of the interval component of $\vartheta_s$ that contains $U_1$ is a size-biased sample from $\mathbf{s}$. An easy induction completes the proof. □

We stress that the construction above also applies for random proper mass-partitions $S$, provided of course that the sequence $U_1, \ldots$ of i.i.d. uniform variables is chosen independently of the interval-representation $\vartheta_S$ of $S$.

We now conclude this section with a useful observation (see Donnelly and Joyce [81] and Gnedin [112]), which is perhaps less obvious than it may look at first sight.

**Proposition 2.3** *Consider for each* $n \in \overline{\mathbb{N}} := \mathbb{N} \cup \{\infty\}$, *a random mass-partition* $S^{(n)}$ *that is proper a.s., and a size-biased reordering* $S^{(n)*}$ *of* $S^{(n)}$. *Then the following two conditions are equivalent:*

(i) *When* $n \to \infty$, $S^{(n)}$ *converges in distribution to* $S^{(\infty)}$ *in* $\mathcal{P}_{\mathrm{m}}$.
(ii) *When* $n \to \infty$, $S^{(n)*}$ *converges in the sense of finite-dimensional distributions to* $S^{(\infty)*}$.

*Proof* (i) ⇒ (ii) Since $\mathcal{P}_{\mathrm{m}}$ is a compact metric space, we may apply Skorokhod representation theorem (see for example Billingsley [56] on its page 287) and assume that $\lim_{n\to\infty} S^{(n)} = S^{(\infty)}$ in $\mathcal{P}_{\mathrm{m}}$ a.s. In particular, $\lim_{n\to\infty} S_k^{(n)} = S_k^{(\infty)}$ a.s. for each $k \in \mathbb{N}$, and the hypothesis that the mass-partitions are proper enables us to apply Scheffé's lemma (see for example [57]). This yields that with probability one,

$$\lim_{n\to\infty} S^{(n)} = S^{(\infty)} \qquad \text{in } \ell^1(\mathbb{N}). \tag{2.1}$$

It follows from (2.1) that a.s., the conditional law of a size-biased sample of $S^{(n)}$ given $S^{(n)}$,

$$\sum_{i=1}^{\infty} S_i^{(n)} \delta_{S_i^{(n)}},$$

converges weakly on the space of probability measures on $[0, 1]$ as $n \to \infty$ to the conditional law of a size-biased sample of $S^{(\infty)}$ given $S^{(\infty)}$,

$$\sum_{i=1}^{\infty} S_i^{(\infty)} \delta_{S_i^{(\infty)}} .$$

This implies that $S_1^{(n)*}$ converges in distributions to $S_1^{(\infty)*}$. An easy iteration of this argument shows that (ii) holds.

(ii) $\Rightarrow$ (i) Recall that size-biased reordered mass-partitions are viewed as random variables with values in the space $\mathcal{S}_{[0,1]}$ of numerical sequences in $[0, 1]$, and that the latter is metric and compact. Furthermore, in this framework, convergence in the sense of finite-dimensional distributions of random variables in $\mathcal{S}_{[0,1]}$ corresponds to weak convergence of probability measures on $\mathcal{S}_{[0,1]}$. By the same argument as above involving Skorokhod's representation theorem, we may and will suppose that $\lim_{n\to\infty} S_k^{(n)*} = S_k^{(\infty)*}$ a.s. for each $k \in \mathbb{N}$.

Again the assumption that the mass-partitions are proper a.s. enables us to invoke Scheffé's lemma, and thus

$$\lim_{n\to\infty} S^{(n)*} = S^{(\infty)*} \qquad \text{in } \ell^1(\mathbb{N}), \text{ a.s.} \tag{2.2}$$

In particular

$$\lim_{n\to\infty} \max_{i\in\mathbb{N}} S_i^{(n)*} = \max_{i\in\mathbb{N}} S_i^{(\infty)*} \qquad \text{a.s.},$$

that is $S_1^{(n)}$ converges to $S_1^{(\infty)}$ as $n \to \infty$, a.s. More generally, we deduce from (2.2) by iteration that $\lim_{n\to\infty} S_k^{(n)} = S_k^{(\infty)}$ a.s. for each $k \in \mathbb{N}$. We conclude from Proposition 2.1 that $S^{(n)}$ converges to $S^{(\infty)}$ in $\mathcal{P}_{\mathrm{m}}$ as $n \to \infty$, with probability 1. $\square$

## 2.2 Random mass-partitions and Poisson measures

In this section, we shall introduce a class of random mass-partitions which are naturally induced by certain random point measures on $]0, \infty[$, and study in details a few important examples. In this direction, we start by presenting some elementary material on a well-known family of probability measures on the $(N-1)$-dimensional simplex related to beta and gamma variables.

### 2.2.1 Multidimensional Dirichlet distributions

To start with, recall that for every $\theta, c > 0$, the gamma law with parameter $(\theta, c)$ is the probability measure on $\mathbb{R}_+$ which has density

$$\frac{c^\theta}{\Gamma(\theta)} x^{\theta-1} e^{-cx}, \qquad x > 0.$$

The parameter $c$ has a very minor role in this section, due to the fact that if a variable $\gamma$ has the gamma$(\theta, c)$ distribution, then $c\gamma$ has the gamma$(\theta, 1)$ distribution.

Next, for every $a, b > 0$, the beta law with parameter $(a, b)$ is the probability measure on $]0, 1[$ which has density

$$\frac{\Gamma(a+b)}{\Gamma(a)\Gamma(b)} x^{a-1}(1-x)^{b-1}, \qquad 0 < x < 1.$$

It is well-known, and easy to check, that if $\gamma$ and $\gamma'$ are two independent gamma variables with respective parameters $(a, c)$ and $(b, c)$, then

$$\frac{\gamma}{\gamma+\gamma'} \quad \text{has the beta}(a, b) \text{ law and is independent of } \gamma+\gamma'. \tag{2.3}$$

Moreover, $\gamma+\gamma'$ has the gamma$(a+b, c)$ distribution.

Throughout this section, $N \geq 2$ denotes a fixed integer. Dirichlet distributions form an important family of probability laws on the simplex

$$\Delta_{N-1} := \left\{ x = (x_1, \dots, x_N) : x_i \geq 0 \text{ for every } i = 1, \dots, N \text{ and } \sum_{i=1}^N x_i = 1 \right\}.$$

**Definition 2.4** *For every $\alpha_1, \dots, \alpha_N > 0$, the probability measure on the simplex $\Delta_{N-1}$ with density*

$$\frac{\Gamma(\alpha_1 + \cdots + \alpha_N)}{\Gamma(\alpha_1)\cdots\Gamma(\alpha_N)} x_1^{\alpha_1-1} \cdots x_N^{\alpha_N-1}, \qquad (x_1, \dots, x_N) \in \Delta_{N-1}$$

*is called the $(N-1)$-**dimensional Dirichlet distribution** with parameter $(\alpha_1, \dots, \alpha_N)$. The special case when $\alpha_1 = \cdots = \alpha_N = 1$ will be referred to as the **uniform distribution** on $\Delta_{N-1}$.*

We now recall a useful representation of Dirichlet distribution based on independent gamma variables, which extends (2.3).

**Lemma 2.2** (i) *Fix $\alpha_1, \dots, \alpha_N, c > 0$, and let $\gamma_1, \dots, \gamma_N$ be independent gamma variables with respective parameters $(\alpha_1, c), \dots, (\alpha_N, c)$. Set*

*$\gamma := \gamma_1 + \cdots + \gamma_N$, so $\gamma$ has a gamma distribution with parameter $(\alpha_1 + \cdots + \alpha_N, c)$. Then the $N$-tuple*

$$(\gamma_1/\gamma, \ldots, \gamma_N/\gamma)$$

*has the $(N-1)$-dimensional Dirichlet distribution with parameter $(\alpha_1, \ldots, \alpha_N)$ and is independent of $\gamma$.*
(ii) *Let $0 < V_1 < \cdots < V_{N-1} < 1$ be the ordered statistics of a family of $N-1$ i.i.d. uniform $[0,1]$ variables. Then the $N$-tuple of the increments*

$$\left(V_1, V_2 - V_1, \ldots, V_{N-1} - V_{N_2}, 1 - V_{N-1}\right)$$

*has the uniform distribution on the simplex $\Delta_{N-1}$.*

*Proof* (i) Let $\mathbf{x} = (\mathbf{x}_1, \ldots, \mathbf{x}_N)$ be a Dirichlet variable with parameter $(\alpha_1, \ldots, \alpha_N)$ and $\gamma$ an independent gamma variable with parameter $(\alpha, c)$ where $\alpha := \alpha_1 + \cdots + \alpha_N$. Set $\gamma_i := \gamma \mathbf{x}_i$ for $i = 1, \ldots, N$, so for every measurable function $f : \mathbb{R}^N \to \mathbb{R}_+$, we have

$$\begin{aligned}
&\mathbb{E}\left(f(\gamma_1, \ldots, \gamma_N)\right) \\
&= \frac{c^\alpha}{\Gamma(\alpha_1)\cdots\Gamma(\alpha_N)} \int_{\Delta_{N-1}} dx_1 \cdots dx_{N-1} x_1^{\alpha_1 - 1} \cdots x_N^{\alpha_N - 1} \\
&\quad \int_0^\infty dy y^{\alpha-1} \mathrm{e}^{-cy} f(yx_1, \ldots, yx_N) \\
&= \frac{c^{\alpha_1} \cdots c^{\alpha_N}}{\Gamma(\alpha_1)\cdots\Gamma(\alpha_N)} \int_{[0,\infty[^N} dz_1 \cdots dz_N z_1^{\alpha_1 - 1} \cdots z_N^{\alpha_N - 1} \mathrm{e}^{-(z_1 + \cdots + z_N)} \\
&\quad f(z_1, \ldots, z_N),
\end{aligned}$$

where at the second equality, we used the change of variables

$$z_1 = yx_1, \ldots, z_{N-1} = yx_{N-1}, z_N = yx_N = y - z_1 - \cdots - z_{N-1}.$$

Thus $\gamma_1, \ldots, \gamma_N$ are independent gamma variables with respective parameters $(\alpha_1, c), \ldots, (\alpha_N, c)$.

(ii) This follows readily from the fact that $(V_1, \ldots, V_{N-1})$ is uniformly distributed on the set $\{(v_1, \ldots, v_{N-1}) : 0 < v_1 < \cdots < v_{N-1} < 1\}$. □

As an application of the representation above of the uniform distribution on the simplex $\Delta_{N-1}$, we present a first simple example of duality between fragmentation and coagulation. Specifically, fragmentation consists of picking a term in a mass-partition by size-biased sampling and splitting this term into two parts using a uniform variable, whereas coagulation consists of merging two distinct terms picked uniformly at random.

**Corollary 2.1** *Let* $\mathbf{x} = (\mathbf{x}_1, \ldots, \mathbf{x}_N)$ *be uniformly distributed on the simplex* $\Delta_{N-1}$.

(i) *Introduce a random variable* $\mathtt{k}$ *whose conditional distribution given* $\mathbf{x}$ *is*

$$\mathbb{P}(\mathtt{k} = k \mid \mathbf{x}) = \mathbf{x}_k, \qquad k = 1, \ldots, N,$$

*so that* $\mathbf{x}_{\mathtt{k}}$ *is a size-biased sample of* $\mathbf{x}$. *Let* $U$ *be an independent uniform variable on* $[0, 1]$. *Then the* $(N+1)$*-tuple*

$$(\mathbf{x}_1, \ldots, \mathbf{x}_{\mathtt{k}-1}, U\mathbf{x}_{\mathtt{k}}, (1-U)\mathbf{x}_{\mathtt{k}}, \mathbf{x}_{\mathtt{k}+1}, \ldots, \mathbf{x}_N)$$

*has the uniform distribution on the simplex* $\Delta_N$.

(ii) *Suppose* $N \geq 3$, *and pick two indices* $\mathtt{j}$ *and* $\mathtt{k}$ *in* $\{1, \ldots, N\}$, *uniformly at random without replacement and independently of* $\mathbf{x}$. *Next, denote by* $\mathbf{x}'$ *the sequence with* $N-1$ *terms which is obtained from* $\mathbf{x}$ *after replacing its* $\mathtt{j}$*-th term* $\mathbf{x}_{\mathtt{j}}$ *by* $\mathbf{x}_{\mathtt{j}} + \mathbf{x}_{\mathtt{k}}$, *and removing its* $\mathtt{k}$*-th term* $\mathbf{x}_{\mathtt{k}}$. *Then* $\mathbf{x}'$ *is uniformly distributed on the simplex* $\Delta_{N-2}$.

*Proof* (i) Thanks to Lemma 2.2(ii), we may suppose that $\mathbf{x}$ is the sequence of the lengths of the interval components (ranked from the left to the right) of a random interval-partition $\vartheta := ]0, 1[\backslash\{U_1, \ldots, U_{N-1}\}$, where $U_1, \ldots, U_{N-1}$ are i.i.d. uniform variables. Let $U_N$ be another uniform variable which is independent of the preceding. By Lemma 2.1, the length of the interval component $I_*$ of $\vartheta$ which contains $U_N$ is a size-biased picked from $\mathbf{x}$. Moreover, it should be plain that conditionally on $\vartheta$ and $I_*$, $U_N$ is uniformly distributed on $I_*$. Hence, the sequence of the lengths of the interval components (ranked from the left to the right) of the interval-partition $]0, 1[\backslash\{U_1, \ldots U_N\}$ is distributed as the $(N+1)$-tuple of the statement. Another application of Lemma 2.2(ii) shows that this variable is uniformly distributed on the simplex $\Delta_N$.

(ii) It is convenient to work with the oriented unit circle $\mathcal{C}$. Let $U_1, \ldots, U_N$ be i.i.d. uniform variables on $\mathcal{C}$ which we use to split $\mathcal{C}$ into $N$ arcs. Specifically, we write $A_i$ for the arc with left-extremity $U_i$; it should be plain from Lemma 2.2(ii) that the sequence $\mathbf{x} := (|A_1|, \ldots, |A_n|)$ is uniformly distributed on $\Delta_{N-1}$.

Next, pick an index $\mathtt{k}$ uniformly at random in $\{1, \ldots, N\}$, independently of the $U_i$ s. Then $U_{\mathtt{k}}$ is the right extremity of some arc, say $A_{\mathtt{j}} = ]U_{\mathtt{j}}, U_{\mathtt{k}}[$. It is easily checked that conditionally on $\mathtt{k}$, $\mathtt{j}$ is uniformly distributed in $\{1, \ldots, N\}\backslash\{\mathtt{k}\}$ and is independent of $\mathbf{x}$. Merging the arcs $A_{\mathtt{j}}$ and $A_{\mathtt{k}}$ amounts to splitting the circle $\mathcal{C}$ using $N-1$ independent uniform variables, namely $\{U_1, \ldots, U_{\mathtt{k}-1}, U_{\mathtt{k}+1}, \ldots, U_N\}$, and the resulting sequence of the arc lengths has the uniform distribution on $\Delta_{N-2}$. This establishes the claim. □

The first part of Corollary 2.1 is of course related to the fact that the Poissonian rain is a self-similar fragmentation chain with index 1 and identifies its dislocation measure as the law of $(1-U/2, U/2, 0, \ldots)$ where $U$ is uniformly distributed on $[0, 1]$. The second part, which is essentially a converse to the first, has a natural interpretation in terms of Kingman's coalescent, as we shall see in Section 4.1.3. We refer to [42] and [78] for recent extensions of Corollary 2.1, which provide further examples of duality between fragmentation and coagulation.

### 2.2.2 Some preliminaries on Poisson random measures

We now recall some basic facts about Poisson random measures; referring to Kingman [140] for background. Let $E$ be a Polish space and $\mu$ a sigma-finite measure on $E$. We call a random measure $\mathtt{M}$ on $E$ *a Poisson measure with intensity* $\mu$ if $\mathtt{M}$ fulfills the following requirements. For every Borel subset $B$ of $E$ with $\mu(B) < \infty$, $\mathtt{M}(B)$ has a Poisson distribution with parameter $\mu(B)$, and if $B_1, \ldots, B_n$ are disjoint Borel sets, the variables $\mathtt{M}(B_1), \ldots, \mathtt{M}(B_n)$ are independent. Plainly, $\mathtt{M}$ is then a sum of Dirac point masses, that is we can express $\mathtt{M}$ in the form

$$\mathtt{M} = \sum_{i\in I} \delta_{\mathsf{a}_i} \tag{2.4}$$

where $\delta_x$ stands for the Dirac point mass at $x$. The $\mathsf{a}_i$ will be referred to as the *atoms* of $\mathtt{M}$. If $\mu(E)$ is finite, then the set of atoms is finite a.s., and more precisely its cardinal $N := \mathtt{M}(E)$ follows the Poisson law with parameter $\mu(E)$. In the representation (2.4), we may then choose $\mathsf{a}_1, \ldots$ to be a sequence of i.i.d. variables with common law $\mu(\cdot)/\mu(E)$ and independent of $N$, and $I = \{1, \ldots, N\}$. If $\mu(E) = \infty$, then there are infinitely many atoms a.s., so we may take $I = \mathbb{N}$. Furthermore, a similar description of the atoms can be obtained using the elementary superposition property of Poisson measure (cf. Lemma 2.4 below) and expressing the intensity measure in the form $\mu = \sum_{n=1}^{\infty} \mu_n$, where each $\mu_n$ is a finite measure.

We now recall three key formulas for the computation of moments, Laplace transforms and distributions related to Poisson point measures. In this direction, we use the convention $\exp(-\infty) = 0$. Consider a measurable map $f : E \to \mathbb{R}$ and write

$$\langle \mathtt{M}, f\rangle := \int_E f(x)\mathtt{M}(dx) = \sum_{i\in I} f(\mathsf{a}_i).$$

**Lemma 2.3** (i) *Suppose either $f \geq 0$ or $f \in L^1(\mu)$. Then we have the* **first moment formula**:

$$\mathbb{E}(\langle \mathtt{M}, f\rangle) = \int_E f(x)\mu(dx)\,.$$

(ii) *Suppose either $f \geq 0$ or $1-\mathrm{e}^{-f} \in L^1(\mu)$. Then we have the* **Campbell formula**:

$$\mathbb{E}(\exp(-\langle \mathtt{M}, f\rangle)) = \exp\left(-\int_E (1-\mathrm{e}^{-f(x)})\mu(dx)\right).$$

(iii) *Let $E_{\mathrm{p}}$ denote the space of point measures on $E$, and $G : E \times E_{\mathrm{p}} \to \mathbb{R}_+$ be some measurable functional. Then we have the* **Palm formula**:

$$\mathbb{E}(\langle \mathtt{M}, fG(\cdot, \mathtt{M})\rangle) := \mathbb{E}\left(\sum_{i\in I} f(\mathsf{a}_i)G(\mathsf{a}_i, \mathtt{M})\right) = \int_E \mathbb{E}(G(x, \delta_x + \mathtt{M}))f(x)\mu(dx)\,.$$

Loosely speaking, the Palm formula can be interpreted as follows. Suppose that the intensity measure $\mu$ gives no mass to some point $x$. Then the conditional distribution of $\mathtt{M}$, given that $\mathtt{M}$ has an atom at $x$, is that of $\delta_x + \mathtt{M}$. Of course, such conditioning is only formal as the probability for $\mathtt{M}$ having an atom at $x$ is 0.

*Proof* The first moment and Campbell formulas are immediate consequences of the description of the distribution of the atoms of $\mathtt{M}$. Let us now establish the Palm formula. Plainly, it suffices to consider functionals $G$ which depend only on the point measure $\mathtt{M}$ and which are of exponential type, that is

$$G(\mathtt{M}) = \exp(-\langle \mathtt{M}, g\rangle)\,,$$

where $g : E \to \mathbb{R}_+$ stands for a generic measurable function. The Campbell formula then gives for every $q \geq 0$

$$\mathbb{E}\left(\exp(-q\sum_{i\in I} f(\mathsf{a}_i))G(\mathtt{M})\right) = \exp\left(-\int_E (1-\mathrm{e}^{-(qf(x)+g(x))})\mu(dx)\right).$$

Taking the derivative in the variable $q$ at 0 yields

$$\begin{aligned}\mathbb{E}\left(G(\mathtt{M})\sum_{i\in I} f(\mathsf{a}_i)\right) &= \int_E \mu(dx)f(x)\mathrm{e}^{-g(x)}\exp\left(-\int_E (1-\mathrm{e}^{-g(y)})\mu(dy)\right)\\ &= \int_E \mu(dx)f(x)\mathbb{E}(G(\delta_x + \mathtt{M}))\,,\end{aligned}$$

where the last equality stems from the Campbell formula. □

Next, we recall the effect of some simple transformations for Poisson random measure.

**Lemma 2.4** • **(Superposition)** *Let $\mu'$ be another sigma-finite measure on $E$ and $\mathtt{M}'$ a Poisson point measure with intensity $\mu'$ which is independent of $\mathtt{M}$. Then $\mathtt{M}+\mathtt{M}'$ is a Poisson point measure with intensity $\mu+\mu'$.*

• **(Image)** *Let $E'$ be another Polish space, $f: E \to E'$ a measurable map, and $\mu'$ the image measure of $\mu$ by $f$. Then the image of $\mathtt{M}$ by $f$ is a Poisson random measure on $E'$ with intensity $\mu'$.*

• **(Marking)** *Let $\rho$ be some probability measure on $E'$, and $\mathtt{z}_1, \ldots$ a sequence of i.i.d. variables with law $\rho$, which is independent of $\mathtt{M}$. Each $\mathtt{z}_i$ is viewed as a mark attached to the atom $\mathtt{a}_i$. Then*

$$\sum_{i\in I} \delta_{(\mathtt{a}_i, \mathtt{z}_i)}$$

*is a Poisson random measure on $E \times E'$ with intensity $\mu \otimes \rho$.*

• **(Change of Probability)** *Let $f: E \to \mathbb{R}$ be a measurable function such that $\mathrm{e}^f - 1 \in L^1(\mu)$. Consider the probability $\tilde{\mathbb{P}}$ which is absolutely continuous with respect to $\mathbb{P}$, with density*

$$\exp\left(\langle \mathtt{M}, f\rangle - \langle \mathrm{e}^f - 1, \mu\rangle\right) .$$

*Then under $\tilde{\mathbb{P}}$, $\mathtt{M}$ is a Poisson random measure with intensity $\mathrm{e}^{f(x)}\mu(dx)$.*

*Proof* The first three statements are straightforward (see for example Kingman [140]); we shall only provide details for the change of probability property. For every measurable function $g: E \to \mathbb{R}_+$, we deduce from the Campbell formula in Lemma 2.3 that

$$\begin{aligned}
\tilde{\mathbb{E}}\left(\exp(-\langle \mathtt{M}, g\rangle)\right) &= \mathbb{E}\left(\exp\left(\langle f-g, \mathtt{M}\rangle - \langle \mathrm{e}^f - 1, \mu\rangle\right)\right) \\
&= \exp\left(-\int_E \left(1-\mathrm{e}^{f(x)-g(x)}\right)\mu(dx)\right. \\
&\qquad \left. -\int_E \left(\mathrm{e}^{f(x)}-1\right)\mu(dx)\right) \\
&= \exp\left(-\int_E \left(1-\mathrm{e}^{-g(x)}\right)\mathrm{e}^{f(x)}\mu(dx)\right) .
\end{aligned}$$

The claim follows. □

### 2.2.3 Mass-partitions induced by Poisson measures

Throughout the rest of this section, we consider a measure $\Lambda$ on $]0, \infty[$ such that

$$\int_{]0,\infty[} (1 \wedge x)\Lambda(dx) < \infty \tag{2.5}$$

and

$$\Lambda(]0, \infty[) = \infty ,$$

and a Poisson random measure $\mathtt{M}$ on $]0, \infty[$ with intensity $\Lambda$. The fact that for every $x > 0$ the tail-intensity $\Lambda(]x, \infty[)$ is finite implies that there are only finitely many atoms in $]x, \infty[$. On the other hand, there are infinitely many atoms in $]0, \infty[$ since $\Lambda(]0, \infty[) = \infty$. We may thus rank these atoms in decreasing order; we denote by

$$\mathsf{a}_1 \geq \mathsf{a}_2 \geq \cdots > 0$$

this ranked sequence.

Next, we point out that condition (2.5) ensures the summability of the series $\sum_1^\infty \mathsf{a}_i$. Indeed, by the first-moment formula of Lemma 2.3, we see that the series $\sum_1^\infty \mathsf{a}_i 1\!\!1_{\{\mathsf{a}_i \leq 1\}}$ converges a.s., and since there are only finitely many atoms in $]1, \infty[$, we have

$$\varsigma := \sum_{i=1}^{\infty} \mathsf{a}_i < \infty \qquad \text{a.s.}$$

Conversely, it can also be checked that the series $\sum_1^\infty \mathsf{a}_i$ diverges a.s. whenever condition (2.5) fails. We also point out that the Campbell formula in Lemma 2.3 enables us to compute the Laplace transform of the series: one gets for every $q \geq 0$ that

$$\mathbb{E}\left(\exp\left(-q\varsigma\right)\right) = \mathbb{E}\left(\exp\left(-q\int_{]0,\infty[} x\mathtt{M}(dx)\right)\right) = \exp(-\Phi(q)) ,$$

where

$$\Phi(q) := \int_{]0,\infty[} (1 - e^{-qx})\Lambda(dx) . \tag{2.6}$$

There is a natural procedure to transform the family of atoms $(\mathsf{a}_i, i \in \mathbb{N})$ of some Poisson random measure on $]0, \infty[$ subject to condition (2.5) into a proper random mass-partition. Namely, the so-called *Poisson-Kingman partition* (see [136] and [187]) is obtained by its normalization by the total sum:

$$S = (\mathsf{a}_i/\varsigma, i \in \mathbb{N}) , \qquad \varsigma = \sum_{i=1}^{\infty} \mathsf{a}_i .$$

Perman [178] and Pitman [187] have obtained formulas for the finite-dimensional distributions of a fairly large class of Poisson-Kingman partitions; unfortunately, the expressions are in general not quite explicit.

Following Perman, Pitman and Yor [179], we will now associate to the Poisson measure $\mathtt{M}$ another random proper mass-partition, by conditioning the Poisson-Kingman partition $S$ on $\varsigma = x$, where $x > 0$ is some fixed real number. In other words, this random mass-partition is given by $(\mathsf{a}_1/x, \mathsf{a}_2/x, \ldots)$ conditionally on $\sum_1^\infty \mathsf{a}_i = x$. Observe that the distribution of the Poisson-Kingman partition can then be recovered as a mixture in the variable $x$ of the latter. However, the conditioning is *a priori* not well-defined as in general $\mathbb{P}(\varsigma = x) = 0$. We shall see that, nonetheless, one can give a rigorous meaning to the conditioning by approximation under a rather mild assumption.

From now on, we shall assume that the intensity measure $\Lambda$ has no atoms, which implies that the atoms $\mathsf{a}_i$ of the Poisson measure $\mathtt{M}$ are all distinct a.s. We shall also suppose that the distribution of the sum of the atoms $\varsigma$ is smooth, in the sense that this random variable possesses a continuous density with respect to Lebesgue measure, which is strictly positive on $]0, \infty[$. This regular density will be denoted by

$$\rho(x) := \mathbb{P}(\varsigma \in dx)/dx\,, \qquad x > 0\,.$$

We refer to Section 28 in Sato [198] for rather mild conditions on the intensity measure $\Lambda$ which ensure the existence of such smooth density.

The idea for making the conditioning rigorous consists of using repeatedly size-biased sampling without replacement in combination with the Palm formula. In this direction, let us denote by $(\mathsf{a}_1^*, \mathsf{a}_2^*, \ldots)$ a size-biased reordering of $(\mathsf{a}_1, \mathsf{a}_2, \ldots)$. This means that conditionally on $(\mathsf{a}_1, \ldots)$ we first pick an index $1^*$ according to the law

$$\mathbb{P}(1^* = k \mid \mathsf{a}_1, \ldots) = \mathsf{a}_k / \sum_{i=1}^{\infty} \mathsf{a}_i = \mathsf{a}_k/\varsigma\,,$$

set $\mathsf{a}_1^* = \mathsf{a}_{1^*}$, then remove $\mathsf{a}_1^*$ from the sequence $(\mathsf{a}_1, \ldots)$, pick $\mathsf{a}_2^*$ by size-biased sampling in this new sequence, and so on. In other words, $(\mathsf{a}_1^*, \mathsf{a}_2^*, \ldots) = \varsigma S^*$ where $S^*$ is a size-biased reordering of the Poisson-Kingman partition $S = (\mathsf{a}_i/\varsigma, i \in \mathbb{N})$. See Section 2.1.3 for details.

**Proposition 2.4** *Let us fix $x > 0$ and assume that the hypotheses above are fulfilled.*

(i) *The conditional law of the size-biased reordering $(\mathsf{a}_1^*, \ldots)$ given $\varsigma \in [x, x+\varepsilon]$ has a weak limit in the sense of convergence of finite-dimensional* *distributions as $\varepsilon \to 0+$, which is denoted by $P_x^*$. More precisely, the law of $(\mathtt{a}_n^*, n \in \mathbb{N})$ under $P_x^*$ is determined by the following Markov-type property:*

$$P_x^*(\mathtt{a}_1^* \in dy) = \frac{y\rho(x-y)}{x\rho(x)} \Lambda(dy), \qquad 0 < y < x,$$

*and the conditional distribution of $(\mathtt{a}_2^*, \mathtt{a}_3^*, \ldots)$ under $P_x^*$ given $\mathtt{a}_1^* = y$ is $P_{x-y}^*$. Moreover, we have*

$$\sum_{i=1}^{\infty} \mathtt{a}_i^* = x, \qquad P_x^*\text{-a.s.}$$

(ii) *As a consequence, the law of the Poisson-Kingman partition $S = (\mathtt{a}_i/\varsigma, i \in \mathbb{N})$ conditioned on $\varsigma \in [x, x+\varepsilon]$ converges weakly as $\varepsilon \to 0$ towards the law of the random mass-partition that results from the rearrangement in decreasing order of the sequence $(\mathtt{a}_i^*/x, i \in \mathbb{N})$ under $P_x^*$.*

*Proof* (i) Let us first check that

$$\mathbb{1}_{\{0 < y < x\}} \frac{y\rho(x-y)}{x\rho(x)} \Lambda(dy)$$

is a probability measure on $]0, x[$. For every $q > 0$, we have

$$\begin{aligned}\int_0^\infty dx \mathrm{e}^{-qx} \int_{]0,x[} \Lambda(dy) y\rho(x-y) &= \int_0^\infty du \mathrm{e}^{-qu} \rho(u) \int_{]0,\infty[} \Lambda(dy) y \mathrm{e}^{-qy} \\ &= \Phi'(q) \mathrm{e}^{-\Phi(q)} \\ &= \int_0^\infty dx \mathrm{e}^{-qx} x\rho(x),\end{aligned}$$

and thus

$$\int_{]0,x[} \Lambda(dy) y\rho(x-y) = x\rho(x).$$

Next, let $f : \mathbb{R}_+ \to \mathbb{R}_+$ be a measurable function; by definition of size-biased sampling and the fact that $\varsigma = \sum_{i=1}^\infty \mathtt{a}_i$, we have that

$$\mathbb{E}\left(f(\mathtt{a}_1^*) \mid \varsigma \in [x, x+\varepsilon]\right) = \frac{\mathbb{E}\left(\varsigma^{-1} \sum_{i=1}^\infty \mathtt{a}_i f(\mathtt{a}_i), \varsigma \in [x, x+\varepsilon]\right)}{\mathbb{P}\left(\varsigma \in [x, x+\varepsilon]\right)}.$$

When $\varepsilon \to 0+$, the denominator in the ratio of the right-hand side is equivalent to $\varepsilon\rho(x)$. On the other hand, the Palm formula (see Lemma 2.3) gives

$$\mathbb{E}\left(\varsigma^{-1}\sum_{i=1}^{\infty} \mathtt{a}_i f(\mathtt{a}_i), \varsigma \in [x, x+\varepsilon]\right)$$

$$= \int_{]0,\infty[} \Lambda(dy) y f(y) \mathbb{E}\left((y+\varsigma)^{-1}, y+\varsigma \in [x, x+\varepsilon]\right)$$

$$\sim x^{-1}\left(\int_{]0,x[} \Lambda(dy) y f(y) \rho(x-y)\right)\varepsilon \qquad \text{as } \varepsilon \to 0+ .$$

This shows that the conditional distribution of $\mathtt{a}_1^*$ given $\varsigma \in [x, x+\varepsilon]$ converges as $\varepsilon \to 0+$ towards the limiting law given in the statement.

We next use an easy induction. Fix an integer $k$ and assume that for every $x > 0$ we have established the convergence in distribution of $(\mathtt{a}_1^*, \ldots, \mathtt{a}_k^*)$ under $\mathbb{P}(\cdot \mid \varsigma \in [x, x+\varepsilon])$ as $\varepsilon \to 0+$, to probability which, by a slight abuse of notation, is denoted by $P_x^*$. This means that for every bounded continuous function $g : \mathbb{R}_+^k \to \mathbb{R}_+$, we have

$$\lim_{\varepsilon\to 0+} \mathbb{E}\left(g(\mathtt{a}_1^*, \ldots, \mathtt{a}_k^*) \mid \varsigma \in [x, x+\varepsilon]\right) = E_x^*(g(\mathtt{a}_1^*, \ldots, \mathtt{a}_k^*)) .$$

Now we have to evaluate $\mathbb{E}\left(f(\mathtt{a}_1^*) g(\mathtt{a}_2^*, \ldots, \mathtt{a}_{k+1}^*) 1\!\!1_{\{x\le\varsigma\le x+\varepsilon\}}\right)$. We can express this quantity in the form

$$\mathbb{E}\left(\sum_{i=1}^{\infty} \mathtt{a}_i f(\mathtt{a}_i) G(\mathtt{a}_i, \mathtt{M})\right) ,$$

where for every $\mathtt{b} > 0$, the functional $G(\mathtt{b}, \mathtt{M})$ is obtained as follows. First we delete the atom at $\mathtt{b}$ in the point measure $\mathtt{M}$, then pick $k$ atoms by size-biased sampling without replacement, say $\mathtt{b}_1^*, \ldots, \mathtt{b}_k^*$, and finally set

$$G(\mathtt{b}, \mathtt{M}) = \varsigma^{-1} g(\mathtt{b}_1^*, \ldots, \mathtt{b}_k^*) 1\!\!1_{\{x\le\varsigma\le x+\varepsilon\}} ,$$

with $\varsigma = \int y\mathtt{M}(dy)$. We then have by the same arguments based on the Palm formula as above that

$$\mathbb{E}\left(\sum_{i=1}^{\infty} \mathtt{a}_i f(\mathtt{a}_i) G(\mathtt{a}_i, \mathtt{M})\right)$$

$$= \int_{]0,\infty[} \Lambda(dy) y f(y) \mathbb{E}\left(G(y, \mathtt{M} + \delta_y)\right)$$

$$= \int_{]0,\infty[} \Lambda(dy) y f(y) \mathbb{E}\left((y+\varsigma)^{-1} g(\mathrm{a}_1^*, \ldots, \mathrm{a}_k^*), y+\varsigma \in [x, x+\varepsilon]\right)$$

$$\sim \varepsilon x^{-1} \int_{]0,x[} \Lambda(dy) y f(y) \rho(x-y) E^*_{x-y}(g(\mathrm{a}_1^*, \ldots, \mathrm{a}_k^*)), \quad \text{as } \varepsilon \to 0+.$$

This establishes the Markov-type description of $P^*_x$.

Finally, under $P^*_x$, the sequence

$$r_n := x - (\mathrm{a}_1^* + \cdots + \mathrm{a}_n^*), \qquad n \in \mathbb{Z}_+$$

is Markovian with values in $[0, x]$. It has decreasing paths and transition probabilities

$$P^*_x(r_{n+1} \in dz \mid r_n = y) = \frac{(y-z)\rho(z)}{y\rho(y)} \Lambda(y - dz), \qquad 0 < z < y.$$

Our assumptions ensure that for every $0 < \varepsilon < x$,

$$\inf_{\varepsilon \le y \le x} P^*_x(r_{n+1} \le \varepsilon \mid r_n = y) > 0,$$

which implies that $r_n$ converges to 0, $P^*_x$-a.s.

(ii) The second assertion follows from (i) and Proposition 2.3. □

We next present a fairly intuitive construction of random mass-partitions defined in the preceding proposition, in the stable case. Specifically, consider an integer-valued random variable $\xi$ which does not live in a strict sub-lattice (i.e. $\mathbb{P}(\xi \in a\mathbb{N}) < 1$ for every integer $a \ge 2$) and such that its probability function fulfills

$$\mathbb{P}(\xi = \ell) \sim c\ell^{-\alpha-1}, \qquad \ell \to \infty,$$

for some $\alpha \in ]0, 1[$ and $c > 0$. Let $\xi_1, \ldots$ be a sequence of i.i.d. copies of $\xi$. For every $k \in \mathbb{N}$, set $\Sigma_k = \xi_1 + \cdots + \xi_k$, and then for every $n \ge k$ let $S^{(k,n)}$ denote a random mass-partition distributed as the rearrangement in decreasing order of $\xi_1/n, \ldots, \xi_k/n$, conditionally on $\Sigma_k = n$.

**Corollary 2.2** *Fix $b > 0$. Under the preceding assumptions, $S^{(k,n)}$ converges in distribution on $\mathcal{P}_{\mathrm{m}}$ as $k, n \to \infty$ with $k \sim bn^\alpha$, to $(\mathrm{a}_1, \ldots)$ conditioned on $\varsigma := \sum_{i=1}^\infty \mathrm{a}_i = 1$, where $(\mathrm{a}_1, \ldots)$ denote the ranked sequence of the atoms of a Poisson random measure on $]0, \infty[$ with intensity $\Lambda(da) = bca^{-1-\alpha} da$, and the singular conditioning is taken in the sense of Proposition 2.4.*

*Proof* Denote for every $n \geq k$ by $(\xi^*_{1,n}, \ldots, \xi^*_{k,n})$ a $k$-tuple distributed as a size-biased reordering of $(\xi_1, \ldots, \xi_k)$ given $\Sigma_k = n$. It is immediately seen that the probability function of $\xi^*_{1,n}$ is

$$\mathbb{P}(\xi^*_{1,n} = \ell) = \frac{k\ell}{n}\mathbb{P}(\xi_1 = \ell \mid \Sigma_k = n) = \frac{k\ell}{n}\mathbb{P}(\xi_1 = \ell) \times \frac{\mathbb{P}(\Sigma_{k-1} = n-\ell)}{\mathbb{P}(\Sigma_k = n)},$$

for $\ell = 1, \ldots, n-k+1$, and then,

$$\text{given } \xi^*_{1,n} = \ell, \ (\xi^*_{2,n}, \ldots, \xi^*_{k,n}) \text{ has the same law as } (\xi^*_{1,n-\ell}, \ldots, \xi^*_{k-1,n-\ell}). \tag{2.7}$$

We shall first establish the weak convergence of $n^{-1}\xi^*_{1,n}$, and then extend to weak convergence of the sequence $n^{-1}(\xi^*_{1,n}, \ldots, \xi^*_{k,n})$ in the sense of finite-dimensional distributions using the Markov type property (2.7). In this direction, recall that $(\mathsf{a}_1, \ldots)$ is the ranked sequence of the atoms of a Poisson random measure on $]0, \infty[$ with intensity $bca^{-1-\alpha}da$, so $\varsigma := \sum_{i=1}^{\infty} \mathsf{a}_i$ is a stable($\alpha$) variable; we write $\rho(x) := \mathbb{P}(\varsigma \in dx)/dx$ for the probability density of the latter.

Now fix $a \in ]0, 1[$ and $b > 0$, and let $\ell, k, n$ tend to $\infty$ with $\ell \sim an$ and $k \sim bn^{\alpha}$. On the one hand, it follows from the hypothesis on the asymptotic behavior of the probability function of $\xi$ that

$$\frac{k\ell}{n}\mathbb{P}(\xi_1 = \ell) \sim bca^{-\alpha}n^{-1}.$$

On the other hand, $n^{-1}\Sigma_k$ converges in distribution to $\varsigma$, and more precisely, we deduce from Gnedenko's local limit theorem (see for example Bingham *et al.* [58] on its page 351) that

$$\frac{\mathbb{P}(\Sigma_{k-1} = n-\ell)}{\mathbb{P}(\Sigma_k = n)} \sim \frac{\rho(1-a)}{\rho(1)}.$$

We conclude that

$$\mathbb{P}(\xi^*_{1,n} = \ell \mid \Sigma_k = n) \sim \frac{a\rho(1-a)}{n\rho(1)} bca^{-1-\alpha}da.$$

This estimate implies that $n^{-1}\xi^*_{1,n}$ converges weakly as $n, k \to \infty$ with $k \sim bn^{\alpha}$ towards the law

$$\frac{a\rho(1-a)}{\rho(1)} bca^{-1-\alpha}da, \qquad 0 < a < 1.$$

By iteration using the Markov-type property (2.7) and comparison with Proposition 2.4, we now see that $(n^{-1}\xi^*_{1,n}, \ldots, n^{-1}\xi^*_{k,n})$ converges in the sense of finite-dimensional distributions as $n, k \to \infty$ with $k \sim bn^{\alpha}$, to $(\mathsf{a}^*_1, \ldots)$, where the latter has the law of a size-biased reordering of $(\mathtt{a}_1, \dots)$ conditioned to have total mass $\varsigma = 1$.

Since we know from Proposition 2.4 that $\sum_{i=1}^{\infty} \mathtt{a}_i^* = 1$ a.s., we can apply Proposition 2.3, which completes the proof of the statement. □

### 2.2.4 Gamma subordinators and Dirichlet processes

In this section and in the next one, it will be convenient to consider certain increasing processes which are naturally related to Poisson random measures $\mathtt{M} = \sum_{i\in I} \delta_{\mathtt{a}_i}$. Recall that the intensity measure $\Lambda$ of $\mathtt{M}$ fulfills (2.5), and introduce an independent sequence $U_1, \dots$ of i.i.d. uniform variables on $[0, 1]$, and then define the increasing process

$$\varsigma(t) := \sum_{i=1}^{\infty} \mathtt{a}_i \mathbb{1}_{\{U_i \le t\}} = \sum_{U_i \le t} \mathtt{a}_i \,, \qquad t \in [0, 1] \,. \tag{2.8}$$

In other words, the collection of the jump times and jump sizes of the purely discontinuous increasing process $\varsigma$ is $\{(U_i, \mathtt{a}_i), i \in \mathbb{N}\}$. Recall also from Lemma 2.4 that the latter is the family of atoms of a marked Poisson measure on $[0, 1]\times]0, \infty[$ with intensity $dt \otimes \Lambda(dx)$.

The classical properties of Poisson random measures that we recalled in Section 2.2.2 imply that $(\varsigma(t), 0 \le t \le 1)$ has independent and stationary increments. This means that for every $0 = t_0 < t_1 < \cdots < t_n < t_{n+1} = 1$, the variables $\varsigma(t_1) - \varsigma(t_0), \dots, \varsigma(t_{n+1}) - \varsigma(t_n)$ are independent, and $\varsigma(t_{i+1}) - \varsigma(t_i)$ has the same law as $\varsigma(t_{i+1} - t_i)$. We say that $(\varsigma(t), 0 \le t \le 1)$ is a *subordinator*[3] on the time interval $[0, 1]$; the intensity measure $\Lambda$ is then referred to as the *Lévy measure*, the formula (2.6) as the *Lévy-Khintchine* formula, the function $\Phi$ there as the *Laplace exponent*, and finally (2.8) as the Lévy-Itô decomposition. Note also that the one-dimensional distributions of the process $(\varsigma(t), 0 \le t \le 1)$ are characterized by

$$\mathbb{E}\left(\exp(-q\varsigma(t))\right) = \exp -t\Phi(q) \,, \qquad q \ge 0 \,.$$

Throughout this section, $\theta, c > 0$ are two fixed real numbers. The subordinator $(\gamma(t), 0 \le t \le 1)$ corresponding to the Lévy measure

$$\Lambda(dx) = \theta x^{-1} \mathrm{e}^{-cx} dx \,, \qquad x > 0$$

[3] More precisely, we have constructed here only a sub-family of subordinators, which have no drift and no killing rate, and which are not compound Poisson. In Chapter 3, we will have to deal with the most general class of subordinators; however, the sub-class we consider here is rich enough for our present purpose.

is called a *gamma subordinator* with parameter $(\theta, c)$. Its Laplace exponent is thus given by

$$\Phi(q) = \theta \int_0^\infty (1 - e^{-qx}) x^{-1} e^{-cx} dx = \theta \ln(1 + q/c), \qquad q \geq 0;$$

note that $\gamma(t)$ has the gamma distribution with parameter $(\theta t, c)$, see Section 2.2.1. Here again, the parameter $c$ will have a very minor role, due to the easy fact that $c\gamma(\cdot)$ is a gamma subordinator with parameter $(\theta, 1)$. In this direction, it might be also interesting to point out that for every $a \in ]0, 1[$, $(\gamma(at), 0 \leq t \leq 1)$ is a gamma subordinator with parameter $(a\theta, c)$.

We start by observing that, thanks to the independence and stationarity of the increments of the subordinator, we may rewrite Lemma 2.2 as follows: For every $0 = t_0 < t_1 < \cdots < t_{n-1} < t_n = 1$, the $n$-tuple

$$\left( \frac{\gamma(t_1) - \gamma(t_0)}{\gamma(1)}, \ldots, \frac{\gamma(t_n) - \gamma(t_{n-1})}{\gamma(1)} \right) \tag{2.9}$$

has the $(n-1)$-dimensional Dirichlet distribution with parameter $(\theta(t_1 - t_0), \ldots, \theta(t_n - t_{n-1}))$. Intuitively, if we let $n$ tend to $\infty$ taking finer and finer subdivisions $\{t_0, t_1, \ldots, t_{n-1}, t_n\}$ of $[0, 1]$, and re-rank the $n$-tuple (2.9) in decreasing order, we should get as limit the ranked sequence of the jumps of $\gamma(\cdot)/\gamma(1)$. This motivates the following definition.

**Definition 2.5** *Let* $(\gamma(t), t \in [0, 1])$ *be a gamma subordinator with parameter* $(\theta, c)$, *and denote by* $\mathsf{a}_1 > \mathsf{a}_2 > \cdots > 0$ *the ranked sequence of its jumps.*

(i) *The process* $(\gamma(t)/\gamma(1), t \in [0, 1])$ *is called a* **Dirichlet process** *with parameter* $\theta$.
(ii) *The distribution of the random mass-partition* $(\mathsf{a}_1/\gamma(1), \ldots)$ *is called the* **Poisson-Dirichlet** *law with parameter* $(0, \theta)$ *and is denoted by* PD$(0, \theta)$.

The Poisson-Dirichlet law is thus an example of a Poisson-Kingman distribution as introduced in the preceding section; see Kingman [140] and Pitman [187]. We now state some elementary properties of Dirichlet processes.

**Proposition 2.5** (i) *The one-dimensional distributions of the Dirichlet process are beta laws. More precisely* $\gamma(t)/\gamma(1)$ *has a beta*$(\theta t, \theta(1-t))$ *law.*
(ii) *The Dirichlet process* $(\gamma(t)/\gamma(1), t \in [0, 1])$ *is independent of* $\gamma(1)$.
(iii) *For every* $x > 0$, *the conditional distribution of* $(x^{-1}\mathsf{a}_1, x^{-1}\mathsf{a}_2, \ldots)$ *given* $\gamma(1) = x$ *is* PD$(0, \theta)$.

*Proof* The first claim merely rephrases (2.3) and the second follows also from (2.3) by an immediate induction using the independence of the increments of the gamma process. The last part of the statement derives readily from the second. □

As an easy consequence, we derive the so-called *residual allocation model* (also sometimes called *stick breaking scheme*) for PD$(0, \theta)$ distributions: Take a sequence $\beta_1, \ldots$ of i.i.d. beta$(1, \theta)$ variables, we break the stick into two smaller sticks with size $\beta_1$ and $1-\beta_1$, keep the former, break the latter into two smaller sticks with size $(1-\beta_1)\beta_2$ and $(1-\beta_1)(1-\beta_2)$, and so on ... Then the random sequence of the lengths is distributed as the size-biased re-ordering of a PD$(0, \theta)$-variable. Here is a formal statement.

**Corollary 2.3** *Let $S$ be a random mass-partition with the* PD$(0, \theta)$ *distribution, $S^* = (S_1^*, S_2^*, \ldots)$ a size-biased reordering of $S$, and set*

$$\beta_1 := S_1^*, \; \beta_2 := S_2^*/(1-\beta_1), \; \ldots, \; \beta_{n+1} := S_{n+1}^* \times \prod_{i=1}^{n} (1-\beta_i)^{-1}, \; \ldots$$

*Then the variables $\beta_1, \ldots$ are i.i.d. with the beta$(1, \theta)$ distribution, that is $\mathbb{P}(\beta_i \in da) = \theta(1-a)^{\theta-1} da$ for $a \in [0, 1]$.*

*Proof* This is an immediate consequence of Propositions 2.4 and 2.5 and the explicit expressions for the Lévy measure of the gamma subordinator and the gamma densities. Indeed, we first find after cancellation of a few terms that

$$\mathbb{P}(S_1^* \in dy) = \theta(1-y)^{\theta-1} dy.$$

Then, in the notation of Proposition 2.4, we know that conditionally on $S_1^* = y$, $(S_2^*, \ldots)$ has the law $P_{1-a}^*$. By Proposition 2.5(ii), the latter coincides with the law of $(1-y)^{-1} S^*$, and we can therefore complete the proof by induction. □

Note that Corollary 2.3 enables us to construct a size-biased reordering of a PD$(0, \theta)$ variable from a sequence of i.i.d. beta variables, since, in the notation used there, we have

$$S_1^* = \beta_1, \; S_2^* = (1-\beta_1)\beta_2, \; \ldots, \; S_{n+1}^* = \beta_{n+1} \prod_{i=1}^{n} (1-\beta_i), \ldots$$

The size-biased reordering $S^*$ of a Poisson-Dirichlet PD$(0, \theta)$ variable $S$ is often called a Griffiths-Engen-McClosey variable with parameter $\theta$ and denoted by GEM$(\theta)$.

The residual allocation model also yields a nice construction due to Ignatov [126] of GEM($\theta$) variables using spacing induced by a simple Poisson random measure. More precisely, let $\mathbf{x}_1 > \mathbf{x}_2 > \cdots > 0$ be the sequence of atoms of a Poisson point measure on $]0,1[$ with intensity $\theta(1-x)^{-1}dx$. These atoms accumulate at 1 and nowhere else; they can be used to split the unit interval into an infinite sequence of intervals ordered from the left to the right. The sequence of spacing is defined as $\mathbf{x}_1, \mathbf{x}_2 - \mathbf{x}_1, \ldots$ Then

$$\mathbb{P}(\mathbf{x}_1 > t) = \exp\left(-\int_0^t \theta x^{-1} dx\right) = t^\theta,$$

so the first spacing has the law beta$(1,\theta)$. The superposition property of Poisson random measures (cf. Lemma 2.4) implies that conditionally on $\mathbf{x}_1 = t$, $\{\mathbf{x}_i, i \geq 2\}$ can be viewed as the family of atoms of a Poisson point measure on $]t,1[$ with intensity $\theta(1-x)^{-1}dx$. Mapping the latter by the dilation $x \to (x-t)/(1-t)$ yields a Poisson random measure on $]0,1[$ with intensity $\theta(1-x)^{-1}dx$. As a consequence, $(\mathbf{x}_2 - \mathbf{x}_1)/(1-\mathbf{x}_1)$ is independent of the first spacing $\mathbf{x}_1$ and also has the beta$(1,\theta)$ distribution. Iterating this argument, we see that the sequence of spacing is given by a residual allocation model based on independent beta$(1,\theta)$ variables, and therefore has the GEM($\theta$) law.

### 2.2.5 Stable subordinators and Poisson-Dirichlet partitions

Throughout this section, $\alpha \in ]0,1[$ and $c > 0$ are fixed parameters. We call a subordinator $\varsigma$ which has a Laplace exponent given by

$$\Phi(q) = cq^\alpha = \frac{c\alpha}{\Gamma(1-\alpha)} \int_0^\infty (1-e^{-qx}) x^{-1-\alpha} dx$$

a stable subordinator with index $\alpha$. Its Lévy measure is given by

$$\Lambda(dx) = \frac{c\alpha}{\Gamma(1-\alpha)} x^{-1-\alpha} dx, \qquad x > 0.$$

The parameter $c > 0$ has a very minor role, as changing $\varsigma$ into $k\varsigma$ merely amounts to change $c$ into $k^\alpha c$. In particular, the following definition does not depend of $c$.

**Definition 2.6** *Let* $\mathsf{a}_1 > \mathsf{a}_2 > \cdots > 0$ *be the sequence of the jumps of a stable subordinator* $(\varsigma(t), t \in [0,1])$. *The law of the random mass-partition* $(\mathsf{a}_1/\varsigma(1), \mathsf{a}_2/\varsigma(1), \ldots)$ *is called the Poisson-Dirichlet distribution with parameter* $(\alpha, 0)$, *and is denoted by* PD$(\alpha, 0)$.

The next proposition states a few basic properties of PD($\alpha$, 0)-distributions. We refer to the survey [191] for much more in this area.

**Proposition 2.6** *Let* $S = (S_1, \dots)$ *be a* PD($\alpha$, 0)*-variable.*

(i) *Put* $R_n = S_{n+1}/S_n$ *for every* $n \in \mathbb{N}$. *Then the variables* $R_1, \dots$ *are independent, and for each* $n \in \mathbb{N}$, $R_n$ *has the beta*($\alpha n$, 1) *distribution, that is* $\mathbb{P}(R_n \le r) = r^{\alpha n}$ *for* $r \in [0, 1]$.
(ii) *The limit*

$$L := \lim_{n\to\infty} n^{1/\alpha} S_n$$

*exists a.s. More precisely, if* $S$ *is given in the form* $S_i = \mathsf{a}_i/\varsigma(1)$ *as in Definition 2.6, then*

$$L = \frac{c}{\Gamma(1-\alpha)\varsigma(1)}.$$

*In particular,* $L > 0$ *a.s. and* $\mathbb{E}(L^a) < \infty$ *for every* $a > -\alpha$.

**Remarks.** • The whole sequence $S$ can be recovered from the independent beta variables $R_1, \dots$ Indeed, since $S_1 + S_2 + \cdots = 1$ a.s., one has

$$S_1 = 1/(1 + R_1 + R_1R_2 + R_1R_2R_3 + \cdots) \quad \text{and} \quad S_{n+1} = S_1R_1 \cdots R_n .$$

• Note that, contrary to the case of gamma subordinators, Proposition 2.6(ii) shows that in the representation of Definition 2.6, $\varsigma(1)$ is *measurable* with respect to the sequence $(\mathsf{a}_1/\varsigma(1), \mathsf{a}_2/\varsigma(1), \dots)$.

*Proof* For the sake of simplicity, we shall suppose that $c = \Gamma(1-\alpha)$, which induces no loss of generality.

(i) The tail of the Lévy measure of $\varsigma$ is $\bar{\Lambda}(x) = \Lambda(]x, \infty[) = x^{-\alpha}$. It follows for the image property of Poisson random measures (see Lemma 2.4) that the family $\{\mathsf{a}_i^{-\alpha}, i \in \mathbb{N}\}$ can be viewed as that of the atoms of a Poisson random measure on $\mathbb{R}_+$ with intensity the Lebesgue measure, ranked in increasing order. Thus the sequence of the increments $\mathbf{e}_1 := \mathsf{a}_1^{-\alpha}, \mathbf{e}_2 := \mathsf{a}_2^{-\alpha} - \mathsf{a}_1^{-\alpha}, \dots$ is formed by i.i.d. standard exponential variables. It then follows that the ratios

$$\frac{\mathsf{a}_1^{-\alpha}}{\mathsf{a}_2^{-\alpha}} = \frac{\mathbf{e}_1}{\mathbf{e}_1 + \mathbf{e}_2}, \quad \frac{\mathsf{a}_2^{-\alpha}}{\mathsf{a}_3^{-\alpha}} = \frac{\mathbf{e}_1 + \mathbf{e}_2}{\mathbf{e}_1 + \mathbf{e}_2 + \mathbf{e}_3}, \dots$$

are mutually independent beta variables with respective parameters (1, 1), (2, 1), ...; see (2.3). We conclude that

$$S_2/S_1 = (\mathsf{a}_1^{-\alpha}/\mathsf{a}_2^{-\alpha})^{1/\alpha}, \; S_3/S_2 = (\mathsf{a}_2^{-\alpha}/\mathsf{a}_3^{-\alpha})^{1/\alpha}, \dots$$

are mutually independent beta variables with respective parameters $(\alpha, 1)$, $(2\alpha, 1), \ldots$

(ii) In the notation of part (i), we have $\mathsf{a}_n^{-\alpha} = \mathbf{e}_1 + \cdots + \mathbf{e}_n$. By the law of large numbers, we deduce that $n^{-1}\mathsf{a}_n^{-\alpha}$ converges to 1 a.s., and we deduce from the representation of $S_n = \mathsf{a}_n/\varsigma(1)$ given in Definition 2.6 that

$$\lim_{n\to\infty} n^{1/\alpha} S_n = 1/\varsigma(1).$$

Then, using for $a > 0$ the identity

$$x^{-a} = \frac{1}{\Gamma(a)} \int_0^\infty \mathrm{e}^{-xt} t^{a-1} dt,$$

we get by Tonelli's theorem

$$\begin{aligned}
\mathbb{E}\left(\varsigma(1)^{-a}\right) &= \frac{1}{\Gamma(a)} \int_0^\infty \mathbb{E}(\mathrm{e}^{-t\varsigma(1)}) t^{a-1} dt \\
&= \frac{1}{\Gamma(a)} \int_0^\infty \exp(-ct^\alpha) t^{a-1} dt \\
&= \frac{1}{\alpha c^{a/\alpha}\Gamma(a)} \int_0^\infty \exp(-s) s^{a/\alpha-1} ds \\
&= \frac{\Gamma(a/\alpha)}{\alpha c^{a/\alpha}\Gamma(a)}.
\end{aligned}$$

We conclude that

$$\mathbb{E}(L^a) = \frac{\Gamma(1+a/\alpha)}{\Gamma(1-\alpha)^{a/\alpha}\Gamma(1+a)}$$

for every $a > 0$, and this formula can then be extended by analytic continuation to any $a > -\alpha$. □

We shall now consider a two-parameter extension of Definition 2.6.

**Definition 2.7** *For every $\alpha \in ]0, 1[$ and $\theta > -\alpha$, we write* PD$(\alpha, \theta)$ *and call Poisson-Dirichlet law with parameters $(\alpha, \theta)$ the probability distribution on $\mathcal{P}_{\mathrm{m}}$ which is absolutely continuous with respect to* PD$(\alpha, 0)$ *with density $L^\theta/\mathbb{E}(L^\theta)$, where $L$ is the variable that appears in Proposition 2.6(ii).*

Probably the most useful tool for the study of this two-parameter family is the following extension of the residual allocation model which was obtained first for PD$(0, \theta)$ random mass partitions (see Corollary 2.3).

**Proposition 2.7** *Fix $\alpha \in ]0,1[$ and $\theta > -\alpha$.*

(i) *Let $S = (S_1, \dots)$ be a random mass partition distributed according to* $\mathrm{PD}(\alpha, \theta)$, *and $N$ a random index with conditional distribution $\mathbb{P}(N = i \mid S) = S_i$, $i \in \mathbb{N}$. Write $S'$ for the random mass partition obtained by the decreasing rearrangement of the terms of the sequence $(S_j/(1-S_N), j \neq N)$. Then $S_N$ and $S'$ are independent, $S_N$ is a beta$(1-\alpha, \alpha)$ variable and $S'$ has the law* $\mathrm{PD}(\alpha, \theta + \alpha)$.

(ii) *Let $\beta_1, \beta_2, \dots$ be a sequence of independent variables such that $\beta_n$ has the beta$(1-\alpha, \theta + n\alpha)$ distribution. Put*

$$S_1^* = \beta_1 \,,\ S_2^* = (1-\beta_1)\beta_2 \,,\ \dots \,,\ S_n^* = \beta_n \times \prod_{i=1}^{n-1}(1-\beta_i), \ \dots$$

*Then the random sequence $S^* = (S_1^*, \dots)$ is distributed as the size-biased reordering of a* $\mathrm{PD}(\alpha, \theta)$ *random mass partition.*

*Proof* Let $\mathtt{M} = \sum_{i=1}^{\infty} \delta_{\mathtt{a}_i}$ be a Poisson point measure on $]0, \infty[$ with intensity $x^{-\alpha-1}dx$, where the atoms $\mathtt{a}_1, \dots$ are ranked in decreasing order. We set $\varsigma(1) = \sum_{i=1}^{\infty} \mathtt{a}_i$, write $P$ for the law of $\mathtt{M}$, and for every $\theta > -\alpha$, denote by $P_\theta$ the absolutely continuous probability measure given by

$$P_\theta = \frac{\varsigma(1)^{-\theta}}{E\left(\varsigma(1)^{-\theta}\right)} P$$

(that this makes sense is plain from Proposition 2.6(ii)). So under $P_\theta$, the random mass-partition $S := (\mathtt{a}_1/\varsigma(1), \dots)$ has the law $\mathrm{PD}(\alpha, \theta)$.

Consider a random index $N$ with conditional distribution

$$P_\theta(N = i \mid \mathtt{M}) = S_i \,, \qquad i \in \mathbb{N},$$

and set $\mathtt{M}' = \sum_{i \neq N} \delta_{\mathtt{a}_i}$.

An application of the Palm formula (cf. Lemma 2.3) yields that for every measurable function $f : [0,1] \to [0, \infty[$ and every non-negative measurable functional $G$ on the space of point measures on $]0, \infty[$

$$\begin{aligned} E_\theta\left(f(S_N)G(\mathtt{M}')\right) &= cE\left(\varsigma(1)^{-\theta-1}\sum_{i=1}^{\infty}\mathtt{a}_i f(\mathtt{a}_i/\varsigma(1))G\left(\sum_{j\neq i}\delta_{\mathtt{a}_j}\right)\right) \\ &= c\int_0^\infty dx x^{-\alpha}E\left((x+\varsigma(1))^{-\theta-1}f\left(\frac{x}{x+\varsigma(1)}\right)G(\mathtt{M})\right), \end{aligned}$$

where $c = 1/E\left(\varsigma(1)^{-\theta}\right)$. Next, we apply Tonelli's theorem and perform the change of variables $y = x/(x+\varsigma(1))$. We get that the preceding quantity equals

$$cE\left(\int_0^1 dy(1-y)^{\alpha-1}y^{-\alpha}f(y)\varsigma(1)^{-\alpha-\theta}G(\mathtt{M})\right)$$
$$= E_{\alpha+\theta}(G(\mathtt{M}))\int_0^1 f(y)\frac{y^{-\alpha}(1-y)^{\alpha-1}}{\Gamma(1-\alpha)\Gamma(\alpha)}dy\,.$$

This shows that under $P_\theta$, $S_N$ and $\mathtt{M}'$ are independent, the former has the beta$(1-\alpha, \alpha)$ distribution, and the latter has the law $P_{\theta+\alpha}$.

The first statement should now be clear, and the second derives from the first by iteration. □

## 2.3 Exchangeable random partitions

In this section, we develop the fundamental observation made by Kingman [139] that mass-partitions are conveniently encoded by certain random partitions of $\mathbb{N}$. This new point of view will have a fundamental importance later in the text; roughly it will provide us with a very powerful method of discretization. The coding can be understood as follows.

Imagine that we have an object with a unit mass, for instance the unit interval endowed with Lebesgue measure, which is split into fragments (i.e. we consider an interval-partition). The ranked sequence of the masses of these fragments is thus given by some mass-partition $\mathbf{s} \in \mathcal{P}_{\mathrm{m}}$. One then introduces a sequence of i.i.d. random points $U_1, \ldots$ which are picked according to the mass distribution of the object (so the $U_i$ are simply i.i.d. uniform variables in the case of the unit interval), and considers the random partition $\pi$ of $\mathbb{N}$, specified by the rule that two indices, say $i$ and $j$, belong to the same block of $\pi$ if and only if the points $U_i$ and $U_j$ belong to the same fragment. An application of the law of large numbers shows that the masses of the fragments can be recovered as the asymptotic frequencies of the blocks of the partition. Conversely, although in general a partition of $\mathbb{N}$ does not necessarily correspond to a mass-partition (because the asymptotic frequencies of blocks may well not exist), there is a bijective correspondence between the laws of *exchangeable* random partitions of $\mathbb{N}$ and probability measures on $\mathcal{P}_{\mathrm{m}}$.

### 2.3.1 Some definition

Recall that $\mathbb{N} = \{1, \dots\}$ stands for the set of positive integers; a *block* is a subset $B \subseteq \mathbb{N}$ . Hereafter, the block formed by the $k$ first integers will play a special role, and it will be convenient to use the notation

$$[k] := \{1, \dots, k\}.$$

In this direction, we also agree that $[\infty] := \mathbb{N}$ for $k = \infty$.

**Definition 2.8** (i) *A partition of $B \subseteq \mathbb{N}$ is a countable collection $\pi = \{\pi_i, i \in \mathbb{N}\}$ of pairwise disjoint blocks such that $\bigcup_{i \in \mathbb{N}} \pi_i = B$, which are always enumerated in increasing order of their least element, that is*

$$\min \pi_i \leq \min \pi_j \qquad \textit{for every } i \leq j,$$

*with the convention that* $\min \emptyset = \infty$.
(ii) *We write $\mathcal{P}_B$ for the set of partitions of $B$. In the special case when $B = [k]$ for some $k \in \overline{\mathbb{N}}$, we simply write $\mathcal{P}_k := \mathcal{P}_{[k]}$; in particular $\mathcal{P}_\infty := \mathcal{P}_{\mathbb{N}}$.*
(iii) *We denote by*

$$\#\pi := \#\{i \in \mathbb{N} : \pi_i \neq \emptyset\} = \max\{i \in \mathbb{N} : \pi_i \neq \emptyset\}$$

*the cardinal of the set of non-empty blocks of a partition $\pi$.*

If $B' \subseteq B$ is a subset of some block $B$ and $\pi \in \mathcal{P}_B$ a partition of $B$, we write $\pi_{|B'}$ for the obvious restriction of $\pi$ to $B'$, that is the partition of $B'$ induced by the sequence of blocks $(\pi_i \cap B', i \in \mathbb{N})$. Restricted partitions naturally yield the notion of compatibility.

**Definition 2.9** *A sequence $\pi^{[1]}, \pi^{[2]}, \dots$ of partitions of $[1], [2], \dots$ is called* **compatible** *if for all integers $k' \leq k$, $\pi^{[k']}$ coincides with the restriction of $\pi^{[k]}$ to $[k']$.*

Plainly, if $\pi \in \mathcal{P}_\infty$ is a partition of $\mathbb{N}$, then the sequence of its restrictions $(\pi_{|[n]}, n \in \mathbb{N})$ is compatible. It is easy to see that the converse holds; here is a formal statement.

**Lemma 2.5** *A sequence of partitions $(\pi^{[n]} : \pi^{[n]} \in \mathcal{P}_n$ and $n \in \mathbb{N})$ is compatible if and only if there exists $\pi \in \mathcal{P}_\infty$ such that $\pi_{|[n]} = \pi^{[n]}$ for every $n \in \mathbb{N}$. Moreover, $\pi$ is then uniquely determined by the sequence $(\pi^{[n]}, n \in \mathbb{N})$.*

*Proof* The compatibility assumption and our rule for labeling blocks of partitions show that for each $i \in \mathbb{N}$, the sequence of blocks $(\pi_i^{[n]}, n \in \mathbb{N})$ increases. If we define

$$\pi_i := \bigcup_{n \in \mathbb{N}} \pi_i^{[n]}, \qquad i \in \mathbb{N},$$

then $(\pi_i, i \in \mathbb{N})$ is a partition of $\mathbb{N}$, say $\pi$, and plainly $\pi_{|[n]} = \pi^{[n]}$ for every $n$. □

Restriction of partitions to $[n]$ also enables us to define a natural metric on $\mathcal{P}_\infty$.

**Lemma 2.6** *The space $\mathcal{P}_\infty$ is endowed with the ultra-metric*

$$d(\pi, \pi') = 1/\max\left\{k \in \mathbb{N} : \pi_{|[k]} = \pi'_{|[k]}\right\}, \qquad \pi, \pi' \in \mathcal{P}_\infty,$$

*with the convention that $1/\max \mathbb{N} = 0$. Then $(\mathcal{P}_\infty, d)$ is compact.*

The terminology *ultra-metric* refers to the fact that the usual triangle inequality can be reinforced into

$$d(\pi, \pi'') \leq \max(d(\pi, \pi'), d(\pi', \pi'')), \qquad \text{for every } \pi, \pi', \pi'' \in \mathcal{P}_\infty.$$

*Proof* Given a sequence $\left(\pi^{(n)}, n \in \mathbb{N}\right)$ in $\mathcal{P}_\infty$, we may extract by the diagonal procedure a subsequence, which is still denoted by $\left(\pi^{(n)}, n \in \mathbb{N}\right)$ for the sake of convenience, such that for every $k \in \mathbb{N}$, the restriction of $\pi^{(n)}$ to $[k]$ is the same for all sufficiently large $n$. More precisely

$$\pi^{(n)}_{|[k]} = \pi^{(n')}_{|[k]} \qquad \text{whenever } n, n' \geq k.$$

This defines a compatible sequence of partitions of $[k]$ when $k$ varies in $\mathbb{N}$, which can thus be expressed in the form $\pi_{|[k]}$ for some $\pi \in \mathcal{P}_\infty$, thanks to Lemma 2.5. It is now straightforward to see that, by construction, $d(\pi, \pi^{(n)}) \leq 1/n$, which establishes the compactness. □

Next, we turn our attention to the notion of asymptotic frequency of blocks, which provides a simple map from $\mathcal{P}_\infty$ to the space $\mathcal{P}_{\mathrm{m}}$ of mass-partitions.

**Definition 2.10** (i) *We say that a block B possesses an* **asymptotic frequency** *if and only if the limit*

$$|B| := \lim_{n\to\infty} \frac{1}{n}\#(B\cap[n]) = \lim_{n\to\infty} \frac{1}{n}\#\{k \in B : k \leq n\}$$

*exists.*
(ii) *If each block of some partition $\pi$ has an asymptotic frequency, then we say that $\pi$ possesses asymptotic frequencies. We then write $|\pi| = (|\pi_1|, \ldots)$, and then $|\pi|^{\downarrow} = (|\pi|_1^{\downarrow}, \ldots)$ for the mass-partition*[4] *given by the decreasing rearrangement of the sequence $|\pi|$.*
(iii) *We say that a partition $\pi$ has* **proper** *asymptotic frequencies if $\pi$ possesses asymptotic frequencies with*

$$\sum_{i=1}^{\infty} |\pi_i| = 1 .$$

When some block of a partition $\pi$ does not have an asymptotic frequency, we decide to write $|\pi| = |\pi|^{\downarrow} = \partial$, where $\partial$ stands for some extra point added to $\mathcal{P}_{\mathrm{m}}$. This allows us to define a natural map $\pi \to |\pi|^{\downarrow}$ from $\mathcal{P}_\infty$ to $\mathcal{P}_{\mathrm{m}} \cup \{\partial\}$ which is measurable but not continuous. Indeed, for any $\pi \in \mathcal{P}_\infty$ and any $\mathbf{s} \in \mathcal{P}_{\mathrm{m}}$, one can easily construct a sequence of partitions $(\pi^{(n)}, n \in \mathbb{N})$ that converges to $\pi$ and such that $|\pi^{(n)}|^{\downarrow}$ tends to $\mathbf{s}$ as $n \to \infty$. This lack of continuity will be a source of some technical difficulties as we continue.

### 2.3.2 Kingman's theory

In this section, we shall consider a natural family of probability measures on $\mathcal{P}_\infty$. To that end, it will be sometimes convenient to identify a partition $\pi$ with an equivalence relation on $\mathbb{N}$, in the sense that $i \overset{\pi}{\sim} j$ if and only if $i$ and $j$ belong to the same block of the partition $\pi$.

For every $n \in \mathbb{N}$, a permutation of $[n]$ is a bijection $\sigma : [n] \to [n]$. For $n = \infty$ (recall that $[\infty] = \mathbb{N}$), we call permutation of $\mathbb{N}$ any bijection $\sigma : \mathbb{N} \to \mathbb{N}$ such that $\sigma(k) = k$ when $k$ is large enough. The group of permutations acts on $\mathcal{P}_n$; more precisely for every permutation $\sigma$ and every partition $\pi$, the relation

$$i \sim j \iff \sigma(i) \overset{\pi}{\sim} \sigma(j), \qquad i, j = 1, \ldots, n$$

is an equivalence relation which can be identified as a partition denoted by $\sigma(\pi)$. In other words, the blocks of $\sigma(\pi)$ are the images of the blocks of $\pi$ by $\sigma^{-1}$, the inverse permutation of $\sigma$.

[4] Fatou's lemma implies that when a partition possesses asymptotic frequencies, then their sum is always bounded from above by 1.

**Definition 2.11** *Let $n \in \overline{\mathbb{N}} := \mathbb{N} \cup \{\infty\}$. A random partition $\pi$ of $[n]$ is called* **exchangeable** *if for every permutation $\sigma$ of $[n]$, $\sigma(\pi)$ has the same law as $\pi$.*

Plainly, a random permutation $\pi$ of $\mathbb{N}$ is exchangeable if and only if the restrictions $\pi_{|[n]}$ are exchangeable for all $n \in \mathbb{N}$. Later in this text, we shall be mainly concerned with the case $n = \infty$; however, for the sake of simplicity, we shall then often omit to specify that we are working with partitions of $\mathbb{N}$. Of course, whenever we will be dealing with random partitions of a finite set, this will be mentioned explicitly.

We now dwell on the informal connection between mass-partitions and partitions of $\mathbb{N}$ which was sketched at the beginning of this section. Let us fix $\mathbf{s} \in \mathcal{P}_{\mathrm{m}}$, and consider an interval representation $\vartheta$ of $\mathbf{s}$. Recall that this means that $\vartheta$ is an open subset of $]0, 1[$ such that the ranked sequence of the lengths of its interval components is given by $\mathbf{s}$. Let $U_1, \ldots$ be an i.i.d. sequence of uniform variables on $[0, 1]$, and consider the random partition $\pi$ of $\mathbb{N}$ induced by the following equivalence relation:

$$i \overset{\pi}{\sim} j \Longleftrightarrow (i = j) \text{ or } (U_i \text{ and } U_j \text{ belong to the same interval component of } \vartheta).$$

The alternative in the right-hand side is needed because when the Lebesgue measure of $\vartheta$ is strictly less than 1, it may happen that $U_i$ does not belong to $\vartheta$ and in that case $\{i\}$ is a singleton of $\pi$.

There is a simple pictorial interpretation of the equivalence relation $\pi$; see the figure below. Think of the unit interval as a paint-box, in which a different color is assigned to each interval component of $\vartheta$. Every integer $i$ then receives the color of the interval to which the variable $U_i$ belongs, and $i$ receives no color if $U_i \notin \vartheta$. The classes of equivalence are given by the families of indices with the same color, where we agree that indices with no color form singletons. We will refer to the random partition $\pi$ defined above as the *paint-box* based on the mass-partition $\mathbf{s}$ (or on the interval-partition $\vartheta$).

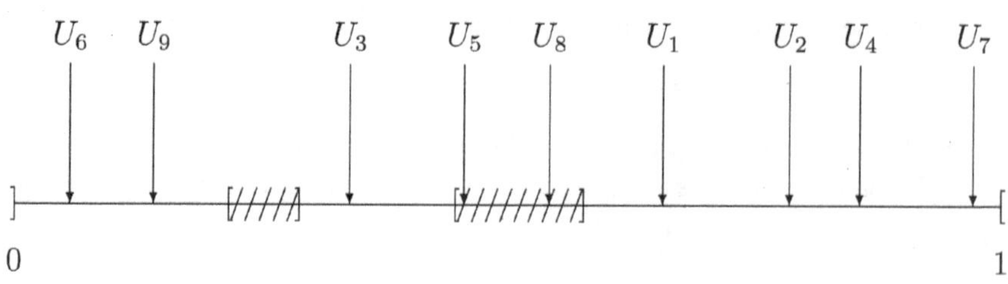


*Paint-box construction:* $\pi = (\{1, 2, 4, 7\}, \{3\}, \{5\}, \{6, 9\}, \{8\})$

**Lemma 2.7** *The paint-box based on* $\mathbf{s} \in \mathcal{P}_m$ *is a random exchangeable partition. Its law does not depend on the choice of the interval representation* $\vartheta$ *of* $\mathbf{s} \in \mathcal{P}_m$*; it will be denoted by* $\varrho_{\mathbf{s}}$.

*Proof* Consider any permutation $\sigma$; the partition $\sigma(\pi)$ is specified by equivalence relation:

$$i \overset{\sigma(\pi)}{\sim} j \iff (i = j) \text{ or } (U_{\sigma(i)} \text{ and } U_{\sigma(j)} \text{ belong to the same interval component of } \vartheta).$$

The variables $U_{\sigma(1)}, \ldots$ are i.i.d. and uniformly distributed on $[0, 1]$; this shows that $\pi$ is exchangeable.

Next, let $\vartheta'$ be another interval representation of $\mathbf{s}$. We can then find a measurable map $f : [0, 1] \to [0, 1]$ which preserves the Lebesgue measure and induces a bijection from $\vartheta$ to $\vartheta'$ such that the image by $f$ of the interval components of $\vartheta$ are precisely the interval components of $\vartheta'$. By construction

$$i \overset{\pi}{\sim} j \iff (i = j) \text{ or } (U'_i \text{ and } U'_j \text{ belong to the same interval component of } \vartheta');$$

which yields our claim, since the variables $U'_n := f(U_n)$ for $n \in \mathbb{N}$, are again i.i.d. and uniformly distributed on $[0, 1]$. □

Lemma 2.7 yields another simple construction which is equivalent to the paint-box: Recall that for every measure $m$, a point $\mathsf{a}$ such that $m(\{\mathsf{a}\}) > 0$ is called an atom of $m$. Given a mass-partition $\mathbf{s}$, consider a probability distribution $F$ on $\mathbb{R}$, such that the ranked sequence of the masses of the atoms of $F$, namely $(F(\mathsf{a}_i) - F(\mathsf{a}_i-), i \in I)$, coincides with the sequence of strictly positive terms in $\mathbf{s}$. Next, let $\xi_1, \ldots$ be a sequence of i.i.d. variables with distribution $F$, and $\pi$ the random partition of $\mathbb{N}$ such that $i$ and $j$ belong to the same block if and only if $\xi_i = \xi_j$. The simple observation that we may choose $\xi_i = F^{-1}(U_i)$ easily yields that $\pi$ is a paint-box based on $\mathbf{s}$.

Next, we state some elementary properties of paint-boxes.

**Proposition 2.8** *Let* $\pi$ *be a paint-box based on some mass-partition* $\mathbf{s} \in \mathcal{P}_m$. *Then the following assertions hold.*

(i) *The paint-box* $\pi$ *possesses asymptotic frequencies. More precisely,* $|\pi|^{\downarrow} = \mathbf{s}$ *and* $|\pi_1|$ *is a size-biased sample of* $\mathbf{s}$ *a.s.*
(ii) *For every* $i \in \mathbb{N}$, *if* $|\pi_i| = 0$, *then* $\pi_i$ *is either a singleton or empty a.s.*
(iii) *More precisely, some blocks of* $\pi$ *are reduced to singletons if and only if* $\mathbf{s}$ *is improper, and in that case, the set of singletons* $\pi_0 := \{i \in \mathbb{N} : i$

*is a singleton of $\pi$\} has an asymptotic frequency given by $|\pi_0| = s_0 = 1 - \sum_{i=1}^{\infty} s_i$ a.s.*

(iv) *If* $\mathbf{s}$ *is proper, then the sequence $|\pi|$ of the asymptotic frequencies of the blocks of $\pi$ is a size-biased reordering of* $\mathbf{s}$.

*Proof* If $A \subseteq ]0, 1[$ is some measurable set and $U_1, \ldots$ a sequence of i.i.d. uniform variables, then $\left(\mathbb{1}_{\{U_i \in A\}}, i \in \mathbb{N}\right)$ is an i.i.d. sequence of Bernoulli variables with mean given by the Lebesgue measure $|A|$ of $A$. It thus follows from the law of large numbers that the random block $B := \{i \in \mathbb{N} : U_i \in A\}$ has an asymptotic frequency $|B| = |A|$.

Assertions (i–iii) now follow immediately from the paint-box construction and the preceding observation. Finally, (iv) is a consequence of Lemma 2.1 and our convention for labeling the blocks of a partition. □

The fundamental result about exchangeable random fragmentations states that the paint-box construction induces a bijection between probability measures on mass-partitions and exchangeable probability measures on $\mathcal{P}_\infty$.

**Theorem 2.1** (Kingman) *Let $\pi$ be an exchangeable random partition of $\mathbb{N}$. Then $\pi$ possesses asymptotic frequencies a.s. More precisely, the law of $\pi$ can be expressed as a mixture of paint-boxes:*

$$\mathbb{P}(\pi \in \cdot) = \int_{\mathcal{P}_m} \mathbb{P}\left(|\pi|^{\downarrow} \in d\mathbf{s}\right) \varrho_{\mathbf{s}}(\cdot),$$

*where $\varrho_{\mathbf{s}}$ stands for the law of the paint-box based on* $\mathbf{s}$, *which is defined in Lemma 2.7.*

In words, the second part of the statement claims that one can construct a version of an arbitrary exchangeable random partition $\pi$ as follows: One first considers some (random) interval representation $\vartheta_S$ of the ranked asymptotic frequencies $|\pi|^{\downarrow} := S$ of $\pi$ and an independent sequence $U_1, \ldots$ of i.i.d. uniform variables on $[0, 1]$. Then the mixture of paint-boxes constructed from $\vartheta_S$ and the $U_i$ has the same law as $\pi$.

Kingman's original proof of Theorem 2.1 uses a martingale argument. We shall present here a simpler approach, due to Aldous [3]. It relies on de Finetti's theorem for sequences of exchangeable variables, which claims that exchangeable sequences of variables are mixtures of i.i.d. sequences. Here is the formal statement.

**Theorem 2.2** (de Finetti) *Let $\xi_1, \ldots$ be an exchangeable sequence of real-valued variables, that is for every permutation $\sigma$ of $\mathbb{N}$, $(\xi_1, \ldots)$ and $(\xi_{\sigma(1)}, \ldots)$ have the same finite-dimensional distributions. Then the sequence of empirical distributions*

$$\mu_n(dx) := \frac{1}{n} \sum_{i=1}^{n} \delta_{\xi_i}(dx)$$

*converges a.s. when $n \to \infty$, in the sense of weak convergence of probability measures on $\mathbb{R}$, to some random probability measure $\mu(dx)$. Moreover, conditionally on $\mu$, the variables $\xi_1, \ldots$ are i.i.d. with joint distribution $\mu$.*

We refer to [3] for a proof of de Finetti's theorem, and now establish Theorem 2.1.

*Proof* For a given partition of $\mathbb{N}$, let us call any function $b : \mathbb{N} \to \mathbb{N}$ that maps all the points of each block of the partition to the same point of this block a selection map. For instance, $b(i)$ can be the smallest element of the block that contains $i$.

Now let $b$ be some selection map for the exchangeable random partition $\pi$ and $U_1, \ldots$ a sequence of i.i.d. uniform variables on $[0, 1]$, which is independent of $\pi$ and $b$, and set $\xi_i = U_{b(i)}$. It should be plain that the law of the sequence $(\xi_i, i \in \mathbb{N})$ does not depend on the choice of the selection map $b$.

The key lies in the observation that the sequence $\xi_1, \ldots$ is exchangeable. Indeed, let $\sigma$ be a permutation of $\mathbb{N}$; we then have

$$\xi_{\sigma(i)} = U_{b(\sigma(i))} = U'_{b'(i)},$$

where $U'_j = U_{\sigma(j)}$ and $b' = \sigma^{-1} \circ b \circ \sigma$. It is immediately seen that $b'$ is a selection map for the partition $\sigma(\pi)$, and that $U'_1, \ldots$ are i.i.d. uniform variables on $[0, 1]$ which are jointly independent of the partition $\sigma(\pi)$ and its selection map $b'$. By assumption, $\pi$ is exchangeable and independent of the $U_i$, so $((U'_i)_{i \in \mathbb{N}}, \sigma(\pi))$ has the same law as $((U_i)_{i \in \mathbb{N}}, \pi)$, and the sequence $\xi_1, \ldots$ is exchangeable.

Next, we observe that we may recover a.s. the partition $\pi$ from the sequence $(\xi_1, \ldots)$, since the blocks of $\pi$ are precisely the sub-families of indices $i$ for which the variables $\xi_i$ take the same value. By de Finetti's theorem, there is some random probability measure $\mu$ on $[0, 1]$ such that conditionally on $\mu$, the variables $\xi_1, \ldots$ are i.i.d. with law $\mu$.

We now work conditionally on $\mu$ and write $q$ for the quantile function of $\mu$, that is the inverse of its distribution function. Introduce the open set of flat points of $q$,

$$\vartheta := \{x \in ]0,1[ : \exists \varepsilon > 0 \text{ such that } q(x) = q(y) \text{ whenever } |y-x| < \varepsilon\},$$

so that the lengths of the intervals components of $\vartheta$ coincide with the masses of the atoms of $\mu$. Introduce an independent sequence $V_1, \ldots$ of i.i.d. uniform variables on $[0,1]$, so that the sequence $(q(V_1), \ldots)$ has the same law as $(\xi_1, \ldots)$ conditionally on $\mu$. Moreover, two distinct indices, say $i$ and $j$, belong to the same block of $\pi$ if and only if $V_i$ and $V_j$ belong to the same interval component of $\vartheta$. This shows that conditionally on $\mu$, $\pi$ is distributed as a paint-box based on $\vartheta$. □

We now present a couple of elementary but useful consequences of Theorem 2.1. In the first, we compute some simple conditional expectations.

**Corollary 2.4** *Let $\pi$ be an exchangeable random partition of $\mathbb{N}$ and $|\pi|$ the sequence of the asymptotic frequencies of its blocks.*

(i) *For every measurable function $f : [0,1] \to \mathbb{R}_+$, we have*

$$\mathbb{E}\left(f(|\pi_1|) \mid |\pi|^{\downarrow}\right) = \sum_{i=1}^{\infty} |\pi|_i^{\downarrow} f(|\pi|_i^{\downarrow}) = \sum_{i=1}^{\infty} |\pi_i| f(|\pi_i|).$$

(ii) *Fix $n \in \mathbb{N}$ and write $\mathbf{1}_{[n]}$ for the partition of $[n]$ which has a unique non-empty block, that is $\mathbf{1}_{[n]} = ([n], \emptyset, \ldots)$. The conditional probability that $\pi_{|[n]} = \mathbf{1}_{[n]}$ given the sequence $|\pi|$ is*

$$\mathbb{P}\left(\pi_{|[n]} = \mathbf{1}_{[n]} \mid |\pi|\right) = |\pi_1|^{n-1}.$$

*Proof* The first claim follows immediately from Proposition 2.8(i), and the second from the paint-box construction. □

The next corollary concerns the asymptotic frequencies of intersections of blocks related to independent exchangeable partitions; it will be useful in the next chapter.

**Corollary 2.5** *Let $B \subseteq \mathbb{N}$ be a block which possesses an asymptotic frequency and $(\pi^{(j)}, j \in \mathbb{N})$ a sequence of independent exchangeable random partitions in $\mathcal{P}_\infty$. Then the block*

$$B' := B \cap \left(\bigcap_{j \in \mathbb{N}} \pi_1^{(j)}\right)$$

*possesses a.s. an asymptotic frequency which is given by the infinite product*

$$|B'| := |B| \times \prod_{j\in\mathbb{N}} |\pi_1^{(j)}| .$$

Later in this text, we shall often apply Corollary 2.5 in the simple situation when $\pi^{(j)} = \mathbf{1}_{\mathbb{N}}$ for $j \geq 2$, in which case the statement yields the following: If $B$ is a block which has an asymptotic frequency and $\pi$ a random exchangeable partition, then the restricted partition $\pi_{|B}$ also possesses asymptotic frequencies, and more precisely, $|B \cap \pi_j| = |B| \times |\pi_j|$ for every $j \in \mathbb{N}$.

*Proof* Thanks to Kingman's theorem, we may assume without loss of generality that each random partition $\pi^{(j)}$ is obtained as the paint-box associated to some interval-partition $\vartheta^{(j)}$ and a sequence $(U_i^{(j)}, i \in \mathbb{N})$ of i.i.d. uniform variables, such that the sequences $(U_i^{(j)}, i \in \mathbb{N})$ are independent as $j$ varies. For each $j \in \mathbb{N}$, write $I^{(j)}$ for the interval component of $\vartheta^{(j)}$ that contains the first uniform variable $U_1^{(j)}$, with the convention that $I^{(j)} = \emptyset$ when $U_1^{(j)} \notin \vartheta^{(j)}$, and work henceforth conditionally on the $I^{(j)}$. Recall that the length $|I^{(j)}|$ coincides with the asymptotic frequency of the first block $\pi_1^{(j)}$. Plainly, if we define for every $i \in \mathbb{N}$

$$\beta_i := \mathbb{1}_{\{U_i^{(j)} \in I^{(j)},\ \forall j \in \mathbb{N}\}} ,$$

then $\beta_2, \beta_3, \ldots$ is a sequence of i.i.d. Bernoulli variables with mean $\prod_{j\in\mathbb{N}} |I^{(j)}|$, and we have

$$B' = \{i \in B : \beta_i = 1\} .$$

Let $b_n$ denote the $n$-th smallest element of $B$; the hypothesis that $B$ possesses an asymptotic frequency reads

$$\lim_{n\to\infty} n/b_n = |B| .$$

It then follows from the law of large numbers that

$$\lim_{n\to\infty} \frac{1}{n} \# \{i \in B : i \leq b_n \text{ and } \beta_i = 1\} = \prod_{j\in\mathbb{N}} |I^{(j)}| \qquad \text{a.s.}$$

By a standard argument of monotonicity, this proves that the block $B'$ possesses an asymptotic frequency which is given by $|B| \times \prod_{j\in\mathbb{N}} |I^{(j)}|$. □

In particular, consider two independent random exchangeable partitions, $\pi$ and $\pi'$. Then it follows from Corollary 2.5 and an easy argument of exchangeability that with probability one, if $C$ is a block of $\pi$ and $C'$ a block of $\pi'$, the block $C \cap C'$ has an asymptotic frequency which is given by $|C \cap C'| = |C||C'|$. This fact is perhaps less obvious that it seems at first sight. Indeed, it might be interesting to point out that if one only assumes that $C$ and $C'$ are two independent random blocks of $\mathbb{N}$ which both possess asymptotic frequencies, then the block $C \cap C'$ may well fail to have asymptotic frequency $|C||C'|$ (simple counterexamples can be constructed taking $C$ and $C'$ deterministic).

We now conclude this section with an important result of continuity in distribution for mixtures of paint-boxes. Indeed, one difficulty related to this representation is that the map $\pi \to |\pi|^{\downarrow}$ that enables one to recover the mass-partition from a paint-box is not continuous. However, this difficulty vanishes when one considers convergence in law.

**Proposition 2.9** *Consider for each $n \in \overline{\mathbb{N}}$, a random exchangeable partition $\pi^{(n)}$, and write $|\pi^{(n)}|^{\downarrow}$ for the mass-partition given by the ranked sequence of the asymptotic frequencies of its blocks. The following conditions are equivalent:*

(i) *When $n \to \infty$, $|\pi^{(n)}|^{\downarrow}$ converges in distribution on $\mathcal{P}_{\mathrm{m}}$ to $|\pi^{(\infty)}|^{\downarrow}$.*
(ii) *When $n \to \infty$, $\pi^{(n)}$ converges in distribution on $\mathcal{P}_{\infty}$ to $\pi^{(\infty)}$.*

*Proof* It is convenient to denote by $S^{(n)}$ a random mass-partition distributed as $|\pi^{(n)}|^{\downarrow}$.

Suppose (i) holds. Since the space of mass-partitions $\mathcal{P}_{\mathrm{m}}$ is metric and compact, we may apply Skorokhod representation theorem (see for example Billingsley [57]) and assume that $\lim_{n\to\infty} S^{(n)} = S^{(\infty)}$ a.s. In particular $\lim_{n\to\infty} S_k^{(n)} = S_k^{(\infty)}$ a.s. for each $k \in \mathbb{N}$. Set for $n \in \overline{\mathbb{N}}$ and $k \in \mathbb{N}$

$$\Sigma_0^{(n)} = 0 \quad \text{and} \quad \Sigma_k^{(n)} = S_1^{(n)} + \cdots + S_k^{(n)},$$

so $\lim_{n\to\infty} \Sigma_k^{(n)} = \Sigma_k^{(\infty)}$ a.s. for every $k \in \mathbb{N}$. Consider the random interval-partition $\vartheta^{(n)} \in \mathcal{P}_{\mathrm{I}}$

$$\vartheta^{(n)} = \bigcup_{k\in\mathbb{N}} ]\Sigma_{k-1}^{(n)}, \Sigma_k^{(n)}[,$$

so each $\vartheta^{(n)}$ is an interval representation of $S^{(n)}$. If we introduce a sequence $U_1, \ldots$ of i.i.d. uniform variables on $[0, 1]$ which is independent of the $S^{(n)}$, we may suppose that each $\pi^{(n)}$ is the mixture of paint-boxes based on $\vartheta^{(n)}$ and $(U_1, \ldots)$.

Now for each $i, k \in \mathbb{N}$, we have that

$$\lim_{n\to\infty} \mathbb{1}_{\{\Sigma_{k-1}^{(n)} < U_i < \Sigma_k^{(n)}\}} = \mathbb{1}_{\{\Sigma_{k-1}^{(\infty)} < U_i < \Sigma_k^{(\infty)}\}} \qquad \text{a.s.},$$

which implies that with probability one, for every $\ell \in \mathbb{N}$, the restrictions of $\pi^{(n)}$ and $\pi^{(\infty)}$ to $[\ell]$ coincide when $n$ is sufficiently large. Thus $\lim_{n\to\infty} \pi^{(n)} = \pi^{(\infty)}$ a.s., which shows that (i) $\Rightarrow$ (ii).

Next, suppose (ii) holds. As $\mathcal{P}_{\mathrm{m}}$ is a compact metric space, the space of probability measures on $\mathcal{P}_{\mathrm{m}}$ is also metric and compact by Prohorov's Theorem (cf. Section 6 in Billingsley [57]), and thus from any subsequence of $(S^{(n)}, n \in \mathbb{N})$ we can extract a subsequence, say $(\tilde{S}^{(n)}, n \in \mathbb{N})$, which converges weakly to some random mass-partition $\tilde{S}^{(\infty)}$. We deduce from the first part of the proof that $\pi^{(\infty)}$ is distributed as a mixture of paint-boxes based on $\tilde{S}^{(\infty)}$, and in particular that the distribution of $\tilde{S}^{(\infty)}$ does not depend on the chosen subsequence (more precisely, $\tilde{S}^{(\infty)}$ has the same law as $|\pi^{(\infty)}|^{\downarrow}$). Thus $S^{(n)}$ converges in distribution to $\tilde{S}^{(\infty)}$, which proves that (ii) $\Rightarrow$ (i). □

### 2.3.3 Exchangeable partition probability functions

Let $\pi$ be an exchangeable random partition. Fix an integer $n \in \mathbb{N}$ and consider a partition $\varphi = (B_1, \ldots, B_k, \emptyset, \ldots)$ of $[n]$ with $B_k \neq \emptyset$. The exchangeability implies that the probability that the restriction of $\pi$ to $[n]$ coincides with $\varphi$ can be expressed in the form

$$\mathbb{P}\left(\pi_{|[n]} = \varphi\right) = \mathrm{p}(\#B_1, \ldots, \#B_k)$$

where $\#B$ denotes the cardinal of the finite block $B$ and $\mathrm{p}$ is called the *exchangeable partition probability function* (in short, EPPF) of $\pi$.

Note that $\mathrm{p}$ is a symmetric function of a finite (but not fixed) number of variables which are all positive integers. By symmetry, one usually writes the argument of EPPF in decreasing order. By compatibility (i.e. $\pi_{|[n]}$ can be obtained as the restriction of $\pi_{|[n+1]}$ to $[n]$), it is immediately seen that an EPPF fulfills the addition rule

$$\mathrm{p}(n_1, \ldots, n_k) = \mathrm{p}(n_1, \ldots, n_k, 1) + \sum_{j=1}^{k} \mathrm{p}(n_1, \ldots, n_{j-1}, n_j + 1, n_{j+1}, \ldots, n_k).$$

The following computations for paint-boxes are straightforward.

**Lemma 2.8** *Let $\mathbf{s} \in \mathcal{P}_{\mathrm{m}}$ be a proper mass-partition such that all the terms of $\mathbf{s}$ are distinct, and $\varrho_{\mathbf{s}}$ the law of the paint-box based on $\mathbf{s}$. Let also $x_1, \ldots, x_k$ be $k$ terms picked without replacement from $\mathbf{s}$, and $(B_1, \ldots, B_k, \emptyset, \ldots)$ be some partition of $[n]$ with $B_k \neq \emptyset$. Then we have*

$$\varrho_{\mathbf{s}}\left(\pi_{|[n]} = (B_1, \ldots, B_k, \emptyset, \ldots), |\pi_1| = x_1, \ldots, |\pi_k| = x_k\right) = x_1^{\#B_1} \cdots x_k^{\#B_k},$$

*and as a consequence*

$$\varrho_{\mathbf{s}}\left(\pi_{|[n]}=(B_1,\ldots,B_k,\emptyset,\ldots)\mid|\pi_1|=x_1,\ldots,|\pi_k|=x_k\right)$$
$$=\left(\prod_{i=1}^{k}x_i^{\#B_i-1}\right)\left(\prod_{j=1}^{k-1}(1-(x_1+\cdots+x_j))\right).$$

*Proof* The first formula is plain from the construction of the paint-box with an interval representation of **s** and i.i.d. uniform variables. The second follows from the fact that the sequence $|\pi|$ is a size-biased reordering of **s** (see Proposition 2.8(ii)), and thus

$$\varrho_{\mathbf{s}}(|\pi_1|=x_1,\ldots,|\pi_k|=x_k)=x_1\times\frac{x_2}{1-x_1}\times\cdots\times\frac{x_k}{1-(x_1+\cdots+x_{k-1})}.$$

□

We can now present a couple of important explicit formulas for the EPPFs related to Poisson-Dirichlet distributions (recall Sections 2.2.4 and 2.2.5). In this direction, for every $\alpha\in[0,1[$ and $\theta>-\alpha$, it is convenient to call any random exchangeable partition such that the ranked sequence of its asymptotic frequencies is distributed according PD$(\alpha,\theta)$ a PD$(\alpha,\theta)$-partition. Alternatively, a PD$(\alpha,\theta)$-partition is a random partition with the same law as a mixture of paint-boxes based on some PD$(\alpha,\theta)$-random mass-partition.

**Theorem 2.3** *For every $\alpha\in[0,1[$ and $\theta>-\alpha$, write $\mathrm{p}_{\alpha,\theta}$ for the EPPF of an $(\alpha,\theta)$-partition. Pick integers $k\le n$ and $n_1,\ldots,n_k$ such that $n_1+\cdots+n_k=n$.*

(i) *For $\alpha=0$, we have* **Ewens sampling formula***:*

$$\mathrm{p}_{0,\theta}(n_1,\ldots,n_k)=\frac{\theta^k}{\theta(\theta+1)\cdots(\theta+n-1)}\prod_{i=1}^{k}(n_i-1)!$$

(ii) *For $\alpha>0$, we have* **Pitman sampling formula***:*

$$\mathrm{p}_{\alpha,\theta}(n_1,\ldots,n_k)=\frac{(\theta/\alpha)_{k\uparrow}}{(\theta)_{n\uparrow}}\prod_{i=1}^{k}-(-\alpha)_{n_i\uparrow},$$

*where for every integer $\ell\ge1$ and real number $a$,*

$$(a)_{0\uparrow}=1\quad\text{and}\quad(a)_{\ell\uparrow}=a(a+1)\cdots(a+\ell-1).$$

*Proof* Recall from Proposition 2.8(iv) that in the proper case, the sequence $|\pi|$ of the asymptotic frequencies of the blocks of an exchangeable random partition $\pi$ is a size-biased reordering of the ranked sequence $|\pi|^{\downarrow}$. When

$\alpha = 0$, we may apply the residual allocation model described in Corollary 2.3 to construct the size-biased reordering of the PD$(0, \theta)$-variable using an i.i.d. sequence $\beta_1, \ldots$ of standard beta variables with parameter $(1, \theta)$. Recall also that the moments of a beta variable with parameter $(a, b)$, say $\beta$, are given by

$$\mathbb{E}\left(\beta^k(1-\beta)^\ell\right) = \frac{\Gamma(a+b)\Gamma(a+k)\Gamma(b+\ell)}{\Gamma(a)\Gamma(b)\Gamma(a+b+k+\ell)}.$$

Then Lemma 2.8 and the preceding observations yield

$$\begin{aligned}&\mathrm{p}_{0,\theta}(n_1, \ldots, n_k)\\ &= \mathbb{E}\left(\left(\beta_1^{n_1-1}(1-\beta_1)^{n_2+\cdots+n_k}\right)\cdots\left(\beta_{k-1}^{n_{k-1}-1}(1-\beta_{k-1})^{n_k}\right)\beta_k^{n_k-1}\right)\\ &= \frac{\theta^k}{\theta(\theta+1)\cdots(\theta+n-1)}\prod_{i=1}^{k}(n_i-1)!\end{aligned}$$

Similar calculations for $\alpha \in ]0, 1[$ and $\theta > -\alpha$ yield the second formula. □

Let us now present a useful consequence of Theorem 2.3, which provides a remarkable recursive construction of PD$(\alpha, \theta)$-partitions due to Dubins and Pitman in the special case $\alpha = 0$ and to Pitman [180] for arbitrary $\alpha \in [0, 1[$. It can be described as follows.

Imagine a *Chinese Restaurant* having an infinite number of tables with infinite capacity, in the sense that each table can accommodate an infinite number of customers. Tables are denoted by $T_1, T_2, \ldots$; initially all the tables are empty. Customers arrive one after the other and pick a table according to a random process that we now explain. Fix $\alpha \in [0, 1[$ and $\theta > -\alpha$. The first customer, denoted by 1, sits at the first table $T_1$. For every $n \geq 1$, if at the time when the $(n+1)$-th customer enters the restaurant there are $k$ non-empty tables, $T_1, \ldots, T_k$, this new customer decides to sit alone at table $T_{k+1}$ with probability $(\theta + k\alpha)/(n+\theta)$, and at table $T_i$ for $1 \leq i \leq k$ with probability $(\#B_i(n) - \alpha)/(n+\theta)$, where $B_i(n)$ is the block of $[n]$ formed by the customers already sat at table $T_i$. So for each $n$, the occupation of the tables yields a partition $\pi(n) = (B_1(n), \ldots)$ of $[n]$. Note that by construction, blocks are labeled according to the increasing order of their least element, in agreement with our convention. Clearly, these partitions are compatible as $n$ varies, so by Lemma 2.5, there exists a unique (random) partition $\pi(\infty)$ of $\mathbb{N}$ such that $\pi(n) = \pi(\infty)_{|[n]}$ for each $n$.

**Corollary 2.6** *The random partition $\pi(\infty)$ constructed above is a* PD$(\alpha, \theta)$-*partition.*

Note that Corollary 2.6 implies in particular that the random partition $\pi(\infty)$ is exchangeable, a property which is not obvious from the construction.

*Proof* An immediate check by iteration can be made that for every partition of $[n]$ with $k$ non-empty blocks, say $(B_1, \ldots, B_k, \emptyset, \ldots)$, the probability that $\pi(n) = (B_1, \ldots, B_k, \emptyset, \ldots)$ is given for $\alpha = 0$ and $\theta > 0$ by

$$\mathbb{P}(\pi(\infty)_{|[n]} = (B_1, \ldots, B_k, \emptyset, \ldots)) = \frac{\theta^k}{\theta(\theta+1)\cdots(\theta+n-1)} \prod_{i=1}^{k} (\#B_i - 1)!\,,$$

and for $0 < \alpha < 1$ and $\theta > -\alpha$ by

$$\mathbb{P}(\pi(\infty)_{|[n]} = (B_1, \ldots, B_k, \emptyset, \ldots)) = \frac{(\theta/\alpha)_{k\uparrow}}{(\theta)_{n\uparrow}} \prod_{i=1}^{k} -(\alpha)_{\#B_i - 1\uparrow}\,.$$

The comparison with Theorem 2.3 establishes the claim. □

In the special case when $\alpha = 0$ and $\theta$ is an integer, the Chinese Restaurant can be interpreted as a variation of Pólya's urn model, see [125]: Let $c_0, c_1, \ldots$ denote a sequence of different colors, and consider an urn which contains initially one ball of color $c_1$ and $\theta$ balls with color $c_0$. At the first step, we pick a ball at random in the urn, note its color $c$ and replace it in the urn together with a new colored ball. More precisely, if $c = c_1$, then the color of the new ball which is added to the urn is $c_1$; whereas if $c = c_0$, then the color of the new ball is $c_2$. We iterate the process in an obvious way. After $n$ steps, there are $\theta + n + 1$ balls in the urn, with colors $c_0, c_1, \ldots, c_k$. We pick a ball at random, uniformly and independently of the preceding drawing. If this ball has color $c_\ell$ for some $\ell = 1, \ldots, k$, then we replace it in the urn together with an additional ball with color $c_\ell$. If the ball has color $c_0$, we replace it in the urn together with an additional ball with the new color $c_{k+1}$. Plainly, the distribution of the numbers of balls with respective colors $c_1, \ldots$ in the urn after $n$ steps is the same as that of the numbers of customers sat at table $T_1, \ldots$ in a Chinese Restaurant process when the total number of customers is $n+1$. We refer to Section 3.1 in [186] for much more on Chinese Restaurants and their applications.

It is also interesting to combine Lemma 2.8 with the formulas obtained in Proposition 2.4 for certain mixtures of paint-boxes related to Poisson random measures. More precisely, let $\mathsf{a}_1 > \mathsf{a}_2 > \cdots > 0$ be the ranked sequence of the atoms of some Poisson point measure on $]0, \infty[$ with intensity $\Lambda$, where $\Lambda$ is an infinite measure on $]0, \infty[$ which has no atoms and fulfills (2.5).

Assume further that the random variable $\varsigma := \sum_{i=1}^{\infty} \mathsf{a}_i$ possesses a continuous density

$$\rho(x) := \mathbb{P}(\varsigma \in dx)/dx$$

which is strictly positive for $x > 0$. Recall from Proposition 2.4 that one can then define a random mass-partition, say $S_{|x}$, by conditioning the Poisson-Kingman partition $S := (\mathsf{a}_1/\varsigma, \mathsf{a}_2/\varsigma, \dots)$ on $\varsigma = x$.

**Corollary 2.7** *In the notation above, write $\mathsf{p}_{|x}$ for the EPPF of a random exchangeable partition given by the mixture of paint-boxes based on $S_{|x}$. Pick integers $k \le n$ and $n_1, \dots, n_k$ such that $n_1 + \dots + n_k = n$. We then have*

$$\begin{aligned}
&\mathsf{p}_{|x}(n_1, \dots, n_k)\\
&= \frac{1}{x^n \rho(x)} \int_{x_1+\dots+x_k<x} \Lambda(dx_1)\cdots\Lambda(dx_k)\,\rho(x-(x_1+\dots+x_k)) \left(\prod_{i=1}^{k} x_i^{n_i}\right).
\end{aligned}$$

*Proof* Let $\pi$ be a random partition obtained by the paint-box construction based on $S_{|x}$. Recall Lemma 2.8, Proposition 2.4, and Proposition 2.8(iv). We deduce that for every partition $(B_1, \dots, B_k, \emptyset, \dots)$ of $[n]$ with $B_k \neq \emptyset$, there is the identity

$$\begin{aligned}
&\mathbb{P}\left(\pi_{|[n]} = (B_1, \dots, B_k, \emptyset, \dots), x|\pi_1| \in dx_1, \dots, x|\pi_k| \in dx_k\right)\\
&\quad = \frac{\rho(x-(x_1+\dots+x_k))}{x^n \rho(x)} \left(\prod_{i=1}^{k} x_i^{\#B_i}\right) \Lambda(dx_1)\cdots\Lambda(dx_k),
\end{aligned}$$

where $x_1, \dots, x_k$ denote generic positive real numbers such that $x_1 + \dots + x_k < x$. Integration with respect to the latter yields the formula of the statement. □

Finally, we mention that one can compute the EPPF, say $\mathsf{p}$, corresponding to the Poisson-Kingman random mass-partition $S$, by integration of the formula of Corollary 2.7 with respect to the distribution of $\varsigma$. Specifically, we immediately get

$$\begin{aligned}
&\mathsf{p}(n_1, \dots, n_k)\\
&= \int_0^\infty \mathbb{P}(\varsigma \in dx)\mathsf{p}_{|x}(n_1, \dots, n_k)\\
&= \int_0^\infty dv\rho(v) \int_{]0,\infty[} \Lambda(dx_1)\dots\int_{]0,\infty[}\\
&\qquad \Lambda(dx_k)(v+x_1+\dots+x_k)^{-n} \left(\prod_{i=1}^{k} x_i^{n_i}\right).
\end{aligned}$$

## 2.4 Comments

The point of view of thinking of a proper mass-partition as the ranked sequence of the masses of the atoms of some discrete probability measure has been motivated originally by problems in non-parametric statistics; see Ferguson [102] and Kingman [136]. The latter introduced the class of random-mass partitions induced by normalized subordinators which is called nowadays Poisson-Kingman partitions; see [187] and references therein. The related family of random-mass partitions discussed in Section 2.2.3, which can be obtained from subordinators by conditioning instead of normalizing, has been considered by Perman, Pitman and Yor [179] to whom Proposition 2.4 is due. In this direction, it is interesting to observe that size-biased reordering often makes the description of the distribution of a random mass-partition much simpler (compare Proposition 2.4 with the formulas obtained by Perman [178] and Pitman [187] for the finite-dimensional distributions of Poisson-Kingman partitions or their conditioned version). We refer to Gnedin [112] and Pitman [181] for some general properties of random mass-partitions and their size-biased permutations.

The Poisson-Dirichlet distribution PD$(0, 1)$ arises in several remarkable limit theorems about the component frequency spectrum of large structures in combinatorics, including the decomposition of permutations in cycles [137, 210], and the decomposition of integers in prime factors [57, 209]. The one-parameter family denoted here by PD$(0, \theta)$ for $\theta > 0$, has been introduced by Kingman [136]. It has appeared in a variety of contexts, such as Bayesian statistics (see in particular Ferguson [102]), population genetics (more precisely to describe the frequency of species in population models with neutral mutations, see Durrett [87], Engen [93], Kingman [137], ...), invariant distributions for certain split and merge transformations (cf. Gnedin and Kerov [113], Pitman [185], Diaconis *et al.* [77],...). We refer to the book [13] by Arratia, Barbour and Tavaré for further examples and references.

The two parameter family PD$(\alpha, \theta)$ was introduced by Perman, Pitman and Yor [179]. It appears naturally in connection with excursion lengths of Brownian motion, Bessel processes and their bridges (see in particular [188, 189, 190]). Section 2.2.5 is mainly borrowed from the survey by Pitman and Yor [191] to which we refer for much more on this topic. It is interesting to stress that for $\theta \geq 0$, PD$(\alpha, \theta)$ distributions are of Poisson-Kingman type, that is they can be constructed as the sequence of jump sizes of some normalized subordinator; see Proposition 21 in [191]. When $\theta \leq -\alpha$, one can still define PD$(\alpha, \theta)$-measures by absolute continuity with respect to

PD($\alpha$, $\theta$)-distributions (cf. Definition 2.7); however, these measures are then only sigma-finite (see [20] and [167, 168] for some applications).

The residual allocation model in Corollary 2.3, which yields the size-biased permutation of a PD(0, $\theta$) random mass-partition (i.e. a GEM($\theta$)-distribution) has been described by Engen [93]; its extension to PD($\alpha$, $\theta$) in Proposition 2.6 is due to Perman, Pitman and Yor [179]. Further related descriptions of PD(0, $\theta$)-distributions in terms of Markovian sequences have been obtained in [126] and [210], and then extended to PD($\alpha$, $\theta$)-distributions in [191].

Exchangeable random partitions and their paint-box representations have been developed by Kingman [139] who was motivated by the construction of the so-called coalescent (cf. the forthcoming Section 4.1); see also Pitman [180]. There is a parallel theory for *compositions*, that is ordered partitions; see Gnedin [111]. More precisely, interval-partitions then play the same role for the representation of exchangeable random compositions as mass-partitions do for the paint-box construction of exchangeable random partitions. It is interesting to mention that subordinators again appear naturally in the framework of exchangeable random compositions; see [114] and [115].

The celebrated sampling formula of Ewens appeared in [101]; it is a cornerstone of the mathematical study of population genetics (see [87, 93, 138]), and has also a variety of applications in combinatorics (see [13] and references therein). Ewens sampling formula is often given in a slightly different form, namely as an expression for the probability that the restriction to $[n]$ of a PD(0, $\theta$)-partition has 'type' $c = (c_1, \dots, c_n)$, that is comprises $c_1$ singletons, $c_2$ doubletons, ... There is a rich literature on asymptotics in Ewens sampling formula as $n \to \infty$, for which we refer to the monograph [13]. The Lecture Notes of Pitman [186] contain a wealthy source of information and references on exchangeable partition probability functions; see also [182] for some slick calculations based on excursions of Brownian motion and stable subordinators.

# 3

# Exchangeable fragmentations

We now resume the study of processes in continuous time that describe the evolution of a unit mass which breaks down randomly into pieces, in such a way that distinct components have independent and self-similar evolutions. The crucial point is that we now allow fragments to split immediately, a situation that could not be handled by the discrete techniques of Chapter 1.

We shall first focus on the homogeneous case when splitting rates are further assumed to be the same for every fragment. The framework of exchangeable random partitions, which has been developed in the preceding chapter, provides a powerful tool for the construction and the analysis of this special class of fragmentation processes. In particular, we shall specify their Poissonian structure and characterize their distributions in terms of an erosion coefficient and rates of sudden dislocations. We shall also point out an important relation between a randomly tagged fragment and a certain subordinator, extending our observations for the evolution of the randomly tagged branch in Chapter 1. Finally, we shall present a transformation of homogeneous fragmentations which enables us to construct general self-similar fragmentations (i.e. with splitting rates proportional to a power function of the mass).

## 3.1 Homogeneous fragmentation processes

We shall now investigate fragmentation processes in the framework of random exchangeable partitions. Let us briefly explain the intuition which will guide us, by presenting an important example that will be used several times to illustrate notions and results in this chapter.

As in Section 2.1.2, consider an interval-representation $\Theta = (\Theta(t), t \geq 0)$ of some homogeneous fragmentation chain $X$ with values in the space $\mathcal{P}_m$ of mass-partitions (so its dislocation measure $\nu$ is conservative or dissipative), started as usual from the configuration $(1, 0 \ldots)$. In other words, $(\Theta(t), t \geq 0)$ is a càdlàg Markovian family of nested open subsets of $]0, 1[$ started from $\Theta(0) = ]0, 1[$, and the process $|\Theta(\cdot)|^{\downarrow}$ of the ranked sequence of spacing (i.e. lengths of the interval components) of $\Theta(\cdot)$ has the same law as $X(\cdot)$. More precisely, the dynamics of $\Theta$ are such that its interval components split independently one of the other and at a rate determined by a finite measure $\nu_{\mathrm{I}}$ on the space $\mathcal{P}_{\mathrm{I}}$ of interval-partitions, such that the image of $\nu_{\mathrm{I}}$ by the map $\vartheta \to |\vartheta|^{\downarrow}$ coincides with the dislocation measure $\nu$ of the fragmentation chain.

Then introduce a sequence $U_1, \ldots$ of i.i.d. uniform variables in $]0, 1[$, independent of the process $\Theta$, and define for each $t \geq 0$ the random partition $\Pi(t)$ such that two different indices $i, j$ belong to the same block of $\Pi(t)$ if and only if $U_i$ and $U_j$ belong to the same interval component of $\Theta(t)$. In other words, for each $t \geq 0$, $\Pi(t)$ is the mixture of paint-boxes based on the random interval-partition $\Theta(t)$ and the same sequence of i.i.d. uniform variables $U_1, \ldots$ For each interval component of $\Theta(t)$, say $I = ]a, b[$, write $B_I = \{i \in \mathbb{N} : U_i \in I\}$ for the block of $\Pi(t)$ corresponding to $I$. We can then observe that conditionally on $\Theta(t)$ and $B_I$, the subsequence $(U_i, i \in B_I)$ consists of i.i.d. variables which are uniformly distributed in $I$, and is independent of the other variables $(U_j, j \notin B_I)$. In particular the image of this subsequence by the normalizing function $x \to (x - a)/(b - a)$ is a sequence of i.i.d. uniform variables on $]0, 1[$. It follows from the dynamics of the interval fragmentation that at time $t + s$, the block $B_I$ is split into $\pi \cap B_I$, where $\pi$ is a random exchangeable partition distributed as $\Pi(s)$. More generally, it should be plain that blocks of $\Pi(t)$ split independently of each other, so that we can obtain $\Pi(t + s)$ by splitting each block of $\Pi(t)$ using a sequence of i.i.d. copies of $\Pi(s)$, which are independent of $\Pi(t)$.

Roughly speaking, this coding of a homogeneous fragmentation chain by a process with values in the space $\mathcal{P}_\infty$ of partitions of $\mathbb{N}$, can be viewed as a dynamical extension of Kingman's paint-box construction. Recall also that $X(t)$ can be recovered from the sequence of the asymptotic frequencies of the blocks of $\Pi(t)$. Such a coding is interesting in that the partition-valued process $\Pi$ can be studied through its restrictions $\Pi_{|[n]}$ to partitions of $[n]$. More precisely, it is easily seen that the latter are Markov chains, so the coding serves as a spatial discretization which enables us to reduce the study of a homogeneous fragmentation to that of a collection of Markov chains.

In this section, we shall first develop some basic material on fragmentation of partitions. This will lead us to introduce homogeneous fragmentations as a

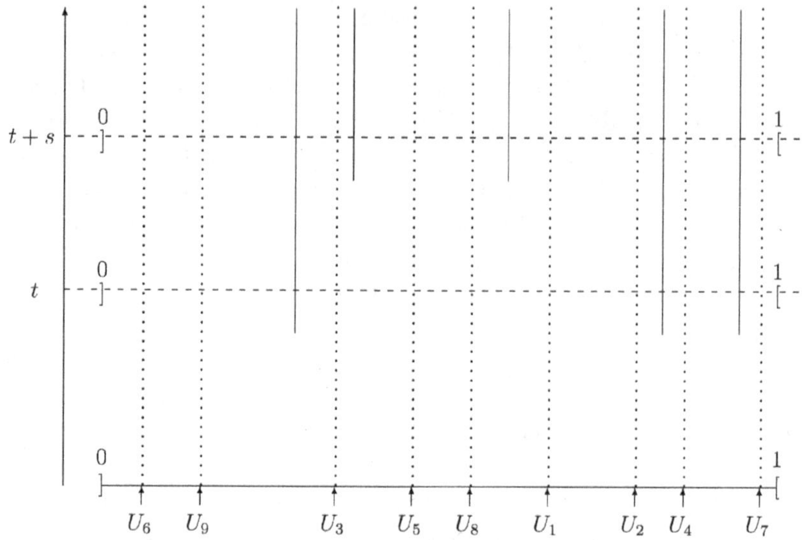


*Interval fragmentation* $\Theta$ *and paint-box process* $\Pi$*:*
$\Pi(t) = (\{1,2,3,5,8\}, \{4\}, \{6,9\}, \{7\})$
$\Pi(t+s) = (\{1,2\}, \{3\}, \{4\}, \{5,8\}, \{6,9\}, \{7\})$

natural family of Markov processes with values in the space $\mathcal{P}_\infty$ of partitions, which we shall then study using Poissonian techniques.

### 3.1.1 Fragmentation of partitions

Motivated by the informal description above, we define a fragmentation operator on partitions of countable sets in the general setting.

**Definition 3.1** *Consider two blocks $B \subseteq B' \subseteq \mathbb{N}$. Let $\pi$ be a partition of $B$ with $\#\pi = n$ non-empty blocks, and $\pi^{(\cdot)} = (\pi^{(i)}, i = 1, \ldots, n)$ be a sequence in $\mathcal{P}_{B'}$. For every integer $i$, we consider the partition of the $i$-th block $\pi_i$ of $\pi$ induced by the $i$-th term $\pi^{(i)}$ of the sequence $\pi^{(\cdot)}$, that is*

$$\pi^{(i)}_{|\pi_i} = \left(\pi^{(i)}_j \cap \pi_i, j \in \mathbb{N}\right).$$

*As $i$ varies in $[n] = \{1, \ldots, n\}$, the collection $\left\{\pi^{(i)}_j \cap \pi_i : i, j \in \mathbb{N}\right\}$ of the blocks of these induced partitions forms a partition of $B$ which we denote by* $\mathrm{Frag}(\pi, \pi^{(\cdot)})$ *and call the fragmentation of $\pi$ by $\pi^{(\cdot)}$.*

For example, take $B = B' = \{1,2,3,4,5,6\}$, $\pi = (\{1,2,5,6\},\{3,4\})$ (for the sake of simplicity we do not write the empty blocks of a partition), $\pi^{(1)} = (\{1,3,5\},\{2,4,6\})$ and $\pi^{(2)} = (\{1,6\},\{2,3,4,5\})$. We split the first block $\{1,2,5,6\}$ of $\pi$ using $\pi^{(1)}$, which yields the partition $(\{1,5\},\{2,6\})$. Similarly, we split the second block $\{3,4\}$ of $\pi$ using $\pi^{(2)}$, which yields the partition $(\{3,4\})$. Putting pieces together, we get $\mathrm{Frag}(\pi, \pi^{(\cdot)}) = (\{1,5\}, \{2,6\},\{3,4\})$.

Observe that the partition $\mathrm{Frag}(\pi, \pi^{(\cdot)})$ is always finer than $\pi$, in the sense that any block of the former is contained in some block of the latter. We also point out that if $\pi'^{(i)} = \pi^{(i)}_{|B}$ is the restriction to $B$, then one has obviously $\mathrm{Frag}(\pi, \pi^{(\cdot)}) = \mathrm{Frag}(\pi, \pi'^{(\cdot)})$; considering a sequence of partitions of a block $B'$ containing $B$ instead simply of $B$ itself in Definition 3.1 is just a matter of convenience for the later discussion.

In the rest of this section, we will use the notation

$$\mathbf{1}_B := (B, \emptyset, \emptyset, \dots)$$

for the partition of $B$ which has a single non-empty block. Note that the sequence $\mathbf{1}_B^{(\cdot)} := (\mathbf{1}_B, \mathbf{1}_B, \dots)$ serves as a neutral element for the fragmentation of partitions of $B$, in the sense that

$$\mathrm{Frag}(\pi, \mathbf{1}_B^{(\cdot)}) = \pi, \qquad \pi \in \mathcal{P}_B.$$

We point out the following elementary continuity result.

**Lemma 3.1** (i) *For every sequence $(\pi^{(i)}, i \in \mathbb{N})$ of partitions of $\mathbb{N}$, the map*

$$\pi \to \mathrm{Frag}(\pi, \pi^{(\cdot)}), \qquad \pi \in \mathcal{P}_\infty$$

*is Lipschitz-continuous.*
(ii) *For each integer $n$, let $\pi^{(\cdot,n)} = (\pi^{(i,n)}, i \in \mathbb{N})$ be a sequence of partitions. Suppose that the limit $\lim_{n\to\infty} \pi^{(i,n)} := \pi^{(i)}$ exists for every $i \in \mathbb{N}$. Then for every partition $\pi \in \mathcal{P}_\infty$, it holds that*

$$\lim_{n\to\infty} \mathrm{Frag}(\pi, \pi^{(\cdot,n)}) = \mathrm{Frag}(\pi, \pi^{(\cdot)}).$$

*Proof* The restriction of partitions to $[n]$ is clearly compatible with the fragmentation operator in the sense that

$$\mathrm{Frag}(\pi, \pi^{(\cdot)})_{|[n]} = \mathrm{Frag}(\pi_{|[n]}, \pi^{(\cdot)}_{|[n]}). \tag{3.1}$$

The very definition of the distance on $\mathcal{P}_\infty$ yields our first claim. The proof of the second assertion is similar. □

We next observe that exchangeability is preserved by fragmentation in the following sense. Pick $n \in \overline{\mathbb{N}}$ and recall that $[\infty] = \mathbb{N}$.

**Lemma 3.2** *Let $\pi \in \mathcal{P}_n$ be a random exchangeable partition and $(\pi^{(i)}, i \in [n])$ a sequence of random partitions of $[n]$ which is independent of $\pi$. Suppose that the sequence $\pi^{(\cdot)}$ is* **doubly-exchangeable**, *in the sense that for every permutation $\sigma$ of $[n]$, the sequences*

$$\left(\sigma(\pi^{(i)}), i \in [n]\right) \quad \textit{and} \quad \left(\pi^{(\sigma(i))}, i \in [n]\right)$$

*both have the same law as $(\pi^{(i)}, i \in [n])$. Then the random partitions $\pi$ and* $\mathrm{Frag}(\pi, \pi^{(\cdot)})$ *are jointly exchangeable, that is their joint distribution is invariant by the action of permutations.*

A typical situation in which Lemma 3.2 will be applied is when $(\pi^{(i)}, i \in [n])$ is given by a sequence of i.i.d. exchangeable random partitions, as obviously the latter are doubly-exchangeable.

*Proof* Write $\pi' = \mathrm{Frag}(\pi, \pi^{(\cdot)})$ and consider some permutation $\sigma$ of $[n]$ with inverse $\sigma^{-1}$. By definition, the blocks of the partition $\sigma^{-1}(\pi')$ are the images by $\sigma$ of the blocks of $\pi'$, so they are given by the family

$$\sigma\left(\pi_j^{(i)} \cap \pi_i\right) = \sigma\left(\pi_j^{(i)}\right) \cap \sigma\left(\pi_i\right), \qquad i, j \in [n].$$

Introduce a bijection $\sigma' : [n] \to [n]$ (dependent on $\pi$ but not on the sequence $\pi^{(\cdot)}$) such that for every $i \in [n]$, $\sigma(\pi_i)$ is the $\sigma'(i)$-th block of $\sigma^{-1}(\pi)$. Then write $\pi'^{(i)}$ for the image of $\pi^{(\sigma'(i))}$ by the permutation $\sigma^{-1}$. The discussion above shows that there is the identity

$$\sigma^{-1}\left(\mathrm{Frag}(\pi, \pi^{(\cdot)})\right) = \mathrm{Frag}(\sigma^{-1}(\pi), \pi'^{(\cdot)}).$$

On the other hand, our assumptions imply that the sequence of $(\pi'^{(i)}, i \in [n])$ is distributed as $(\pi^{(i)}, i \in [n])$ and is independent of $\pi$. Since $\sigma^{-1}(\pi)$ has the same law as $\pi$, this completes the proof. □

We now conclude this section by presenting a fairly general procedure for constructing functions $\Pi : t \to \Pi(t)$ with values in $\mathcal{P}_\infty$, such that the partition $\Pi(t)$ gets finer as $t$ increases. Typically, for every fixed $n \in \mathbb{N}$, the function $t \to \Pi^{[n]}(t)$ is defined recursively by splitting exactly one block of the partition at certain discrete times. Then one checks that for every $t \geq 0$, the sequence $(\Pi^{[n]}(t), n \in \mathbb{N})$ is compatible, and thus can be identified as a unique partition $\Pi(t) \in \mathcal{P}_\infty$. The fundamental feature of this construction is that, even though each restricted function $t \to \Pi^{[n]}(t) = \Pi_{|[n]}(t)$ has a discrete evolution (i.e. is piecewise constant), the set formed by its jump-times as $n$ varies in $\mathbb{N}$ may be everywhere dense, and then the unrestricted function $t \to \Pi(t)$ has a continuous evolution (i.e. is nowhere piecewise constant).

Specifically, call '*discrete point measure* on $\mathbb{R}_+ \times \mathcal{P}_\infty \times \mathbb{N}$' any measure $\mathtt{m}$ which can be expressed in the form

$$\mathtt{m} = \sum_{(t,\pi,k)\in\mathcal{D}}^{\infty} \delta_{(t,\pi,k)},$$

where $\mathcal{D}$ is a subset of $\mathbb{R}_+ \times \mathcal{P}_\infty \times \mathbb{N}$ such that the following two conditions hold. First, for every real number $t' \geq 0$ and integer $n \geq 1$

$$\#\left\{(t, \pi, k) \in \mathcal{D} : t \leq t', \pi_{|[n]} \neq \mathbf{1}_{[n]}, k \leq n\right\} < \infty,$$

where $\mathbf{1}_{[n]} = ([n], \emptyset, \ldots)$ stands for the partition of $[n]$ which has a single non-empty block (recall that $\mathbf{1}_{[n]}$ plays the role of a neutral element for the fragmentation operator in the space of partitions of $[n]$). Second, $\mathtt{m}$ has at most one atom on each fiber $\{t\} \otimes \mathcal{P}_\infty \otimes \mathbb{N}$, that is

$$\mathtt{m}(\{t\} \otimes \mathcal{P}_\infty \otimes \mathbb{N}) = 0 \text{ or } 1.$$

Starting from an arbitrary discrete point measure $\mathtt{m}$ on $\mathbb{R}_+ \times \mathcal{P}_\infty \times \mathbb{N}$, we first describe informally the construction of a family of nested partitions $(\Pi(t), t \geq 0)$. For every fixed $n \in \mathbb{N}$, the assumption that the point measure $\mathtt{m}$ is discrete enables us to define a càdlàg step-path $\Pi^{[n]} : t \to \Pi^{[n]}(t)$ with values in the space of partitions of $[n]$, which can only jump at times $t$ at which the fiber $\{t\} \times \mathcal{P}_\infty \times \mathbb{N}$ carries an atom of $\mathtt{m}$, say $(t, \pi, k)$, such that $\pi_{|[n]} \neq \mathbf{1}_{[n]}$ and $k \leq n$. In that case, $\Pi^{[n]}(t)$ is the partition obtained by replacing the $k$-th block of $\Pi^{[n]}(t-)$, namely $\Pi_k^{[n]}(t-)$, by the restriction of $\pi$ to this block, and leaving the other blocks unchanged. To give an example, take for instance $n = 6$, and suppose that $\mathtt{m}$ has an atom at $(t, k, \pi)$ with $k = 2$ and $\pi_{|[6]} = (\{1, 2, 3\}, \{4, 5, 6\})$. Assume for instance that $\Pi^{[6]}(t-) = (\{1, 6\}, \{2, 3, 4\}, \{5\})$. Then at time $t$, we split the second block of $\Pi^{[6]}(t-)$, that is $\{2, 3, 4\}$, using the partition $\pi_{|[6]}$. This produces two new blocks, $\{2, 3\}$ and $\{4\}$, and thus $\Pi^{[6]}(t) = (\{1, 6\}, \{2, 3\}, \{4\}, \{5\})$.

To make this construction more formal in the framework of this section, we introduce the following notation. For every $n \in \mathbb{N}$ and every pair $(\pi, k) \in \mathcal{P}_\infty \times \mathbb{N}$, we write $\Delta_n^{(\cdot)}(\pi, k)$ for the (finite) sequence of partitions in $\mathcal{P}_n$ given for $i = 1, \ldots, n$ by

$$\Delta_n^{(i)}(\pi, k) = \begin{cases} \mathbf{1}_{[n]} & \text{if } i \neq k, \\ \pi_{|[n]} & \text{if } i = k. \end{cases}$$

Next, let $\mathtt{m}^{[n]}$ be the point measure on $[0, \infty[\times(\mathcal{P}_n)^n$ whose atoms are the images by the map

$$(t, \pi, k) \to (t, \Delta_n^{(\cdot)}(\pi, k))$$

of the atoms $(t, \pi, k)$ of $\mathtt{m}$ such that $\pi_{|[n]} \neq \mathbf{1}_{[n]}$ and $k \leq n$. We write $t_0 = 0$ and $(t_1, \Delta_n^{(\cdot)}(1)), (t_2, \Delta_n^{(\cdot)}(2)), \ldots$ for the sequence of the atoms of $\mathtt{m}^{[n]}$, ranked in increasing order of their first coordinate. We then set $\Pi^{[n]}(t) = \mathbf{1}_{[n]}$ for every $t \in [t_0, t_1[$ and define recursively

$$\Pi^{[n]}(t) = \mathrm{Frag}\left(\Pi^{[n]}(t_{i-1}), \Delta_n^{(\cdot)}(i)\right), \qquad \text{for every } t \in [t_i, t_{i+1}[ .$$

It should be plain that this rigorous construction merely rephrases the informal one above.

We now check the compatibility of the sequence of partitions $(\Pi^{[n]}(t), n \in \mathbb{N})$.

**Lemma 3.3** *In the notation above, for every $t \geq 0$, the sequence $(\Pi^{[n]}(t), n \in \mathbb{N})$ is compatible, and thus there exists a unique partition $\Pi(t) \in \mathcal{P}_\infty$ such that $\Pi_{|[n]}(t) = \Pi^{[n]}(t)$ for every $n \in \mathbb{N}$. Moreover the function $\Pi : t \to \Pi(t)$ is càdlàg.*

*Proof* Fix $n \geq 2$ and consider the first atom $(t_1, \Delta_n^{(\cdot)}(1))$ of $\mathtt{m}^{[n]}$. Plainly, $\Pi^{[n-1]}(t) = \Pi^{[n]}_{|[n-1]}(t) = \mathbf{1}_{[n-1]}$ for every $t \in [0, t_1[$. The atom $(t_1, \Delta_n^{(\cdot)}(1))$ corresponds to some atom $(t_1, \pi, k)$ of the discrete point measure $\mathtt{m}$ with $k \leq n$ and $\pi_{|[n]} \neq \mathbf{1}_{[n]}$. Consider first the case when $\pi_{|[n-1]} \neq \mathbf{1}_{[n-1]}$ and $k \leq n-1$. Then the first atom of $\mathtt{m}^{[n-1]}$ is $(t_1, \Delta_{n-1}^{(\cdot)}(1))$, where for $i \in [n-1]$, $\Delta_{n-1}^{(i)}(1)$ is simply the restriction of $\Delta_n^{(i)}(1)$ to $[n-1]$. It follows immediately from (3.1) that $\Pi^{[n-1]}(t) = \Pi^{[n]}_{|[n-1]}(t)$ for every $t \in [t_1, t_2[$.

Next, consider the case when $\pi_{|[n-1]} = \mathbf{1}_{[n-1]}$ or $k = n$. Then $\mathtt{m}^{[n-1]}$ has no atoms on $[0, t_2[\times(\mathcal{P}_{n-1})^{n-1}$, and it follows again from (3.1) that $\Pi^{[n-1]}(t) = \Pi^{[n]}_{|[n-1]}(t) = \mathbf{1}_{[n-1]}$ for every $t \in [0, t_2[$. By iteration, this shows that the restriction of $\Pi^{[n]}$ to $[n-1]$ coincides with $\Pi^{[n-1]}$ (observe that the convention for enumerating the blocks of a partition of $[n]$, say $\gamma$, is compatible with the operation of restriction to $[n]$, in the sense that the sequence of the blocks of $\gamma_{|[n-1]}$ is simply $\gamma_1 \cap [n-1], \gamma_2 \cap [n-1], \ldots$). By the Compatibility Lemma 2.5, there exists a unique partition $\Pi(t)$ of $\mathcal{P}_\infty$ such that $\Pi_{|[n]}(t) = \Pi^{[n]}(t)$ for every $n \in \mathbb{N}$. The very construction shows that for each integer $n$, the function $t \to \Pi_{|[n]}(t)$ is càdlàg, and thus $t \to \Pi(t)$ is càdlàg as well. □

### 3.1.2 Homogeneous fragmentation as Markov processes

Recall the notion of homogeneous fragmentation chains (i.e. self-similar with index $\alpha = 0$) which has been developed in Chapter 1. We are interested here in Markov processes with values in the space of partitions in which, roughly speaking, blocks split independently of each other and with the same intensity. The notion of fragmentation operator leads naturally to the following rigorous definition.

**Definition 3.2** *Fix $n \in \overline{\mathbb{N}}$, and let $\Pi = (\Pi(t), t \geq 0)$ be a Markov process with values in $\mathcal{P}_n$ with càdlàg sample paths.*

(i) *$\Pi$ is called a* **homogeneous fragmentation process** *if its semigroup can be described as follows. For every $t, t' \geq 0$, the conditional distribution of $\Pi(t+t')$ given $\Pi(t) = \pi$ is the law of* $\mathrm{Frag}(\pi, \pi^{(\cdot)})$, *where $\pi^{(\cdot)} = (\pi^{(1)}, \ldots)$ is an i.i.d. sequence of exchangeable random partitions (whose law only depends on $t'$).*

(ii) *A homogeneous fragmentation process $\Pi$ is called* **standard** *if it starts from $\mathbf{1}_{[n]}$, the partition of $[n]$ into a single non-empty block, $([n], \emptyset, \ldots)$.*

The description of the semigroup implies that blocks in the partition $\Pi(t)$ split independently of each other, which should be viewed as the branching property in the setting of partition-valued processes. The term *homogeneous* in Definition 3.2 refers to the fact that, roughly speaking, in the transition from $\Pi(t)$ to $\Pi(t+t')$, all the blocks of $\Pi(t)$ play the same role, in the sense that they are split according to the same random procedure, independently of the composition of these blocks. This is a much stronger property than the invariance of the law of $\Pi$ under the action of permutations; see the forthcoming Proposition 3.1.

Motivated by applications to processes with values in the space of mass-partitions that we have in mind, we shall implicitly deal with the case $n = \infty$ of Definition 3.2 later in the text, except when it is explicitly mentioned otherwise. Of course homogeneous fragmentations in $\mathcal{P}_n$ for $n \in \mathbb{N}$ are much easier to study as the state space is then finite, and in this direction, the following elementary lemma will be quite useful.

**Lemma 3.4** *Let $\Pi = (\Pi(t), t \geq 0)$ be a process with values in $\mathcal{P}_\infty$ and for every integer $n$, write $\Pi_{|[n]} = (\Pi_{|[n]}(t), t \geq 0)$ for its restriction to $\mathcal{P}_n$. Then $\Pi$ is a homogeneous fragmentation process in $\mathcal{P}_\infty$ if and only if $\Pi_{|[n]}$ is a homogeneous fragmentation process in $\mathcal{P}_n$ for all $n \in \mathbb{N}$.*

*Proof* If $\Pi$ is a homogeneous fragmentation process, then identity (3.1) shows that each restriction $\Pi_{|[n]}$ also fulfills the requirements of Definition 3.2. The converse is immediate. □

We stress the importance of the compatibility property (3.1) of the restriction for the fragmentation operator. Indeed, the restriction map $\pi \to \pi_{|[n]}$ is not injective, and in general the image of a Markov process by a map which is not injective may fail to be Markovian.

We also point out that the exchangeable random partition $\pi^{(1)}$ which arises in the description of the semigroup in Definition 3.2, has the law of $\Pi(t')$ when $\Pi$ is standard (take $t = 0$ in Definition 3.2(i)). In this direction, we note that if $(\Pi^{(i)}(t), t \geq 0)$, $i \in \mathbb{N}$, is a sequence of i.i.d. copies of a standard homogeneous fragmentation $\Pi$, then for every $\pi \in \mathcal{P}_\infty$, the process

$$\operatorname{Frag}(\pi, \Pi^{(\cdot)}(t)), \qquad t \geq 0$$

is a version of the same homogeneous fragmentation started from $\pi$. This is the reason why we shall focus on standard fragmentations later in the text, except when it is explicitly mentioned otherwise. We now develop a couple of simple observations in this vein.

**Proposition 3.1** (i) *The semigroup of a homogeneous fragmentation enjoys the Feller property, that is for every continuous function $\varphi : \mathcal{P}_\infty \to \mathcal{P}_\infty$, the map*

$$\pi \to \mathbb{E}\left(\varphi(\operatorname{Frag}(\pi, \Pi^{(\cdot)}(t))\right), \qquad \pi \in \mathcal{P}_\infty,$$

*is continuous for each $t \geq 0$, and we have*

$$\lim_{t\to 0} \mathbb{E}\left(\varphi(\operatorname{Frag}(\pi, \Pi^{(\cdot)}(t))\right) = \varphi(\pi), \qquad \pi \in \mathcal{P}_\infty.$$

(ii) *The distribution of a standard homogeneous fragmentation is invariant by the natural action of permutations of $\mathbb{N}$, that is for every permutation $\sigma$ of $\mathbb{N}$, the processes $(\sigma(\Pi(t)), t \geq 0)$ and $(\Pi(t), t \geq 0)$ have the same distribution.*

*Proof* The first assertion follows immediately from Lemma 3.1. Next, we deduce from the Markov property and Lemma 3.2 that for every permutation $\sigma$ of $\mathbb{N}$, the processes $(\sigma(\Pi(t)), t \geq 0)$ and $(\Pi(t), t \geq 0)$ have the same finite-dimensional distribution. Since both processes have càdlàg paths a.s. and take values in a compact metric space, they have the same distribution. □

The Feller property ensures that the (completed) natural filtration $(\mathcal{F}_t)_{t\geq 0}$ of $\Pi$ is right-continuous, that is for every $t \geq 0$ there is the identity

$$\mathcal{F}_t = \mathcal{F}_{t+} := \bigcap_{\varepsilon>0} \mathcal{F}_{t+\varepsilon} .$$

Furthermore, $\Pi$ enjoys the *strong Markov property*, in the sense that for every a.s. finite $(\mathcal{F}_t)$-stopping time, say $T$, the conditional distribution of $\Pi(T+t')$ given $\mathcal{F}_T$ is the law of $\mathrm{Frag}(\pi, \pi^{(\cdot)})$, where $\pi = \Pi(t)$ and $\pi^{(\cdot)} = (\pi^{(1)}, \ldots)$ is a sequence of i.i.d. copies of $\Pi(t')$. See Section III.2 in Revuz and Yor [192] for details.

We now turn our attention to the dynamics of homogeneous fragmentations. Since for every $n \in \mathbb{N}$, the space $\mathcal{P}_n$ of partitions of $[n]$ is finite, Lemma 3.4 shows that the restricted processes $\Pi_{|[n]}$ are Markov *chains*. This leads us to consider the jump rates for the standard process

$$q_\pi := \lim_{t\to 0+} \frac{1}{t} \mathbb{P}\left(\Pi_{|[n]}(t) = \pi\right) , \qquad \pi \in \mathcal{P}_n \backslash \{\mathbf{1}_{[n]}\} ;$$

see (1.7). Obviously, these jump rates inherit an exchangeability property from $\Pi$, and it is easily seen that they entirely characterize the law of $\Pi$. More precisely, we have the following:

**Lemma 3.5** *For every permutation $\sigma$ of $[n]$ and every $\pi \in \mathcal{P}_n\backslash\{\mathbf{1}_{[n]}\}$, there is the identity*

$$q_\pi = q_{\sigma(\pi)} .$$

*Moreover, the family of jump rates $(q_\pi, \pi \in \mathcal{P}_n\backslash\{\mathbf{1}_{[n]}\}$ and $n \in \mathbb{N})$ determines the law of $\Pi$.*

*Proof* The first assertion is immediate from Proposition 3.1(ii). To establish the second, we have to check for every $k \in \mathbb{N}$ that the jump rates of the restricted process $\Pi_{|[k]}$ are entirely determined by the collection $(q_\pi, \pi \in \mathcal{P}_n\backslash\{\mathbf{1}_{[n]}\}$ and $n \in \mathbb{N})$. The argument relies on exchangeability and the branching property of the semigroup.

More precisely, consider some partition $\pi' \in \mathcal{P}_n$, and let $\pi'' \in \mathcal{P}_n$ be another partition with $\pi'' \neq \pi'$, which can be obtained from $\pi'$ by the fragmentation of a single block. This means there is an index $k \leq \#\pi'$ and a sequence $\pi^{(\cdot)} = (\pi^{(i)}, i \leq \#\pi')$ such that the following holds. First $\pi^{(i)} = \mathbf{1}_{[n]}$ for $i \neq k$, second the restriction of $\pi^{(k)}$ to the $k$-th block $\pi'_k$ of $\pi'$ is not trivial, that is $\pi^{(k)}_{|\pi'_k} \neq \mathbf{1}_{\pi'_k}$, and third $\pi'' = \mathrm{Frag}(\pi', \pi^{(\cdot)})$. Let $j \geq 2$ denote the cardinal of $\pi'_k$, and consider some permutation $\sigma$ of $[n]$ which maps $[j]$ on the block $\pi'_k$. Finally, introduce the partition $\pi \in \mathcal{P}_j$ given by the image of the partition

$\pi^{(k)}_{|\pi'_k}$ of $\pi'_k$ by $\sigma$. An application of Lemma 3.2 shows that the jump rate from $\pi'$ to $\pi''$ coincides with the jump rate $q_\pi$ from $\mathbf{1}_{[j]}$ to $\pi$, that is

$$\lim_{t\to 0}\frac{1}{t}\mathbb{P}\left(\Pi_{|[n]}(t)=\pi'' \mid \Pi_{|[n]}(0)=\pi'\right)=q_\pi\,.$$

Finally, we point out that all the other jump rates are zero. This is obvious if $\pi''$ cannot be expressed in the form $\pi''=\mathrm{Frag}(\pi',\pi^{(\cdot)})$ for any sequence $\pi^{(\cdot)}$ in $\mathcal{P}_n$. Next, suppose that $\pi''$ can be obtained from $\pi'$ by the fragmentation of two or more of its blocks. The fact that blocks split independently of each others then implies that

$$\mathbb{P}\left(\Pi_{|[n]}(t)=\pi'' \mid \Pi_{|[n]}(0)=\pi'\right)=O(t^2) \qquad \text{as } t\to 0\,,$$

and thus the jump rate from $\pi'$ to $\pi''$ must be zero. We conclude that the family $\left(q_\pi,\ \pi\in\mathcal{P}_n\backslash\{\mathbf{1}_{[n]}\} \text{ and } n\in\mathbb{N}\right)$ characterizes the law of the homogeneous fragmentation. □

The fundamental result about the collection of jump rates in Lemma 3.5 is that it can be described by a single measure on $\mathcal{P}_\infty$. In this direction, it is convenient to introduce the notation

$$\mathcal{P}_{n',\pi}=\left\{\pi'\in\mathcal{P}_{n'}:\pi'_{|[n]}=\pi\right\}\,,$$

where $\pi\in\mathcal{P}_n$ and $n'\in\{n,n+1,\ldots,\infty\}$. Recall that for every block $B$, $\mathbf{1}_B$ denotes the partition $(B,\emptyset,\ldots)$.

**Proposition 3.2** *Let $(q_\pi:\pi\in\mathcal{P}_n\backslash\{\mathbf{1}_{[n]}\}$ and $n\in\mathbb{N})$ be the family of jump rates of some homogeneous fragmentation $\Pi$. There exists a unique measure $\mu$ on $\mathcal{P}_\infty$ such that $\mu(\{\mathbf{1}_{\mathbb{N}}\})=0$ and*

$$\mu\left(\mathcal{P}_{\infty,\pi}\right)=q_\pi$$

*for every $n\in\mathbb{N}$ and every partition $\pi\in\mathcal{P}_n\backslash\{\mathbf{1}_{[n]}\}$. More precisely, the measure $\mu$ assigns a finite mass to the sets $\left\{\pi\in\mathcal{P}_\infty:\pi_{|[n]}\neq\mathbf{1}_{[n]}\right\}$ for all $n\in\mathbb{N}$, and is exchangeable, that is invariant by the action of permutations of $\mathbb{N}$.*

The measure $\mu$ in Proposition 3.2 which enables us to represent the jump rates of a homogeneous fragmentation $\Pi$ will be referred to as the *splitting rate* of $\Pi$. Before proving Proposition 3.2, let us consider again the example at the beginning of Section 3.1. □

**Example** Let $\Pi$ be the homogeneous fragmentation associated to an interval representation $\Theta$ of some homogeneous fragmentation chain with dislocation measure $\nu$. We suppose that we start from $\Theta(0) = ]0,1[$, that is $\Pi$ is standard, and that $\nu$ has no atom at $\mathbf{1} = (1,0,\ldots)$. Denote by $T$ the first splitting time of $\Theta$, so $T$ has an exponential distribution with parameter $\nu(\mathcal{P}_{\mathrm{m}})$, and the ranked sequence of the lengths of the interval components of $\Theta(T)$ is distributed according to $\nu(\cdot)/\nu(\mathcal{P}_{\mathrm{m}})$. By construction, $T$ is the first jump time of $\Pi$, and $\Pi(T)$ has the law of the mixture of paint-boxes

$$\int_{\mathcal{P}_{\mathrm{m}}} \varrho_{\mathbf{s}} \nu(d\mathbf{s})/\nu(\mathcal{P}_{\mathrm{m}}),$$

independently of $T$. It follows readily that the jump rates of $\Pi_{|[n]}$ can be expressed in the form

$$\begin{aligned} q_\pi &= \lim_{t\to 0} \tfrac{1}{t}\mathbb{P}(T \le t, \Pi_{|[n]}(T) = \pi) \\ &= \mathbb{P}(\Pi_{|[n]}(T) = \pi) \lim_{t\to 0} \tfrac{1}{t}\mathbb{P}(T \le t) \\ &= \int_{\mathcal{P}_{\mathrm{m}}} \varrho_{\mathbf{s}}(\mathcal{P}_{\infty,\pi})\nu(d\mathbf{s}), \end{aligned}$$

and we conclude that the splitting rate $\mu$ of $\Pi$ is given by the mixture of paint-boxes

$$\mu = \int_{\mathcal{P}_{\mathrm{m}}} \varrho_{\mathbf{s}} \nu(d\mathbf{s}).$$

We now establish Proposition 3.1.

*Proof* Take any $n' \ge n$, and note the identity

$$\mathcal{P}_{\infty,\pi} = \bigcup_{\pi' \in \mathcal{P}_{n',\pi}} \mathcal{P}_{\infty,\pi'}, \tag{3.2}$$

where, in the right-hand side, the union is over disjoint subsets of $\mathcal{P}_\infty$. Because the Markov chain $\Pi_{|[n]}$ can be obtained as the restriction of $\Pi_{|[n']}$ to $[n]$, its jump rate $q_\pi$ from $\mathbf{1}_{[n]}$ to $\pi \in \mathcal{P}_n \backslash \{\mathbf{1}_{[n]}\}$ coincides with the total jump rate for $\Pi_{|[n']}$ from $\mathbf{1}_{[n']}$ to $\mathcal{P}_{n',\pi}$. In other words, we have

$$q_\pi = \sum_{\pi' \in \mathcal{P}_{n',\pi}} q_{\pi'}. \tag{3.3}$$

This shows that the function

$$\mu : \mathcal{P}_{\infty,\pi} \to \mu(\mathcal{P}_{\infty,\pi}), \qquad \pi \in \mathcal{P}_n \backslash \{\mathbf{1}_{[n]}\} \text{ for some } n \in \mathbb{N}$$

is additive, and we conclude by an easy application of Caratheodory's extension theorem that $\mu$ has a unique extension to a measure on $\mathcal{P}_\infty \backslash \{\mathbf{1}_{\mathbb{N}}\}$. The exchangeability is immediate from Lemma 3.5. $\square$

### 3.1.3 Poissonian structure

Our purpose now is to construct a homogeneous fragmentation $\Pi$ with a given splitting rate $\mu$. In this direction, consider an exchangeable[1] measure $\mu$ on $\mathcal{P}_\infty$ such that

$$\mu(\{\mathbf{1}_{\mathbb{N}}\}) = 0 \quad \text{and} \quad \mu\left(\{\pi \in \mathcal{P}_\infty : \pi_{|[n]} \neq \mathbf{1}_{[n]}\}\right) < \infty \qquad \text{for every } n \geq 2. \tag{3.4}$$

Recall the construction of a family $(\Pi(t), t \geq 0)$ of nested partitions from a discrete point measure $\mathtt{m}$ presented at the end of Section 3.1.1. From now on, we shall consider the situation when $\mathtt{m} := \mathtt{M}$ is random and, more precisely, distributed as a Poisson random measure on $[0, \infty[ \times \mathcal{P}_\infty \times \mathbb{N}$ with intensity $dt \otimes \mu(d\pi) \otimes \#$, where $\#$ stands for the counting measure on $\mathbb{N}$ and $\mu$ is a splitting rate on $\mathcal{P}_\infty$. It can be immediately checked from the assumptions on $\mu$ that with probability one, $\mathtt{M}$ fulfills the requirements of Section 3.1.1 for discrete point measures.

**Proposition 3.3** *In the notation above, the process $\Pi = (\Pi(t), t \geq 0)$ is a standard homogeneous fragmentation with splitting rate $\mu$.*

*Proof* We start by checking that for every $t \geq 0$, the random partition $\Pi(t)$ is exchangeable. In the notation of Section 3.1.1, for every $n \in \mathbb{N}$, the random measure $\mathtt{M}^{[n]}$ on $[0, \infty[ \times (\mathcal{P}_n)^n$ is a Poisson random measure with intensity $dt \otimes \tilde{\mu}^{[n]}$, where $\tilde{\mu}^{[n]}$ is the image by the map $(\pi, k) \to \Delta_n^{(\cdot)}(\pi, k)$ of the restriction of $\mu \otimes \mathbb{N}$ to pairs $(\pi, k) \in \mathcal{P}_\infty \times \mathbb{N}$ with $\pi_{|[n]} \neq \mathbf{1}_{[n]}$ and $k \leq n$. Our assumptions on $\mu$ ensure that $\tilde{\mu}^{[n]}$ is a finite measure on $(\mathcal{P}_n)^n$, which is proportional to the law of a doubly exchangeable sequence of random partitions in $\mathcal{P}_n$, in the sense of Lemma 3.2.

Then, observe that $\Delta_n^{(\cdot)}(1), \Delta_n^{(\cdot)}(2), \ldots$ are i.i.d. doubly exchangeable sequences of random partitions of $\mathcal{P}_n$. By conditioning on the number of atoms of $\mathtt{M}^{[n]}$ on $[0, t] \times (\mathcal{P}_n)^n$, we deduce from Lemma 3.2 and an easy induction that $\Pi^{[n]}(t)$ is an exchangeable partition of $[n]$, and therefore $\Pi(t)$ is an exchangeable partition. Repeating this argument shows that the pair of random partitions $(\Pi(t), \Pi(t+t'))$ is jointly exchangeable for every $t, t' \geq 0$.

We now finish the proof by checking that $\Pi$ is a homogeneous fragmentation with the desired splitting rate. In this direction, we see from the very construction of $\Pi^{[n]} = \Pi_{|[n]}$ that the latter is a continuous time Markov chain, hence all that is needed is to identify the jump rates. So fix $j \in \{2, \ldots, n\}$

[1] We may observe that, thanks to exchangeability, it suffices to require that (3.4) is fulfilled for $n = 2$, as then it is automatically fulfilled for all $n \geq 2$.

and consider $\pi' \in \mathcal{P}_n$, a partition such that $\pi'_1 = [j]$. Let $\pi \in \mathcal{P}_j \backslash \{\mathbf{1}_{[j]}\}$ and denote by $\pi''$ the partition of $[n]$ obtained from $\pi'$ by the fragmentation of its first block $[j]$ according to $\pi$, and leaving all the other blocks unchanged. The Poissonian construction shows that the jump rate of $\Pi^{[n]}$ from $\pi'$ to $\pi''$ is given by $\mu(\mathcal{P}_{\infty,\pi}) = q_\pi$. By an easy argument of exchangeability similar to that in the proof of Lemma 3.5, we conclude that $\Pi^{[n]}$ has the same jump rates as the homogeneous fragmentation of $\mathcal{P}_n$ with splitting rate $\mu$, which completes the proof. □

**Example** Let $\nu$ be a finite measure on $\mathcal{P}_{\mathrm{m}}$ which has no atom at $\mathbf{1} = (1, 0, \ldots)$, and define the mixture of laws of paint-boxes (cf. Section 2.3.2)

$$\varrho_\nu(d\pi) = \int_{\mathrm{s} \in \mathcal{P}_{\mathrm{m}}} \varrho_{\mathrm{s}}(d\pi)\nu(d\mathrm{s}), \qquad \pi \in \mathcal{P}_\infty.$$

Plainly, $\varrho_\nu$ is a finite exchangeable measure on $\mathcal{P}_\infty$ with no atom at $\mathbf{1}_{\mathbb{N}}$; and we can thus construct as above a standard homogeneous fragmentation $\Pi$ with splitting rate $\varrho_\nu$. The first atom $(T, \Gamma)$ of a Poisson random measure on $[0, \infty[ \times \mathcal{P}_\infty$ with intensity $dt \otimes \varrho_\nu(d\pi)$ has the law

$$\mathbb{P}(T \in dt, \Gamma \in d\pi) = \varrho_\nu(d\pi) \exp(-\varrho_\nu(\mathcal{P}_\infty)t)dt,$$

and we see from the Poissonian construction of $\Pi$ that the latter specifies the distribution of the first jump of $\Pi$. In particular, the position $\Pi(T)$ after the first jump is a random exchangeable partition such that the random mass-partition $|\Pi(T)|^\downarrow$ given by the ranked sequence of its asymptotic frequencies has the law $\nu(\cdot)/\nu(\mathcal{P}_{\mathrm{m}})$. Further, it is independent of the time $T$ of the first jump which follows the exponential distribution with parameter $\nu(\mathcal{P}_{\mathrm{m}})$. On the other hand, recall from the dynamics of $\Pi$ that its blocks split independently of each other and at the same rate governed by $\varrho_\nu$. It follows easily that the homogeneous fragmentation $\Pi$ possesses asymptotic frequencies at all times, and that the process $|\Pi(T)|^\downarrow$ is a homogeneous fragmentation chain with dislocation measure $\nu$. Of course, this agrees with the observations made in the discussion at the beginning of Section 3.1.1 and after Proposition 3.2.

## 3.2 Asymptotic frequencies

The fundamental Theorem 2.1 of Kingman gives a representation of random exchangeable partitions based on the paint-box construction, which provides a bijection with laws of random mass-partitions. In this section, we shall gain an insight into homogeneous fragmentations by considering asymptotic frequencies of certain involved partitions. We shall first obtain a representation of splitting rates by extending Kingman's theorem to certain infinite measures on $\mathcal{P}_\infty$, and then describe the process of the asymptotic frequencies of the first block $|\Pi_1(\cdot)|$.

### 3.2.1 Erosion and dislocation

We have seen that the distribution of a homogeneous fragmentation is determined by its splitting rate, which is an exchangeable measure on $\mathcal{P}_\infty$ such that (3.4) holds. Conversely, the Poissonian construction shows that any such measure can be viewed as the splitting rate of some homogeneous fragmentation. This leads us to investigate exchangeable measures which fulfill (3.4) in further detail. In this direction, we first give two fundamental examples.

First, for every $n \in \mathbb{N}$, we write $\epsilon^{(n)}$ for the partition of $\mathbb{N}$ that has exactly two non-empty blocks, $\{n\}$ and $\mathbb{N}\backslash\{n\}$. If $\delta_\pi$ stands for the Dirac point mass at $\pi \in \mathcal{P}_\infty$, then the measure

$$\epsilon := \sum_{n=1}^{\infty} \delta_{\epsilon^{(n)}}$$

is an exchangeable measure which fulfills (3.4). We shall refer to $\epsilon$ as the *erosion rate*.

To construct the second example, recall that for every mass-partition $\mathbf{s} \in \mathcal{P}_{\mathrm{m}}$, $\varrho_{\mathbf{s}}$ stands for the distribution of the paint-box based on $\mathbf{s}$. By a slight abuse of notation, we shall write $\mathbf{1}$ for the mass-partition $\mathbf{1} = (1, 0, \ldots)$. Consider some sigma-finite measure $\nu$ on $\mathcal{P}_{\mathrm{m}}$ such that

$$\nu(\{\mathbf{1}\}) = 0 \quad \text{and} \quad \int_{\mathcal{P}_{\mathrm{m}}} (1 - s_1)\, \nu(d\mathbf{s}) < \infty, \tag{3.5}$$

where $s_1$ stands for the first term of the mass-partition $\mathbf{s}$.

**Lemma 3.6** *Given a measure $\nu$ on $\mathcal{P}_{\mathrm{m}}$ such that (3.5) holds, define a measure $\varrho_\nu$ on $\mathcal{P}_\infty$ by*

$$\varrho_\nu(d\pi) = \int_{\mathbf{s} \in \mathcal{P}_{\mathrm{m}}} \varrho_{\mathbf{s}}(d\pi)\nu(d\mathbf{s}).$$

*Then $\varrho_\nu$ is an exchangeable measure on $\mathcal{P}$ which fulfills (3.4), and can thus be viewed as the splitting rate of some homogeneous fragmentation.*

Later in the text, we shall refer to $\varrho_\nu$ as a *dislocation rate* and to $\nu$ as a *dislocation measure*.

*Proof* Each $\varrho_{\mathbf{s}}$ is an exchangeable probability measure on $\mathcal{P}_\infty$. Exchangeability is preserved by mixing, so $\varrho_\nu$ is an exchangeable measure on $\mathcal{P}_\infty$. As for all $\mathbf{s} \in \mathcal{P}_{\mathrm{m}}$, the measures $\varrho_{\mathbf{s}}$ assign zero mass to the trivial partition $\mathbf{1}_{\mathbb{N}}$, the same holds for the mixture $\varrho_\nu$.

For every $n \in \mathbb{N}$ and $\mathbf{s} \in \mathcal{P}_{\mathrm{m}}$, the paint-box construction shows that

$$\varrho_{\mathbf{s}}\left(\left\{\pi \in \mathcal{P}_\infty : \pi_{|[n]} \neq \mathbf{1}_{[n]}\right\}\right) = 1 - \sum_{k=1}^{\infty} s_k^n \leq 1 - s_1^n \leq n(1 - s_1).$$

Hence

$$\varrho_\nu\left(\left\{\pi \in \mathcal{P}_\infty : \pi_{|[n]} \neq \mathbf{1}_{[n]}\right\}\right) \leq n \int_{\mathcal{P}_{\mathrm{m}}} (1 - s_1)\nu(d\mathbf{s})$$

and the right-hand side is finite by (3.5). □

The main result of this section, which is essentially a consequence of Kingman's representation of exchangeable random partitions as a mixture of paint-boxes, is that every splitting rate of a homogeneous fragmentation can be expressed as the linear combination of the erosion rate $\epsilon$ and a dislocation rate.

**Theorem 3.1** *Let $\mu$ be an exchangeable measure on $\mathcal{P}_\infty$ which fulfills (3.4). Then there exists a unique $\mathtt{c} \geq 0$ and a unique measure $\nu$ on $\mathcal{P}_{\mathrm{m}}$ that fulfills (3.5) such that*

$$\mu = \mathtt{c}\epsilon + \varrho_\nu .$$

*Specifically, the following holds:*

(i) *$\mu$-almost every partition $\pi \in \mathcal{P}_\infty$ possesses asymptotic frequencies.*

(ii) *The restriction of $\mu$ to the subset of partitions $\pi$ with $|\pi|^\downarrow \neq \mathbf{1}$ is a dislocation rate. More precisely, let $|\mu|^\downarrow$ be the image measure of $\mu$ by the mapping $\pi \to |\pi|^\downarrow$. The restriction*

$$\nu(d\mathbf{s}) := \mathbb{1}_{\{\mathbf{s}\neq\mathbf{1}\}} |\mu|^\downarrow(d\mathbf{s})$$

*of $|\mu|^\downarrow$ to $\mathcal{P}_{\mathrm{m}} \backslash \{\mathbf{1}\}$ fulfills (3.5), and*

$$\mathbb{1}_{\{|\pi|^\downarrow \neq \mathbf{1}\}} \mu(d\pi) = \varrho_\nu(d\pi) .$$

(iii) *The restriction of $\mu$ to the subset of partitions $\pi$ with $|\pi|^\downarrow = \mathbf{1}$ is proportional to the erosion rate, that is there is a real number $\mathtt{c} \geq 0$ such that*

$$\mathbb{1}_{\{|\pi|^\downarrow = \mathbf{1}\}} \mu(d\pi) = \mathtt{c}\epsilon(d\pi) .$$

Later in the text, we shall refer to $\mathrm{c}$ as the *erosion coefficient* and to $\nu$ as the *dislocation measure* of the homogeneous fragmentation $\Pi$. One can compute the jump rates $q_\pi$ of the restricted chains $\Pi_{|[n]}$ explicitly in terms of the erosion coefficient $\mathrm{c}$ and the dislocation measure $\nu$ (in this direction, recall the calculations in Lemma 2.8); however, the expressions than can be obtained are in general rather involved.

*Proof* (i) For every integer $n$, write $\mu_n$ for the restriction of $\mu$ to $\{\pi \in \mathcal{P}_\infty, \pi_{|[n]} \neq \mathbf{1}_{[n]}\}$. Then $\mu_n$ is a finite measure on $\mathcal{P}_\infty$, and is invariant by the action of permutations that coincide with the identity on $[n]$. This leads us to define the $n$-shift $\vec{\pi}$ of a partition $\pi$ by

$$i \overset{\vec{\pi}}{\sim} j \Longleftrightarrow i+n \overset{\pi}{\sim} j+n\,, \qquad i, j \in \mathbb{N}\,,$$

and then to write $\vec{\mu}_n$ for the image of $\mu_n$ by this shift.

Then $\vec{\mu}_n$ is an exchangeable finite measure on $\mathcal{P}_\infty$, and by Theorem 2.1, $\vec{\mu}_n$-almost every partition has asymptotic frequencies. More precisely,

$$\vec{\mu}_n(d\pi) = \int_{\mathcal{P}_\mathrm{m}} \varrho_\mathbf{s}(d\pi)\vec{\mu}_n(|\pi|^\downarrow \in d\mathbf{s}) \tag{3.6}$$

is a regular disintegration of $\vec{\mu}_n$. As shift does not affect asymptotic frequencies, $\mu_n$-almost every partition has asymptotic frequencies. This establishes the first claim.

(ii) Note that if we write $\{i \not\sim j\}$ for the event that $i$ and $j$ do not belong to the same block, then by (3.6), for every $\mathbf{s} \in \mathcal{P}_\mathrm{m}$

$$\begin{aligned}\mu_n(n+1 \not\sim n+2 \mid |\pi|^\downarrow = \mathbf{s}) &= \varrho_\mathbf{s}(1 \not\sim 2)\\ &= 1 - \textstyle\sum_{k=1}^\infty s_k^2\\ &\geq 1 - s_1\left(\textstyle\sum_{k=1}^\infty s_k\right)\\ &\geq 1 - s_1\,.\end{aligned}$$

Hence, if we denote by $\nu_n(d\mathbf{s}) = \mathbb{1}_{\{s\neq 1\}}|\mu_n|^\downarrow(d\mathbf{s})$ the restriction to $\mathcal{P}_\mathrm{m}\backslash\{\mathbf{1}\}$ of the image measure of $\mu_n$ by the map $\pi \to |\pi|^\downarrow$, then

$$\mu_n(n+1 \not\sim n+2) \geq \int_{\mathcal{P}_\mathrm{m}} (1-s_1)\nu_n(d\mathbf{s})\,.$$

On the one hand, the finite measure $\nu_n$ increases as $n \uparrow \infty$ to the measure $\nu$ defined in the statement, so

$$\lim_{n\to\infty} \int_{\mathcal{P}_\mathrm{m}} (1-s_1)\nu_n(d\mathbf{s}) = \int_{\mathcal{P}_\mathrm{m}} (1-s_1)\nu(d\mathbf{s})\,.$$

On the other hand,

$$\mu_n(n+1 \not\sim n+2) \leq \mu(n+1 \not\sim n+2) = \mu(1 \not\sim 2) < \infty .$$

We thus see that $\nu$ fulfills (3.5).

Finally, fix $k \in \mathbb{N}$ and pick a partition $\pi^{[k]} \neq \mathbf{1}_{[k]}$ of $[k]$. We have by monotone convergence

$$\mu(\pi_{|[k]} = \pi^{[k]}, |\pi|^{\downarrow} \neq \mathbf{1}) = \lim_{n\to\infty} \mu(\pi_{|[k]} = \pi^{[k]},$$
$$\pi_{|\{k+1,\dots,k+n\}} \neq \mathbf{1}_{\{k+1,\dots,k+n\}}, |\pi|^{\downarrow} \neq \mathbf{1}) .$$

In the notation introduced in (i), we see from an obvious permutation that

$$\mu(\pi_{|[k]} = \pi^{[k]}, \pi_{|\{k+1,\dots,k+n\}} \neq \mathbf{1}_{\{k+1,\dots,k+n\}},$$
$$|\pi|^{\downarrow} \neq \mathbf{1}) = \vec{\mu}_n(\pi_{|[k]} = \pi^{[k]}, |\pi|^{\downarrow} \neq \mathbf{1}) .$$

Applying (3.6) and then letting $n$ tend to $\infty$, we conclude that

$$\mu(\pi_{|[k]} = \pi^{[k]}, |\pi|^{\downarrow} \neq \mathbf{1}) = \int_{\mathcal{P}_m} \varrho_{\mathbf{s}}(\pi_{|[k]} = \pi^{[k]}) \nu(d\mathbf{s}) .$$

This establishes (ii) as $k$ is arbitrary and the restriction $|\pi|^{\downarrow} \neq \mathbf{1}$ excludes the trivial partition $\mathbf{1}_{\mathbb{N}}$.

(iii) Consider $\tilde{\mu}$, the restriction of $\mu$ to the event $\{1 \not\sim 2, |\pi|^{\downarrow} = (1, 0 \dots)\}$, which has finite mass. Its image by the 2-shift as defined in (i) is an exchangeable finite measure on $\mathcal{P}_\infty$ for which almost every partition has asymptotic frequencies $\mathbf{1} = (1, 0, \dots)$, and hence it must be proportional to the Dirac mass at the trivial partition. Let us denote by $\pi$ the partition with non-void blocks $\{1\}$ and $\{2, \dots\}$, $\pi'$ the partition with non-void blocks $\{2\}$ and $\{1, 3, \dots\}$, and $\pi''$ the partition with non-void blocks $\{1\}$ , $\{2\}$ and $\{3, \dots\}$. We thus have that

$$\tilde{\mu} = \mathtt{c}\delta_{\pi} + \mathtt{c}'\delta_{\pi'} + \mathtt{c}''\delta_{\pi''}$$

where $\mathtt{c}, \mathtt{c}', \mathtt{c}'' \geq 0$ are some real numbers and $\delta$ stands for the Dirac point mass. By exchangeability, we see that the assumption $\mu(1 \not\sim 2) < \infty$ forces $\mathtt{c}'' = 0$. It is also immediate that $\mathtt{c} = \mathtt{c}'$, and the fact that $\mu$ restricted to the event $\{|\pi|^{\downarrow} = \mathbf{1}\}$ coincides with $\mathtt{c}\epsilon$ is now clear again by exchangeability. □

It is interesting to observe that condition (3.5) allows the dislocation measure $\nu$ (and thus the dislocation rate $\varrho_\nu$ also) to be infinite. It is then seen from the Poissonian construction of the homogeneous fragmentation $\Pi$, that when the dislocation measure $\nu$ is infinite, $\Pi$ immediately leaves the initial state $\mathbf{1}_{\mathbb{N}}$; in other words the initial block $\mathbb{N}$ splits instantaneously.

Clearly, such phenomena never occur for a fragmentation chain. Indeed, homogeneous fragmentation processes with no erosion and finite dislocation measures correspond exactly to homogeneous fragmentation chains in $\mathcal{P}_{\mathrm{m}}$ and, more precisely, the dislocation measure $\nu$ of $\Pi$ then coincides with the dislocation measure of the chain.

Recall also that the trivial partition $\mathbf{1}_{\mathbb{N}}$ serves as a neutral element for the fragmentation operator, so a partition $\pi \in \mathcal{P}_\infty$ which is closed to $\mathbf{1}_{\mathbb{N}}$ should be thought of as small from the point of view of fragmentation. Similarly, a mass-partition $\mathbf{s}$ for which the first term $s_1$ is close to 1 is small, in the sense that a small mass-partition produces one large fragment $s_1$ and all the remaining ones are small. Informally, we see from the Poissonian construction of homogeneous fragmentation that condition (3.5) allows infinitely many small dislocations to occur, but guarantees that their accumulation does not instantaneously reduce the initial block $\mathbb{N}$ into singletons.

This discussion shows obvious similarities with subordinators; see Chapter 1 in [30]. Indeed, the latter are constructed by the summation of the atoms of a certain Poisson random measure on $\mathbb{R}_+$, whose intensity is given by the so-called Lévy measure $\Lambda$ of the subordinator. The hypothesis (3.5) for dislocation measures then bears the same role as the condition $\int_{\mathbb{R}_+}(1 \wedge x)\Lambda(dx) < \infty$ for a measure on $\mathbb{R}_+$ to be the Lévy measure of some subordinator, as the latter is the necessary and sufficient condition for the summability of the atoms.

To conclude this section, we shall discuss the effect of erosion in homogeneous fragmentations. In this direction, we first provide a simple construction of a pure erosion process. Let $\mathbf{e}_1, \ldots$ be a sequence of i.i.d. exponential variables with parameter 1, and for every $t \geq 0$, denote by

$$S(t) := \{i \in \mathbb{N} : \mathbf{e}_i \leq t\} .$$

Fix $\mathtt{c} \geq 0$ and denote by $\mathcal{E}^{(\mathtt{c})}(t)$ the partition of $\mathbb{N}$ such that each point of $S(\mathtt{c}t)$ is a singleton and $\mathbb{N}\backslash S(\mathtt{c}t)$ is a block of $\mathcal{E}^{(\mathtt{c})}(t)$.

**Lemma 3.7** *In the notation above, $\mathcal{E}^{(\mathtt{c})} := \left(\mathcal{E}^{(\mathtt{c})}(t), t \geq 0\right)$ is a pure erosion with coefficient $\mathtt{c}$, that is it is a standard homogeneous fragmentation with erosion coefficient $\mathtt{c}$ and dislocation measure $\nu \equiv 0$. In particular, the ranked asymptotic frequencies are simply given by*

$$|\mathcal{E}^{(\mathtt{c})}(t)|^{\downarrow} = \mathrm{e}^{-\mathtt{c}t}\mathbf{1} = (\mathrm{e}^{-\mathtt{c}t}, 0, \ldots) .$$

*Proof* It should be plain that for every $t \geq 0$, $\mathcal{E}^{(\mathtt{c})}(t)$ is an exchangeable random partition and, by the lack of memory property of the exponential law, that $\mathcal{E}^{(c)}$ is a homogeneous fragmentation. Obviously, $\mathcal{E}^{(c)}(t)$ has exactly one block that is not reduced to a singleton, and it is immediate from the law of large numbers that the asymptotic frequency of this block is $e^{-ct}$.

In order to complete the proof, we simply need to compute the jump rates of the continuous Markov chain $\mathcal{E}^{(c)}_{|[n]}$. So consider $\pi^{[n]} \in \mathcal{P}_n \backslash \{\mathbf{1}_{[n]}\}$, and suppose first that $\pi^{[n]}$ is of the type $([n]\backslash\{k\}, \{k\}, \emptyset, \ldots)$ for some $k = 1, \ldots, n$. In this case, we have

$$\lim_{t\to 0+} \frac{1}{t}\mathbb{P}\left(\mathcal{E}^{(c)}_{|[n]}(t) = \pi^{[n]}\right) = \lim_{t\to 0+} \frac{1}{t}\mathbb{P}\left(\mathbf{e}_k \le ct\right) = c\,.$$

Next, if $\pi^{[n]}$ is not of the preceding type, then clearly

$$\lim_{t\to 0+} \frac{1}{t}\mathbb{P}\left(\mathcal{E}^{(c)}_{|[n]}(t) = \pi^{[n]}\right) = 0\,.$$

Recall the definition of the erosion rate $\epsilon$ in Section 3.2.1. We have thus checked that the jump rates of the restriction $\mathcal{E}^{(c)}_{|[n]}$ can be expressed as

$$c\epsilon\left(\{\pi \in \mathcal{P}_\infty : \pi_{|[n]} = \pi^{[n]}\}\right),$$

which enables us to conclude that the splitting rate of $\mathcal{E}^{(c)}$ is indeed $c\epsilon$. □

Next, we combine the pure erosion process constructed above with a fragmentation. Specifically, let $\Pi^{(0)}$ be a homogeneous fragmentation with zero erosion coefficient and dislocation measure $\nu$, and $|\Pi^{(0)}|^{\downarrow}$ the process of its ranked asymptotic frequencies. Assume that $\Pi^{(0)}$ is independent of the sequence of exponential variables $\mathbf{e}_i$ and, for a fixed $c \ge 0$, write $\Pi^{(c)}(t)$ for the unique partition such that each $i \in S(ct)$ is a singleton of $\Pi^{(c)}(t)$ and the restrictions of $\Pi^{(c)}(t)$ and $\Pi^{(0)}(t)$ to $\mathbb{N}\backslash S(ct)$ coincide.

**Proposition 3.4** *In the notation above, $\Pi^{(c)} := \left(\Pi^{(c)}(t), t \ge 0\right)$ is a homogeneous fragmentation with erosion coefficient $c$ and dislocation measure $\nu$. Moreover, for every $t \ge 0$, the partitions $\Pi^{(c)}(t)$ and $\Pi^{(0)}(t)$ possess asymptotic frequencies a.s., and there is the identity*

$$|\Pi^{(c)}(t)|^{\downarrow} = e^{-ct}|\Pi^{(0)}(t)|^{\downarrow}\,, \qquad a.s.$$

*Proof* It might be intuitively clear from the Markov property of $\Pi^{(0)}$ and the lack of memory of the exponential variables that $\Pi^{(c)}$ is indeed a homogeneous fragmentation; however, the rigorous argument is somewhat heavy.

First, we observe that for every $t \ge 0$, $\Pi(t)$ is an exchangeable partition (this follows from the exchangeability of $\Pi^{(0)}(t)$ and the fact that the variables $\mathbf{e}_i$ are i.i.d. and independent of $\Pi^{(0)}(t)$). Then, we need some notation. For every $\pi \in \mathcal{P}_\infty$ and $r \geq 0$, we write $\mathcal{E}^{(r)}(\pi)$, and call $\pi$ eroded with rate $r$ the random partition obtained as follows. Recall that $S(r)$ corresponds to the set of singletons induced by a pure erosion process at time $r$, in the sense specified above. Then $\mathcal{E}^{(r)}(\pi)$ is the partition such that each $i \in S(r)$ is a singleton of $\mathcal{E}^{(r)}(\pi)$ and the restrictions of $\mathcal{E}^{(r)}(\pi)$ and $\pi$ to $\mathbb{N}\backslash S(r)$ coincide. In particular, $\Pi^{(\mathtt{c})}(t) = \mathcal{E}^{(\mathtt{c}t)}(\Pi^{(0)}(t))$.

The key lies in the following identity in distribution, which is readily checked. Consider $\pi \in \mathcal{P}_\infty$ and a sequence $\pi^{(\cdot)}$ in $\mathcal{P}_\infty$. Then

$$\mathcal{E}^{(r)}\left(\mathrm{Frag}\left(\pi, \pi^{(\cdot)}\right)\right) \overset{(\mathrm{d})}{=} \mathrm{Frag}\left(\pi, \gamma^{(\cdot)}\right),$$

where, in the right-hand side, $\gamma^{(\cdot)} = (\gamma^{(i)}, i \in \mathbb{N})$ is a sequence of independent random partitions such that $\gamma^{(i)}$ is distributed as $\mathcal{E}^{(r)}\left(\pi^{(i)}\right)$ for each $i \in \mathbb{N}$. We apply this identity when $r = \mathtt{c}t$ and the random partitions $\pi^{(i)}$ are i.i.d. distributed as $\Pi^{(0)}(t)$ (so that $\mathcal{E}^{(\mathtt{c}t)}\left(\pi^{(i)}\right)$ is a version of $\Pi^{(\mathtt{c})}(t)$), and combine this with the Markov property of $\Pi^{(0)}$ and the lack of memory of exponential variables. We get that $\Pi^{(\mathtt{c})}$ is a homogeneous fragmentation.

Next, we need to calculate the jump rates of $\Pi$, so we fix some integer $n \in \mathbb{N}$ and consider the restricted Markov chain $\Pi_{|[n]}$. Pick $\pi \in \mathcal{P}_n$ with $\pi \neq \mathbf{1}_{[n]}$, and first suppose that $\pi$ has exactly two non-empty blocks, one of which is a singleton. Then it should be plain from the construction that the jump rate $q_\pi^{(\mathtt{c})}$ of $\Pi_{|[n]}$ from $\mathbf{1}_{[n]}$ to $\pi$ is given by $q_\pi^{(\mathtt{c})} = q_\pi^{(0)} + \mathtt{c}$, where $q_\pi^{(0)}$ stands for the jump rate of $\Pi_{|[n]}^{(0)}$ from $\mathbf{1}_{[n]}$ to $\pi$. Next, when $\pi$ is not of the preceding type, then plainly $q_\pi = q^{(0)}$. In other words, there is the identity

$$q_\pi^{(\mathtt{c})} = q_\pi^{(0)} + \mathtt{c} q_\pi^{\mathcal{E}},$$

where $q_\pi^{\mathcal{E}}$ stands for the jump rates of the pure erosion with unit coefficient from $\mathbf{1}_{[n]}$ to $\pi$. This shows that the splitting rate of $\Pi^{(\mathtt{c})}$ is the sum of that of $\Pi^{(0)}$ and $\mathtt{c}\epsilon$, that is $\Pi^{(\mathtt{c})}$ has erosion coefficient $\mathtt{c}$ and dislocation measure $\nu$.

Finally, it is immediately seen from the construction of $\Pi^{(\mathtt{c})}(t)$ from $\Pi^{(0)}(t)$ and the independent exponential variables $\mathbf{e}_i$ and the law of large numbers, that $|\Pi^{(\mathtt{c})}(t)|^\downarrow = \mathrm{e}^{-\mathtt{c}t}|\Pi^{(0)}(t)|^\downarrow$ a.s. □

### 3.2.2 Subordinator representation of the tagged fragment

We next turn our attention to the process of the asymptotic frequency of the first block $|\Pi_1(\cdot)|$ in a homogeneous fragmentation process; the motivation stems from the following interpretation. Consider an object with a unit mass that falls in part randomly as time passes. In order to investigate its evolution, we pick at random a sequence $U_1, \ldots$ of i.i.d. points in the object according to the mass distribution, and we consider the process of nested partitions $(\Pi(t), t \geq 0)$ such that two distinct integers, say $i$ and $j$, belong to the same block of $\Pi(t)$ if and only if the points $U_i$ and $U_j$ belong to the same component of the object at time $t$. In this setting, the law of large numbers implies that the asymptotic frequency of the first block $|\Pi_1(t)|$ coincides with the mass of the component of the object which contains $U_1$ at time $t$. In other words, the process $|\Pi_1(\cdot)|$ describes the evolution of the mass of the component which contains a point tagged at random according to the mass distribution of the object; it will therefore often be referred to as the process of the *tagged fragment*. Properties of the tagged fragment will provide a most important tool for the study of homogeneous fragmentations.

Loosely speaking, given that at time $t$ the tagged point $U_1$ belongs to some component, $U_1$ is again distributed according to the mass-distribution of this component. Therefore, one expects that in the situation in which the evolution of the object is governed by a homogeneous fragmentation, the process of the mass of the tagged component should fulfill the identity in distribution

$$|\Pi_1(t+t')| \stackrel{(d)}{=} |\Pi_1(t)||\Pi'_1(t')|,$$

where $\Pi'_1(t')$ is a copy of $\Pi_1(t')$ that is independent of $\Pi_1(t)$. In other words, the process $\ln|\Pi_1(\cdot)|$ should have independent and stationary increments.

The informal argument above can be made rigorous. But before doing so, we need to recall some basic material on *subordinators*. Remember that a large class of subordinators has already been introduced in Section 2.2.4; however, we will have to consider here the most general family. We refer to Chapter III in [29] or [30] for details.

Let $(\mathcal{F}_t)_{t\geq 0}$ denote the natural filtration of the homogeneous fragmentation $\Pi(\cdot)$. An $(\mathcal{F}_t)$-subordinator is an $(\mathcal{F}_t)$-adapted right-continuous process $(\xi(t), t \geq 0)$ with values in $[0, \infty]$ such that $\xi_0 = 0$ a.s. and given $\xi(t) < \infty$, the increment $\xi(t+t') - \xi(t)$ is independent of $\mathcal{F}_t$ and has the same law as $\xi(t')$. The point $\infty$ is an absorbing state for $\xi$, and its hitting time $\zeta := \inf\{t \geq 0 : \xi(t) = \infty\}$ is usually referred to as the *lifetime* of $\xi$. The lifetime follows an exponential law whose parameter $\mathtt{k} \geq 0$ is known as the *killing rate* ($\zeta = \infty$ a.s. when $\mathtt{k} = 0$). Any subordinator $\xi$ with killing rate $\mathtt{k} > 0$ can be obtained from a subordinator $\xi'$ with zero killing rate and an independent exponential variable $\zeta$ with parameter $\mathtt{k}$ by killing $\xi'$ at time $\zeta$, that is

$$\xi(t) = \xi'(t) \text{ for } t < \zeta \text{ and } \xi(t) = \infty \text{ for } t \geq \zeta\,.$$

Note in particular that $\xi(\zeta-) < \infty$ a.s. when the killing rate is $\mathtt{k} > 0$.

The one-dimensional distributions of a subordinator are specified by the so-called *Laplace exponent* $\Phi$ that is given by the identity

$$\mathbb{E}\left(\exp(-q\xi(t))\right) = \exp(-t\Phi(q)), \qquad t, q > 0,$$

where we implicitly use the convention that $e^{-\infty} = 0$. In turn, $\Phi$ is given by the celebrated *Lévy-Khintchine formula*

$$\Phi(q) = \mathtt{k} + \mathtt{d}q + \int_{]0,\infty[} (1 - e^{-qx})\Lambda(dx), \tag{3.7}$$

where $\mathtt{d} \geq 0$ is the so-called *drift coefficient* and $\Lambda$ a measure on $]0, \infty[$ with $\int (1 \wedge x)\Lambda(dx) < \infty$, is known as the *Lévy measure* of $\xi$. In other words, the Lévy-Khintchine formula enables us to calculate the Laplace transform of the one-dimensional distribution of a subordinator in terms of its drift and killing coefficients and its Lévy measure. Observe that, by the independence and stationarity of the increments, this characterizes the law of the entire process.

Finally, we recall the celebrated Lévy-Itô decomposition of subordinators into the continuous part and the jump part. Specifically, there is a random measure $N$ on $[0, \infty[\times]0, \infty]$ which has a Poisson distribution with intensity $dt \otimes (\Lambda(dx) + \mathtt{k}\delta_\infty(dx))$, such that

$$\xi(t) = \mathtt{d}t + \int_{[0,t]\times]0,\infty]} xN(dt, dx), \qquad t \geq 0.$$

In this direction, we stress that the lifetime $\zeta = \inf\{t \geq 0 : \xi(t) = \infty\}$ can be identified as the first instant when an infinite atom arises in the Poisson measure,

$$\zeta = \inf\{t \geq 0 : N(\{(t, \infty)\}) = 1\}.$$

We can now resume the study of homogeneous fragmentations. The Poissonian construction of Section 3.1.3 is reminiscent of the Lévy-Itô decomposition of subordinators. If we intend to develop further this informal analogy, we may wish to link the erosion coefficient (respectively, the dislocation measure $\nu$) of a fragmentation to the drift coefficient (respectively, the Lévy measure) of a subordinator. In this direction, the crucial Lévy-Khintchine formula for subordinators, which enables one to identify the distribution of the subordinator, has so far no analog for homogeneous fragmentations.

We shall now partly fill this gap. More precisely, consider a homogeneous fragmentation $\Pi$ with splitting rate $\mu$. Recall from Theorem 3.1 that $\mu = \mathtt{c}\epsilon + \varrho_\nu$ is the canonical decomposition into the sum of an erosion rate with coefficient $\mathtt{c} \geq 0$, and a dislocation rate $\varrho_\nu$, where $\nu$ is a measure on $\mathcal{P}_\mathrm{m}$ which fulfills (3.5). Since for each fixed $t \geq 0$, the random partition $\Pi(t)$ is exchangeable, we know from Kingman's Theorem 2.1 that the first block $\Pi_1(t)$ has an asymptotic frequency a.s. We stress that this event depends on $t$; nonetheless we shall see below that the existence of asymptotic frequencies for $\Pi_1(t)$ holds in fact simultaneously for all $t$, with probability one. We are now able to state the main result of this section.

**Theorem 3.2** *Let $\Pi$ be a homogeneous fragmentation with erosion coefficient $\mathtt{c}$ and dislocation measure $\nu$. With probability one, the blocks $\Pi_1(t)$ have an asymptotic frequency simultaneously for all $t \geq 0$, and the process $\xi = (\xi(t), t \geq 0)$ defined by*

$$\xi(t) := -\ln(|\Pi_1(t)|), \qquad t \geq 0,$$

*is an $(\mathcal{F}_t)$-subordinator. The Laplace exponent of $\xi$,*

$$\Phi(q) = -\ln \mathbb{E}\left(e^{-q\xi(1)}\right) = -\ln \mathbb{E}\left(|\Pi_1(1)|^q\right), \qquad q > 0,$$

*is given by*

$$\Phi(q) = \mathtt{c}(q+1) + \int_{\mathcal{P}_m} \left(1 - \sum_{n=1}^{\infty} s_n^{q+1}\right) \nu(d\mathbf{s}).$$

*Alternatively, the drift coefficient $\mathtt{d}$ coincides with the erosion coefficient $\mathtt{c}$, the killing rate is given by*

$$\mathtt{k} = \mathtt{c} + \int_{\mathcal{P}_m} s_0 \nu(d\mathbf{s}), \qquad \text{with } s_0 := 1 - \sum_{j=1}^{\infty} s_j,$$

*and the Lévy measure by*

$$\Lambda(dx) = e^{-x} \sum_{j=1}^{\infty} \nu(-\ln s_j \in dx), \qquad x \in ]0, \infty[.$$

Before tackling the proof, let us make a couple of comments on Theorem 3.2.

The resemblance between Theorem 3.2 and the Lévy-Khintchine formula for subordinators is quite appealing; however, there is a major difference. In the case of subordinators, the Laplace exponent determines the distribution of the entire process, whereas in general the function $\Phi$ alone does not enable one to recover the characteristics of the fragmentation. To be more precise, the Laplace transform (3.7) determines uniquely the drift and killing coefficients and the Lévy measure of the subordinator, but in general one cannot recover[2]

[2] There is however, one important situation in which $\Phi$ determines $\mathtt{c}$ and $\nu$: it is when dislocations are binary a.s., in the sense that $\nu(s_1 + s_2 \neq 1) = 0$. Then one can identify $\nu$ as a measure on $[1/2, 1]$, namely the image of $\nu$ by the map $\mathbf{s} \to s_1$, and the latter is then entirely characterized by $\Phi$.

the dislocation measure $\nu$ from the expression for $\Phi$ in Theorem 3.2. Indeed, it is not hard to construct two homogeneous fragmentations, say $\Pi^{(1)}$ and $\Pi^{(2)}$ with different dislocation measures, such that $|\Pi_1^{(1)}(\cdot)|$ and $|\Pi_1^{(2)}(\cdot)|$ have the same distribution (in this direction, recall that in general, the law of a size-biased sample of a random mass-partition does not determine the distribution of the latter).

Theorem 3.2 should be compared with Proposition 1.6 for homogeneous fragmentation chains. More precisely, recall the framework and notation of Section 1.2.3, and suppose that the erosion coefficient $\mathtt{c}$ is zero, and that the dislocation measure $\nu$ is both finite and conservative (i.e. $\nu(\mathcal{P}_{\mathrm{m}}) < \infty$ and $\nu\{\mathbf{s} : \sum_{i=1}^{\infty} s_i \neq 1\} = 0$). In particular, the Malthusian parameter is $p^* = 1$ and the intrinsic martingale is trivial, $\mathcal{M} \equiv 1$. Then Proposition 1.6 shows that the processes of the so-called tagged branch and of the tagged fragment have the same distribution, an identity which should of course be intuitively obvious. In this direction, we also stress that this identity fails when the dislocation measure is dissipative, because then the intrinsic martingale is no longer trivial. In the latter situation, the tagged fragment reaches 0 in a finite time (which has an exponential distribution with parameter $\mathtt{k} = \Phi(0) > 0$), and the tagged branch in Section 1.2.3 rather corresponds to the tagged fragment conditioned to remain strictly positive forever.

The proof of Theorem 3.2 is broken into several lemmas. To start with, we set

$$\Phi(q) = \mathtt{c}(q+1) + \int_{\mathcal{P}_{\mathrm{m}}} \left(1 - \sum_{n=1}^{\infty} s_n^{q+1}\right) \nu(d\mathbf{s}), \qquad q \geq 0.$$

Note the similarity with (1.17); more precisely we have $\Phi(q) = \kappa(q+1)$ whenever the erosion coefficient $\mathtt{c}$ is zero and the dislocation measure $\nu$ finite. It can be immediately checked that $\Phi$ can also be expressed by a Lévy-Khintchine formula (3.7). More precisely, we have the following.

**Lemma 3.8** *Introduce the real number*

$$\mathtt{k} = \mathtt{c} + \int_{\mathcal{P}_{\mathrm{m}}} \left(1 - \sum_{j=1}^{\infty} s_j\right) \nu(d\mathbf{s}),$$

*and the measure*

$$\Lambda(dx) = \mathrm{e}^{-x} \sum_{j=1}^{\infty} \nu(-\ln s_j \in dx), \qquad x \in ]0, \infty[.$$

*Then* $\int(1\wedge x)\Lambda(dx)<\infty$ *and there is the identity*

$$\Phi(q)=\mathtt{k}+\mathtt{c}q+\int_{]0,\infty[}(1-\mathrm{e}^{-qx})\Lambda(dx)\,,\qquad q\geq 0\,.$$

Next, we observe that for each fixed $t\geq 0$, $\Pi(t)$ is an exchangeable partition, and thus the first block $\Pi_1(t)$ has an asymptotic frequency a.s. To start with, we point at the following obvious property of the increments of $|\Pi_1(\cdot)|$. Recall that $(\mathcal{F}_t)_{t\geq 0}$ denotes the natural filtration of $\Pi(\cdot)$.

**Lemma 3.9** *For every* $t, t'\geq 0$, *conditionally on* $|\Pi_1(t)|>0$, *the ratio* $|\Pi_1(t+t')|/|\Pi_1(t)|$ *is independent of* $\mathcal{F}_t$ *and has the same law as* $|\Pi_1(t')|$.

*Proof* The very construction of $\Pi(\cdot)$ implies that $\Pi_1(t+t')=\Pi_1(t)\cap\Pi'_1(t')$, where $\Pi'(t')$ is a random exchangeable partition which is independent of $\mathcal{F}_t$ and has the same law as $\Pi(t')$. When we compute asymptotic frequencies of first blocks using Corollary 2.5, this yields $|\Pi_1(t+t')|=|\Pi_1(t)||\Pi'_1(t')|$ a.s., which in turn yields our claim. □

The rest of our analysis relies on the following elementary formula for the entire moments of $|\Pi_1(t)|$.

**Lemma 3.10** *For every* $t\geq 0$, *we have*

$$\mathbb{E}\left(|\Pi_1(t)|^k\right)=\exp(-t\Phi(k))\,,\qquad k\in\mathbb{N}.$$

*Proof* Observe first the fact that $\Pi(t)$ is an exchangeable random partition and Corollary 2.4 yield the identity

$$\mathbb{E}\left(|\Pi_1(t)|^k\right)=\mathbb{P}\left(\Pi(t)_{|[k+1]}=\mathbf{1}_{[k+1]}\right)\,,\qquad k\in\mathbb{N}\,.$$

Recall now the construction in Section 3.1.1 of $\Pi(\cdot)$ in terms of the Poisson random measure $\mathtt{M}$ on $[0,\infty[\times\mathcal{P}_\infty\times\mathbb{N}$ with intensity $dt\otimes\mu\otimes\#$, where $\mu$ stands for the splitting rate. Write $\mathtt{M}_1$ for the random measure on $[0,\infty[\times\mathcal{P}_\infty$ derived from $\mathtt{M}$ by retaining only the atoms on the fiber $[0,\infty[\times\mathcal{P}_\infty\times\{1\}$. Then $\mathtt{M}_1$ is a Poisson random measure with intensity $dt\otimes\mu$, and for all $t\geq 0$, the first block $\Pi_1(t)$ is constructed using only the atoms of $\mathtt{M}_1$ on $[0,t]\times\mathcal{P}_\infty$. More precisely, we have that the restriction $\Pi_{|[k+1]}(t)$ of $\Pi(t)$ to $[k+1]$ coincides with the partition $\mathbf{1}_{[k+1]}$ of $[k+1]$ into a single non-empty block if and only if $[k+1]\subseteq\Pi_1(t)$, and this holds if and only if

$$\mathtt{M}_1\left([0,t]\times\left\{\pi\in\mathcal{P}_\infty:\pi_{|[k+1]}\neq\mathbf{1}_{[k+1]}\right\}\right)=0\,.$$

From an elementary result on Poisson random measures, we get that the probability of this event equals

$$\exp\left(-t\mu\left(\left\{\pi\in\mathcal{P}_\infty:\pi_{|[k+1]}\neq\mathbf{1}_{[k+1]}\right\}\right)\right).$$

On the other hand, recall from Theorem 3.1 that the splitting rate $\mu$ is given in terms of the erosion coefficient $\mathtt{c}$ and the dislocation measure $\nu$ by $\mu=\mathtt{c}\epsilon+\varrho_\nu$. We now see from the paint-box construction that

$$\mu\left(\left\{\pi\in\mathcal{P}_\infty:\pi_{|[k+1]}\neq\mathbf{1}_{[k+1]}\right\}\right)=\mathtt{c}(k+1)+\int_{\mathcal{P}_{\mathrm{m}}}\left(1-\sum_{n=1}^{\infty}s_n^{k+1}\right)\nu(ds).$$

By definition, the right-hand side above equals $\Phi(k)$, which completes the proof. □

As each random variable $|\Pi_1(t)|$ takes values in [0, 1], and since $|\Pi_1(t)|\geq|\Pi_1(t')|$ a.s. whenever $0\leq t\leq t'$, we may define a right-continuous increasing process $(\xi(t),t\geq0)$ with values in $[0,\infty]$ by

$$\xi(t)=\lim_{\mathbb{Q}\ni t'\downarrow t}\ln\left(1/|\Pi_1(t')|\right),\qquad t\geq0.$$

The notation indicates that we consider the decreasing limit as the rational number $t'$ decreases to $t$.

**Lemma 3.11** (i) *The process* $\exp(-\xi(\cdot))$ *is a version of* $|\Pi_1(\cdot)|$, *that is for each* $t\geq0$,

$$\mathbb{P}\left(|\Pi_1(t)|=\exp(-\xi(t))\right)=1.$$

(ii) *The process* $\xi$ *is an* $(\mathcal{F}_t)$*-subordinator.*
(iii) *The Laplace exponent of* $\xi$ *is* $\Phi$.

*Proof* (i) Lemma 3.10 implies that $\mathbb{E}(|\Pi_1(t)|)$ tends to 1 when $t\to0+$, and thus $|\Pi_1(t)|$ converges in probability to 1. It follows from Lemma 3.9 that the process $|\Pi_1(\cdot)|$ is right-continuous in probability, and thus $\exp(-\xi(\cdot))$ is a version of $|\Pi_1(\cdot)|$.

(ii) The process $\xi$ is increasing and right-continuous by construction. It also has independent and stationary increments by Lemma 3.9 and (i). Thus $\xi$ is a subordinator.

(iii) By Lemma 3.8, we know that $\Phi$ can be viewed as the Laplace exponent of some subordinator, say $\xi'$. On the other hand, we known from Lemma 3.10 that for every $k\in\mathbb{N}$,

$$\mathbb{E}\left(\exp(-k\xi(t))\right)=\mathbb{E}\left(\exp(-k\xi'(t))\right)=\exp(-t\Phi(k)).$$

The random variables $\exp(-\xi(t))$ and $\exp(-\xi'(t))$ thus have the same integer moments, and since they take values in $[0, 1]$, they must have the same distribution, and in particular the same Laplace transform. □

We finally complete the proof of Theorem 3.2 by showing that asymptotic frequencies of the first blocks exist simultaneously for all times.

**Lemma 3.12** *The probability that for all $t > 0$ the blocks $\Pi_1(t)$ have an asymptotic frequency which are given by*

$$|\Pi_1(t)| = \exp(-\xi(t)),$$

*equals one.*

*Proof* It will be convenient to introduce the following notation for the upper and lower asymptotic frequencies of a block $B$:

$$|B|^+ = \limsup_{n\to\infty} \frac{1}{n}\#(B \cap [n]),$$

$$|B|^- = \liminf_{n\to\infty} \frac{1}{n}\#(B \cap [n]).$$

Using the fact that for $t' < t < t''$, the partition $\Pi(t')$ is coarser than $\Pi(t)$ and $\Pi(t'')$ finer than $\Pi(t)$, we see that with probability one, it holds for all $t > 0$ that

$$\exp(-\xi(t)) \le |\Pi_1(t)|^- \le |\Pi_1(t)|^+ \le \exp(-\xi(t-)).$$

Hence, whenever $\xi$ is continuous at time $t$, $\Pi_1(t)$ has an asymptotic frequency given by $\exp(-\xi(t))$. Observe also that, since $\xi_0 = 0$ and $\xi$ is right-continuous at 0,

$$\lim_{t\to 0} |\Pi_1(t)|^- = 1 \qquad \text{a.s.}$$

We next turn our attention to discontinuity points of $\xi$. In this direction, fix an arbitrary integer $k \in \mathbb{N}$ and a real number $\varepsilon > 0$, and consider the instant $T$ when $\xi$ makes its $k$-th jump of size at least $\varepsilon$. Then $T$ is an $(\mathcal{F}_t)$-stopping time (recall that the filtration $(\mathcal{F}_t)_{t\ge 0}$ is right-continuous), so the strong Markov property easily implies that conditionally on $\{T < \infty\}$, for every $t > 0$

$$|\Pi_1(T)|^- - (1 - |\Pi_1'(t)|^-) \le |\Pi_1(T+t)|^- \le |\Pi_1(T)|^-,$$

$$|\Pi_1(T)|^+ - (1 - |\Pi_1'(t)|^-) \le |\Pi_1(T+t)|^+ \le |\Pi_1(T)|^+,$$

where $|\Pi'_1(\cdot)|^-$ denotes the process of the lower asymptotic frequency of the first block in some homogeneous fragmentation $\Pi'(\cdot)$ which has the same law as $\Pi(\cdot)$. Recall that $|\Pi'_1(t)|^-$ converges to 1 a.s. when $t \to 0+$, we deduce that

$$\lim_{t\to 0+} |\Pi_1(T+t)|^- = |\Pi_1(T)|^-$$

and

$$\lim_{t\to 0+} |\Pi_1(T+t)|^+ = |\Pi_1(T)|^+$$

a.s. on $\{T < \infty\}$. Since the family of positive $t$ for which $\xi$ is continuous at time $T+t$ has 0 as an accumulation point, and since $\xi$ is right-continuous, we conclude that almost surely on $\{T < \infty\}$

$$|\Pi_1(T)|^- = |\Pi_1(T)|^+ = \exp\{-\xi_T\} .$$

In other words, we have shown that with probability one, at the instant $T$ when $\xi$ makes its $k$-th jump of size at least $\varepsilon$, the first block $\Pi_1(T)$ has an asymptotic frequency which is given by $\exp(-\xi(T))$. This is valid simultaneously for all integers $k \in \mathbb{N}$ and all $\varepsilon \in \{1/n, n \in \mathbb{N}\}$, so for all the jump times of $\xi$. □

### 3.2.3 Lévy-Itô decomposition of the tagged fragment

Theorem 3.2 describes the dynamics of the tagged fragment in terms of some subordinator. We now would like to enlighten this representation by connecting the Poissonian construction of homogeneous fragmentation $\Pi$ in Section 3.1.3 to the Lévy-Itô decomposition of the subordinator $\xi = -\ln|\Pi_1(\cdot)|$.

The process of the first block $\Pi_1(\cdot)$ of $\Pi(\cdot)$ can be obtained from the restriction of the Poisson random measure $\mathtt{M}$ to the fiber $\mathbb{R}_+ \times \mathcal{P}_\infty \times \{1\}$. This leads us to introduce the random set

$$\mathcal{A} = \{t \geq 0 : \mathtt{M}(\{t\} \times \mathcal{P}_\infty \times \{1\}) = 1\} ,$$

that is $\mathcal{A}$ is the set of times $t$ at which $\mathtt{M}$ has an atom, say $(t, \pi, 1)$, on this fiber. Then, for every $t \in \mathcal{A}$, we write $\pi(t) = \pi$ for the second coordinate of this atom. The random point measure

$$\mathtt{M}_1 := \sum_{t\in\mathcal{A}} \delta_{(t,\pi(t))}$$

is Poisson on $\mathbb{R}_+ \times \mathcal{P}_\infty$ with intensity $dt \otimes \mu(d\pi)$, where $\mu$ is the splitting rate of $\Pi$. From the very construction of $\Pi$, we have the identity

$$\Pi_1(t) = \bigcap_{r\in\mathcal{A}\cap[0,t]} \pi_1(r) .$$

Recall also that $\Pi$ has càdlàg paths, and that for every $t > 0$, if we denote by $\Pi(t-)$ the left-limit of $\Pi$ at time $t$, then

$$\Pi_1(t-) = \bigcap_{r\in\mathcal{A}\cap[0,t[} \pi_1(r) .$$

We know from Theorem 3.1 that the splitting rate $\mu$ can be expressed as the sum of an erosion rate and of a dislocation rate, $\mu = \mathtt{c}\epsilon + \varrho_\nu$, where $\mathtt{c} \geq 0$ is the erosion coefficient and $\nu$ the dislocation measure of $\Pi$. This leads us to decompose the set $\mathcal{A}$ in the form $\mathcal{A} = \mathcal{A}_e \cup \mathcal{A}_d$, where $\mathcal{A}_e$ and $\mathcal{A}_d$ are the sets of times of apparition of atoms corresponding respectively to erosion and to dislocation,

$$\mathcal{A}_e := \left\{t \in \mathcal{A} : |\pi(t)|^{\downarrow} = \mathbf{1}\right\} , \quad \mathcal{A}_d := \left\{t \in \mathcal{A} : |\pi(t)|^{\downarrow} \neq \mathbf{1}\right\} ,$$

and set

$$\mathtt{M}_e := \sum_{t\in\mathcal{A}_e} \delta_{(t,\pi(t))} , \quad \mathtt{M}_d := \sum_{t\in\mathcal{A}_d} \delta_{(t,\pi(t))} .$$

Then $\mathtt{M}_e$ and $\mathtt{M}_d$ are two independent Poisson point measures with respective intensity $c\epsilon$ and $\varrho_\nu$ (see Lemma 2.4) and $\mathtt{M}_1 = \mathtt{M}_e + \mathtt{M}_d$.

We may now state:

**Proposition 3.5** *The following assertions hold with probability one:*

(i) *For every $t \in \mathcal{A}$, the partition $\pi(t)$ possesses asymptotic frequencies.*

(ii) *The lifetime of the subordinator $\xi = -\ln|\Pi_1(\cdot)|$,*

$$\zeta := \inf\{t \geq 0 : \xi(t) = \infty\} = \inf\{t \geq 0 : |\Pi_1(t)| = 0\} ,$$

*is given by*

$$\zeta = \inf\{t \in \mathcal{A} : \pi_1(t) = \{1\}\} = \inf\{t \in \mathcal{A} : |\pi_1(t)| = 0\} .$$

(iii) *For every $0 \leq t < \zeta$, there is the identity*

$$|\Pi_1(t)| = \mathrm{e}^{-\mathtt{c}t} \prod_{r\in\mathcal{A}_d\cap[0,t]} |\pi_1(r)| ,$$

*where $\mathtt{c} \geq 0$ is the erosion coefficient.*

(iv) *For every $0 < t < \zeta$, the block $\Pi_1(t-)$ has an asymptotic frequency which is given by*

$$|\Pi_1(t-)| = \lim_{t'\uparrow t} |\Pi_1(t')| = \mathrm{e}^{-\mathtt{c}t} \prod_{r\in\mathcal{A}_d\cap[0,t[} |\pi_1(r)| .$$

Before establishing this result, let us make a few comments.

First, part (ii) of the statement explains the formula for the killing rate $\mathtt{k}$ given in Theorem 3.2. Indeed, Proposition 3.5(ii) identifies $\mathtt{k}$ as the rate at which partitions $\pi$ with $\pi_1 = \{1\}$ occur in the Poisson point measure $\mathtt{M}_1 = \mathtt{M}_e + \mathtt{M}_d$. Thus we can express $\mathtt{k}$ as the sum of the rates of occurrence of such partitions in $\mathtt{M}_e$ and in $\mathtt{M}_d$, respectively. Plainly, the first term in this sum is simply the erosion coefficient. In order to compute the second, observe that for every improper mass-partition $\mathbf{s} \in \mathcal{P}_{\mathrm{m}}$ (i.e. such that $s_0 := 1 - \sum_{j=1}^{\infty} s_j > 0$), the probability that 1 is a singleton in a paint-box based on $\mathbf{s}$ equals $s_0$. So putting the pieces together, we recover the identity $\mathtt{k} = \mathtt{c} + \int_{\mathcal{P}_{\mathrm{m}}} s_0 \nu(d\mathbf{s})$.

Second, part (iii) of the statement can be viewed as the Lévy-Itô decomposition of the tagged fragment $|\Pi_1(t)| = \exp(-\xi_t)$. Indeed, taking logarithms yields

$$\xi_t = \mathtt{c}t - \sum_{r \in \mathcal{A}_d \cap [0,t]} \ln |\pi_1(r)| , \qquad 0 \le t < \zeta ,$$

which shows in particular that the drift coefficient of $\xi$ is $\mathtt{c}$. On the other hand, since $\mathtt{M}_d$ is a Poisson random measure with intensity $dt \otimes \varrho_\nu$, the image property of Lemma 2.4 implies that the point measure

$$\sum_{t \in \mathcal{A}_d} \delta_{(t, -\ln |\pi_1(t)|)}$$

is Poisson with intensity $dt \otimes \Lambda$, where $\Lambda$ is the image of $\varrho_\nu$ by the map $\pi \to -\ln |\pi_1|$. It is then easy to deduce from Proposition 2.8(ii) that

$$\Lambda(dx) = \mathrm{e}^{-x} \sum_{j=1}^{\infty} \nu(-\ln s_j \in dx) , \qquad x \in ]0, \infty] ,$$

which agrees with Theorem 3.2.

Last, we stress that the identity (iv) is by no mean obvious, as the map $\pi \to |\pi_1|$ is not continuous on $\mathcal{P}_\infty$.

We now tackle the proof of Proposition 3.5.

*Proof* By the very definition of the erosion, $\epsilon$-almost every partition $\pi$ possesses asymptotic frequencies which are given either by $|\pi| = (0, 1, 0, \ldots)$ or by $|\pi| = (1, 0, 0, \ldots)$, depending on whether the singleton of $\pi$ is $\{1\}$ or not. Similarly, $\varrho_\nu$-almost every partition $\pi$ possesses asymptotic frequencies such that $|\pi|^\downarrow \neq \mathbf{1}$. By the construction of Poisson measures, this proves (i).

Next, for every $t \ge 0$, introduce

$$B_e(t) = \bigcap_{r \in \mathcal{A}_e \cap [0,t]} \pi_1(r) , \quad B_d(t) = \bigcap_{r \in \mathcal{A}_d \cap [0,t]} \pi_1(r) ,$$

with the convention that such an intersection equals $\mathbb{N}$ in the situation when the set of indices is empty. By the construction of the nested family of partitions, we thus have

$$\Pi_1(t) = B_e(t) \cap B_d(t) .$$

When the total mass of the dislocation measure $\nu(\mathcal{P}_m)$ is finite, $\mathcal{A}_d \cap [0, t]$ has finitely many points a.s. and, conditionally on the number of points, $(\pi(r) : r \in \mathcal{A}_d \cap [0, t])$ forms a finite sequence of the i.i.d. partitions which are exchangeable (because $\varrho_\nu$ is a finite exchangeable measure). It then follows from Corollary 2.5 that the block $B_d(t)$ has an asymptotic frequency a.s. which is given by

$$|B_d(t)| = \prod_{r \in \mathcal{A}_d \cap [0,t]} |\pi_1(r)| , \qquad \text{a.s.} \tag{3.8}$$

When $\nu$ is infinite, we consider the restrictions of $\mathtt{M}_d$ to the sets

$$\Gamma_n := \left\{ (t, \pi) \in \mathbb{R}_+ \times \mathcal{P}_\infty : 1 - \frac{1}{n} \leq |\pi|_1^\downarrow < 1 - \frac{1}{n+1} \right\} , \qquad n \in \mathbb{N} .$$

We obtain a sequence of independent Poisson random measures with intensity, say $dt \otimes \varrho_n$, where $\varrho_n$ is a finite exchangeable measure on $\mathcal{P}_\infty$. Plainly, $\mathtt{M}_d = \sum_{n=1}^\infty \mathbb{1}_{\Gamma_n} \mathtt{M}_d$, and combining the preceding result and Corollary 2.5, we see that (3.8) still holds.

We now turn our attention to the block $B_e(t)$. First recall that the partitions $\pi(r)$ for $r \in \mathcal{A}_e$ are of the type $\epsilon^{(n)}$, where the latter denotes the partition of $\mathbb{N}$ which has two non-empty blocks, $\{n\}$ and $\mathbb{N}\backslash\{n\}$. In particular, $|\pi_1(r)| = 0$ or 1. On the other hand, $B_e(t)$ corresponds to the first block of a pure erosion process with coefficient $\mathtt{c}$ at time $t$. By Lemma 3.7, we know that either this block is reduced to $\{1\}$, or its asymptotic frequency is $\mathrm{e}^{-\mathtt{c}t}$, so with probability one

$$|B_e(t)| = \mathbb{1}_{\{B_e(t) \neq \{1\}\}} \mathrm{e}^{-\mathtt{c}t} = \mathbb{1}_{\{t < \zeta_e\}} \mathrm{e}^{-\mathtt{c}t} ,$$

where $\zeta_e := \inf \{t \in \mathcal{A}_e : \pi_1 = \{1\}\}$. We conclude from another application of Corollary 2.5 that for every fixed $t \geq 0$, there is the identity

$$|\Pi_1(t)| = |B_e(t)||B_d(t)| = \mathbb{1}_{\{t < \zeta_e\}} \mathrm{e}^{-\mathtt{c}t} \prod_{r \in \mathcal{A}_d \cap [0,t]} |\pi_1(r)| \quad \text{a.s.}$$

Since the processes above are càdlàg, these identities hold simultaneously for all $t \geq 0$ with probability one, which establishes both (ii) and (iii).

The assertion (iv) is now plain from an argument of monotonicity whenever $|\Pi_1(\cdot)|$ is continuous at time $t$. Next, fix an integer $n \in \mathbb{N}$ and consider the stopping time

$$T := \inf \left\{ t \in \mathcal{A}_d : |\pi(r)|_1^\downarrow < 1 - 1/n \right\} .$$

A slight variation of the argument used above to establish (iii) shows that $\Pi_1(T-)$ has an asymptotic frequency given by the identity (iv). By an easy iteration based on the Markov property, we deduce that more generally, the same holds for all $t \in \mathcal{A}_d$ such that $|\pi(r)|_1^\downarrow < 1 - 1/n$. Letting $n$ tend to $\infty$, we get that (iv) holds with probability one for all $t > 0$ such that either $t \in \mathcal{A}_d$ or $|\Pi(\cdot)|$ is continuous at time $t$. Since we know from (iii) that the discontinuity points of $|\Pi_1(\cdot)|$ are times in $\mathcal{A}_d$, this establishes (iv). □

We now conclude this section with a slight reinforcement of Theorem 3.2. More precisely we shall show that all the random partitions $\Pi(t)$ possess asymptotic frequencies simultaneously for all $t \geq 0$, a.s. Maybe more surprisingly, even though the map $\pi \to |\pi|^\downarrow$ is not continuous, the process $t \to |\Pi(t)|^\downarrow$ has, loosely speaking, the same regularity as $t \to \Pi(t)$.

**Proposition 3.6** *The following assertions hold with probability one:*

(i) *For all $t > 0$, the partitions $\Pi(t)$ and $\Pi(t-)$ possess asymptotic frequencies.*
(ii) *The function $t \to |\Pi(t)|^\downarrow$ is càdlàg, and for every $t > 0$, there is the identity*

$$\lim_{t' \to t-} |\Pi(t')|^\downarrow = |\Pi(t-)|^\downarrow .$$

*Proof* We already know that with probability one, $\Pi_1(t)$ has an asymptotic frequency for every $t \geq 0$ and that $t \to |\Pi_1(t)|$ is càdlàg. Recall from Proposition 3.1(ii) that $\Pi$ is exchangeable, that is for every permutation $\sigma$, the process $(\sigma(\Pi(t)), t \geq 0)$ has the same distributions as $(\Pi(t), t \geq 0)$. By letting transpositions act on $\Pi$, we get that with probability one, for every $t \geq 0$ and every $n \in \mathbb{N}$, the block of $\Pi(t)$ which contains $n$, say $B_n(t)$, has an asymptotic frequency, and moreover $t \to |B_n(t)|$ is càdlàg. It follows readily that with probability one, the partitions $\Pi(t)$ possess asymptotic frequencies for all $t \geq 0$, and that $t \to |\Pi(t)|^\downarrow$ is càdlàg.

The same argument, combined with Proposition 3.5(iv), establishes the existence of asymptotic frequencies of the blocks of $\Pi(t-)$ simultaneously for all $t > 0$ with probability one, as well as the identity in (ii). □

## 3.3 Self-similar fragmentations

The preceding sections have underlined the crucial role of Kingman's theory of exchangeable random partitions for the construction and study of extensions of homogeneous fragmentation chains to the situation when fragments can split immediately. In the present section, this approach will be developed further to the self-similar case. Roughly, the main technical result is a transformation of self-similar fragmentations extending Corollary 1.2, which enables us to change the index of self-similarity, and thus to reduce the general self-similar case to the homogeneous one. In particular, this implies that the distribution of such processes is determined by the index of self-similarity $\alpha$ ($\alpha = 0$ in the homogeneous case), an erosion coefficient and a dislocation measure, and also yields a useful characterization of the law of the tagged fragment.

### 3.3.1 Definition and first properties

Throughout this section, we consider a càdlàg process $\Pi = (\Pi(t), t \geq 0)$ with values in $\mathcal{P}_\infty$ and started from the trivial partition $\Pi(0) = \mathbf{1}_{\mathbb{N}}$ a.s. We suppose that $\Pi$ is exchangeable, that is its law is invariant by the natural action of permutations. We further assume that the following requirements are fulfilled with probability 1:

$$\Pi(t) \text{ possesses asymptotic frequencies } |\Pi(t)| \text{ for all } t \geq 0, \tag{3.9}$$

and for every $i \in \mathbb{N}$, if we denote by $B_i(t)$ the block of $\Pi(t)$ which contains $i$, then

$$\text{the process } t \to |B_i(t)| \text{ has right-continuous paths.} \tag{3.10}$$

We shall write $(\mathcal{F}_t)_{t\geq 0}$ for the natural filtration of $\Pi$ after the completion by null events and $\mathbb{P}$ for the law of $\Pi$. Proposition 1.2 leads us to make the following definition.

**Definition 3.3** *Let* $\Pi$ *be a càdlàg process satisfying (3.9) and (3.10). We call* $\Pi$ *a* **self-similar fragmentation process** *with index* $\alpha \in \mathbb{R}$ *if and only if, for every* $t, t' \geq 0$*, the conditional distribution of* $\Pi(t+t')$ *given* $\mathcal{F}_t$ *is that of the law of* $\mathrm{Frag}(\pi, \pi^{(\cdot)})$*, where* $\pi = \Pi(t)$ *and* $\pi^{(\cdot)} = \left(\pi^{(i)}, i \in \mathbb{N}\right)$ *is a family of independent random partitions, such that for each* $i \in \mathbb{N}$*,* $\pi^{(i)}$ *has the same distribution as* $\Pi(t'|\pi_i|^\alpha)$.

We stress that, since $\pi = \Pi(t)$ is exchangeable, if a block of $\pi$ has zero asymptotic frequency, then it is necessarily either empty or a singleton; see Proposition 2.8(ii). Because the fragmentation of a singleton or the empty set is necessarily a trivial operation, the definition above makes sense for $\alpha < 0$ even when $|\pi_i| = 0$ for some indices $i$.

Let us briefly discuss a couple of examples.

**Example 1** Assume that $\Pi$ is a standard homogeneous fragmentation. Then we know from Theorem 3.2 that the conditions (3.9) and (3.10) are fulfilled, and Definition 3.2 shows that $\Pi$ is a self-similar fragmentation with index $\alpha = 0$.

**Example 2** The second example is merely an extension of that at the beginning of Section 3.1.1. Specifically, consider as in Section 2.1.2, a self-similar interval-fragmentation chain $\Theta = (\Theta(t), t \geq 0)$. As previously, we associate to $\Theta$ a paint-box process $\Pi = (\Pi(t), t \geq 0)$ by introducing an independent sequence $U_1, \ldots$ of i.i.d. uniform variables in $]0, 1[$, and then defining for each $t \geq 0$ the random partition $\Pi(t)$ such that two different indices $i, j$ are in the same block of $\Pi(t)$ if and only if $U_i$ and $U_j$ belong to the same interval component of $\Theta(t)$. It is then easily checked that the requirements of Definition 3.3 are fulfilled, and thus $\Pi(\cdot)$ is a self-similar fragmentation process with index $\alpha$.

By definition, self-similar fragmentations are Markov processes; however, we stress that (except for the homogeneous case $\alpha = 0$), their semigroups are only defined for partitions which possess asymptotic frequencies. More precisely, consider an arbitrary partition $\pi \in \mathcal{P}_\infty$ which has asymptotic frequencies and such that each block with zero frequency is either empty or a singleton. Then denote by $\mathbb{P}_\pi$ the law of the process

$$\mathrm{Frag}(\pi, \Pi^{(\cdot)}(t|\pi_\cdot|^\alpha)), \qquad t \geq 0, \tag{3.11}$$

where $\Pi^{(\cdot)} = \left(\Pi^{(i)}, i \in \mathbb{N}\right)$ is a family of independent copies of $\Pi$. One can rephrase Definition 3.3 by saying that for every $t \geq 0$, the conditional distribution of the shifted process $\Pi(t+\cdot)$ given $\mathcal{F}_t$ is $\mathbb{P}_\pi$, where $\Pi(t) = \pi$. We further point to the following weak continuity property, which will serve as a substitute for the Feller property.

**Lemma 3.13** *Let $(\pi^{(n)}, n \in \mathbb{N})$ be a sequence in $\mathcal{P}_\infty$ which converges to some partition $\pi^{(\infty)}$. Suppose that for every $n \in \overline{\mathbb{N}}$, the partition $\pi^{(n)}$ possesses asymptotic frequencies and that each of its blocks with zero frequency is either empty or a singleton. Suppose further that if $B_i^{(n)}$ stands for the block of $\pi^{(n)}$ that contains the integer i, then*

$$\lim_{n\to\infty} |B_i^{(n)}| = |B_i^{(\infty)}|, \qquad \text{for all } i \in \mathbb{N}.$$

*Then $\mathbb{P}_{\pi(n)}$ converges to $\mathbb{P}_\pi$ as $n \to \infty$ in the sense of weak convergence of finite-dimensional distributions.*

The proof is straightforward and therefore omitted.

In order to study self-similar fragmentations, we have to introduce some more notation. Recall that for each integer $i \in \mathbb{N}$ and real number $s \in [0, \infty]$, we denote by $B_i(s)$ the block of the partition $\Pi(s)$ which contains $i$, with the convention that $B_i(\infty) = \{i\}$. Then we write

$$\mathcal{G}_i(t) = \sigma(B_i(s), s \le t)$$

for the sigma-field generated by that block up to time $t$, and completed as usual with events with zero probability. Plainly, there is the identity

$$\mathcal{F}_t = \bigvee_{i \in \mathbb{N}} \mathcal{G}_i(t)\,.$$

**Definition 3.4** *We call* **stopping line** *a family $L = (L_i, i \in \mathbb{N})$ of random variables with values in $[0, \infty]$ such that for each $i \in \mathbb{N}$:*

(i) *$L_i$ is a $(\mathcal{G}_i(t))$-stopping time,*
(ii) *$L_i = L_j$ for every $j \in B_i(L_i)$.*

For example, first passage times such as $L_i := \inf\{t \ge 0 : |B_i(t)| \le a\}$ for some fixed level $a \in ]0, 1[$ define a stopping line. Let us look at some immediate properties. Observe from Definition 3.4(ii) that if $L$ is a stopping line, then for all integers $i, j$, the blocks $B_i(L_i)$ and $B_j(L_j)$ are either identical or disjoint. Thus the collection $\{B_i(L_i), i \in \mathbb{N}\}$ induces a partition of $\mathbb{N}$ which we denote by $\Pi(L)$; further we know from (3.9) that $\Pi(L)$ possesses asymptotic frequencies a.s. Note also that for every $t \ge 0$, both $L + t := (L_i + t, i \in \mathbb{N})$ and $L \wedge t := (L_i \wedge t, i \in \mathbb{N})$ are also stopping lines. This leads us to define for every $t \ge 0$ the random partition

$$\Pi \circ \theta_L(t) := \Pi(L + t)$$

and the sigma-field $\mathcal{F}_L := \sigma(\Pi(L \wedge t), t \ge 0)$. Note that there is the identity

$$\mathcal{F}_L = \bigvee_{i \in \mathbb{N}} \mathcal{G}_i(L_i)\,.$$

Loosely speaking, stopping lines play the same role for fragmentation processes as stopping times do for Markov processes, in the sense that the branching property can be extended to stopping lines. Here is a precise statement.

**Lemma 3.14** (Extended branching property) *Let $L = (L_i, i \in \mathbb{N})$ be a stopping line. Then the conditional distribution of the process $\Pi \circ \theta_L : t \to \Pi(L + t)$ given $\mathcal{F}_L$ is $\mathbb{P}_\pi$, where $\pi = \Pi(L)$.*

*Proof* We shall first check by induction Lemma 3.13 when $L$ only takes finitely many values. When $L_i \equiv t$ for all $i \in \mathbb{N}$, the statement merely rephrases Definition 3.3, so let us assume that the extended fragmentation property has been proved for every stopping line taking at most $n$ values, and consider a stopping line $L$ taking values in $\{t_1, \ldots, t_{n+1}\}$ where $0 \leq t_1 < \cdots < t_{n+1} \leq \infty$. We may apply the extended branching property to the stopping line $L \wedge t_n$, so conditionally on $\Pi(L \wedge t_n) = \pi'$, $\Pi \circ \theta_{L \wedge t_n}$ is independent of $\mathcal{F}_{L \wedge t_n}$ and has the law $\mathbb{P}_{\pi'}$.

We introduce the random partition of $\mathbb{N}$ into two non-empty blocks,

$$B := \{i \in \mathbb{N} : L(i) \leq t_n\} \quad \text{and} \quad B' := \{i \in \mathbb{N} : L(i) > t_n\} .$$

We stress that $B$ and $B'$ are measurable with respect to the sigma-field $\mathcal{F}_{L \wedge t_n}$, so the extended branching property for the stopping line $L \wedge t_n$ implies that the restricted processes $\left(\Pi \circ \theta_{L \wedge t_n}\right)_{|B}$ and $\left(\Pi \circ \theta_{L \wedge t_n}\right)_{|B'}$ are conditionally independent given $\mathcal{F}_{L \wedge t_n}$. By an application of the branching property at time $t_{n+1} - t_n$ for the second one, we now easily conclude that the extended branching property holds for $L$.

We now complete the proof for a general stopping line $L$. We may approximate $L$ by a decreasing sequence $(L^{(n)}, n \in \mathbb{N})$ of stopping lines taking only finitely many values. For instance, one may choose

$$L_i^{(n)} = \begin{cases} 2^{-n}[2^n L_i + 1] & \text{if } L_i \leq 2^n , \\ \infty & \text{otherwise.} \end{cases}$$

Next, consider a bounded random variable $Y$ which is measurable with respect to $\mathcal{F}_L$, a bounded measurable functional $\varphi : \mathcal{P}_\infty \to \mathbb{R}$ and an arbitrary time $t \geq 0$. On the one hand, the continuity of $\varphi$ and the right-continuity of $\Pi$ give

$$\mathbb{E}\left[Y\varphi(\Pi \circ \theta_L(t))\right] = \lim_{n\to\infty} \mathbb{E}\left[Y\varphi(\Pi \circ \theta_{L_n}(t))\right] .$$

On the other hand, as $\mathcal{F}_L \subseteq \mathcal{F}_{L_n}$ for each $n$, the extended branching property for $L_n$ yields

$$\mathbb{E}\left[Y\varphi(\Pi \circ \theta_{L^{(n)}}(t))\right] = \mathbb{E}\left[Y\mathbb{E}_{\pi^{(n)}}(\varphi(\Pi(t)))\right] ,$$

where in the right-hand side, $\pi^{(n)} = \Pi(L^{(n)})$. Using (3.10) and Lemma 3.13, we see that

$$\lim_{n\to\infty} \mathbb{E}_{\pi^{(n)}}[\varphi(\Pi(t))] = \mathbb{E}_{\pi}[\varphi(\Pi(t))] \qquad \text{a.s.}$$

where $\pi = \Pi(L)$, and we conclude by dominated convergence that

$$\mathbb{E}\left[Y\varphi(\Pi \circ \theta_L(t))\right] = \mathbb{E}\left[Y\mathbb{E}_{\pi}(\varphi(\Pi(t)))\right] .$$

Thus the extended branching property holds for the stopping line $L$. □

### 3.3.2 Changing the index of self-similarity

We shall now see that Lemma 3.14 provides a simple transformation for self-similar fragmentations which enables us to change the index of self-similarity. The interest of such transformation is that it reduces the study to the homogeneous case $\alpha = 0$, which has been treated in the preceding sections. Roughly, the idea is to perform a change of the time-scale in order to modify the rate of fragmentations. Clearly, this change has to affect adequately each fragment, which requires the extended branching property stated in Lemma 3.14. Specifically, recall that for every $i \in \mathbb{N}$ and $r \geq 0$, $|B_i(r)|$ denotes the asymptotic frequency of the block of $\Pi(r)$ that contains $i$. Then introduce for an arbitrary $\beta \in \mathbb{R}$ the family of stopping lines

$$L_i^{\{\beta\}}(t) = \inf\left\{u \geq 0 : \int_0^u |B_i(r)|^{-\beta}\, dr > t\right\}, \qquad t \geq 0,$$

and consider the random partition

$$\Pi^{\{\beta\}}(t) := \Pi(L^{\{\beta\}}(t)).$$

We are now able to state the main result of this section.

**Theorem 3.3** *If $\Pi$ is a self-similar fragmentation with index $\alpha$, then the process $\Pi^{\{\beta\}} = \left(\Pi^{\{\beta\}}(t), t \geq 0\right)$ is a self-similar fragmentation with index $\alpha + \beta$. Moreover $\Pi$ can be recovered from $\Pi^{\{\beta\}}$; more precisely $\Pi = (\Pi^{\{\beta\}})^{\{-\beta\}}$ in the obvious notation.*

Before proving Theorem 3.3, we point out that the result is straightforward in the case of fragmentation chains. Indeed, if we use the genealogical coding of the chain as in Section 1.2.1, then we see that the transformation $\Pi \to \Pi^{\{\beta\}}$ amounts to changing the lifetime $\zeta$ of a block (i.e. a particle) $B$ into $|B|^{-\beta}\zeta$, and thus the statement essentially rephrases Corollary 1.2.

*Proof* Note that for every integer $i$, the process $t \to L_i^{\{\beta\}}(t)$ is right-continuous and that the block $B_i^{\{\beta\}}(t)$ of $\Pi^{\{\beta\}}(t)$ that contains $i$ is given by $B_i^{\{\beta\}}(t) = B_i(L_i^{\{\beta\}}(t))$. It is then immediate that $\Pi^{\{\beta\}}$ is a càdlàg process in $\mathcal{P}_\infty$ started from the trivial partition $\mathbf{1}_{\mathbb{N}}$, and which is exchangeable. By assumptions (3.9) and (3.10), we see that the random partitions $\Pi^{\{\beta\}}(t)$ possess asymptotic frequencies for all $t \geq 0$, and that the processes $t \to |B_i^{\{\beta\}}(t)|$ are right-continuous a.s.

Next, let $\mathcal{F}_t^{\{\beta\}} := \mathcal{F}_{L^{\{\beta\}}(t)}$ denote the sigma-field generated by the process $\Pi^{\{\beta\}}$ up to time $t$. We now need to check that $\Pi^{\{\beta\}}$ fulfills the requirements of Definition 3.3 in the filtration $(\mathcal{F}_t^{\{\beta\}})_{t\geq 0}$ and with index of self-similarity $\alpha+\beta$ instead of $\alpha$. Fix $t, t' \geq 0$ and write $\pi = \Pi^{\{\beta\}}(t) = \Pi(L^{\{\beta\}}(t))$. We apply the extended branching property of Lemma 3.14 for the stopping line $L^{\{\beta\}}(t)$, denoting by $\Pi^{(\cdot)}$ the sequence of independent copies of $\Pi$ which arises in Definition (3.11) of the law $\mathbb{P}_\pi$.

Pick $i \in \mathbb{N}$, and let $j \in \mathbb{N}$ denote the index of the block of $\pi$ which contains $i$; in other words, $\pi_j = B_i^{\{\beta\}}(t) = B_i(L_i^{\{\beta\}}(t))$. For every $r \geq 0$, let $C_i(r)$ denote the block of $\Pi^{(j)}(r)$ that contains $i$. Then, by construction of the fragmentation operator, we have

$$B_i(L_i^{\{\beta\}}(t)+r) = B_i^{\{\beta\}}(t) \cap C_i(r),$$

and we deduce the identity

$$L_i^{\{\beta\}}(t+t') = L_i^{\{\beta\}}(t) + \lambda_i(t'),$$

where

$$\lambda_i(t') := \inf\left\{s \geq 0 : \int_0^s |B_i^{\{\beta\}}(t) \cap C_i(r)|^{-\beta} dr > t'\right\}.$$

Then recall from Corollary 2.5 that

$$|B_i^{\{\beta\}}(t) \cap C_i(r)| = |B_i^{\{\beta\}}(t)||C_i(r)|,$$

so

$$\lambda_i(t') = \inf\left\{s \geq 0 : \int_0^s |C_i(r)|^{-\beta} dr > t'|B_i^{\{\beta\}}(t)|^\beta\right\}.$$

The extended branching property thus shows that the conditional distribution of $\Pi^{\{\beta\}}(t+t')$ given $\mathcal{F}_t^{\{\beta\}}$ is that of the law of Frag$(\pi, \pi^{(\cdot)})$, where $\pi = \Pi^{\{\beta\}}(t)$ and $\pi^{(\cdot)} = \left(\pi^{(j)}, j \in \mathbb{N}\right)$ is a family of independent random partitions, such that for each $j \in \mathbb{N}$, $\pi^{(j)}$ has the same distribution as $\Pi^{\{\beta\}}(t'|\pi_j|^{\alpha+\beta})$. Thus $\Pi^{\{\beta\}}$ is a self-similar fragmentation with index $\alpha+\beta$. Finally the identity $\Pi = (\Pi^{\{\beta\}})^{\{-\beta\}}$ is immediate. □

An obvious consequence of Theorem 3.3, combined with the characterization of homogeneous fragmentations in Theorem 3.1, is that the law of a self-similar fragmentation is determined by its index of self-similarity $\alpha \in \mathbb{R}$, and the erosion coefficient $\mathtt{c} \geq 0$ and the dislocation measure $\nu$ of the homogeneous fragmentation $\Pi^{\{-\alpha\}}$. We call $(\alpha, \mathtt{c}, \nu)$ the characteristics of $\Pi$.

Another useful consequence of Theorem 3.3 concerns the so-called tagged fragment, that is the process $|\Pi_1(\cdot)|$ which gives the asymptotic frequency of the block that contains 1 as time passes. It is easy to see from the branching and self-similarity properties of self-similar fragmentation that the process of the tagged fragment is Markovian and fulfills the scaling property. Specifically, for every $a > 0$, the law of the rescaled process $(a|\Pi_1(a^\alpha t)|, t \geq 0)$ coincides with the law of $|\Pi_1(\cdot)|$ started from $|\Pi_1(0)| = a$ (in the sense of the Markov property). Lamperti [148] has established an important and useful representation of general self-similar Markov processes with values in $[0, \infty[$ in terms of Lévy processes. In the present situation, Theorem 3.2 immediately yields Lamperti's representation for the tagged fragment as a simple transformation of a subordinator.

**Corollary 3.1** *Let $\Pi$ be a self-similar fragmentation with characteristics $(\alpha, \mathtt{c}, \nu)$. Consider a subordinator $\xi = (\xi_t, t \geq 0)$ distributed as in Theorem 3.2. Introduce the time-change*

$$T(t) = \inf\left\{u : \int_0^u \exp(\alpha \xi_r) dr > t\right\}, \qquad t \geq 0,$$

*and set $Z_t = \exp(-\xi_{T(t)})$ (with the convention that $Z_t = 0$ if $T(t) = \infty$). Then the processes $(Z_t, t \geq 0)$ and $(|\Pi_1(t)|, t \geq 0)$ have the same law.*

*Proof* The homogeneous fragmentation $\Pi^{\{-\alpha\}}$ has erosion coefficient $\mathtt{c}$ and dislocation measure $\nu$, so by Theorem 3.2, there exists a subordinator $\xi$ distributed as in the statement, such that $|\Pi_1^{\{-\alpha\}}(t)| = \exp(-\xi_t)$ for all $t \geq 0$. Recall that $\Pi_1(t) = \Pi_1^{\{-\alpha\}}(T(t))$ where

$$T(t) = \inf\left\{s \geq 0 : \int_0^s |\Pi_1^{\{-\alpha\}}(r)|^{-\alpha} dr > t\right\}.$$

Our claim follows. □

We now conclude this section by presenting an easy application of the preceding results.

**Corollary 3.2** *Let $\Pi$ be a self-similar fragmentation with characteristics $(\alpha, \mathtt{c}, \nu)$ and $(Z_t, t \geq 0)$ the process defined in Corollary 3.1. Then one has*

$$\mathbb{E}\left(\sum_{i=1}^{\infty} |\Pi_i(t)|\right) = \mathbb{P}(Z_t > 0), \qquad t \geq 0.$$

*In particular, for every $t > 0$, the random partition $\Pi(t)$ has proper asymptotic frequencies a.s. (i.e. $\sum_{i=1}^\infty |\Pi_i(t)| = 1$ a.s.) if and only if the index of self-similarity $\alpha$ is non-negative, the erosion coefficient is $\mathtt{c} = 0$ and the dislocation measure conservative, that is $s_0 := 1 - \sum_{i=1}^\infty s_i = 0$, for $\nu$-a.e. $\mathbf{s} \in \mathcal{P}_{\mathrm{m}}$.*

*Proof* Recall that $|\Pi_1(t)|$ is a size-biased pick from the sequence $|\Pi(t)|$ (cf. Corollary 2.4), so the first claim follows from Corollary 3.1. In particular, $\Pi(t)$ is proper a.s. if and only if $Z_t > 0$ a.s., or equivalently $\xi_r < \infty$ for all $r > 0$ and $T(t) < \infty$ a.s. The latter holds if and only if the killing rate of the subordinator $\xi$ is $\mathtt{k} = 0$ and the index of self-similarity is non-negative (because for every subordinator $\xi$ with infinite lifetime, $\int_0^\infty e^{\alpha\xi_t} dt = \infty$ a.s. if and only if $\alpha \geq 0$). By Theorem 3.2, this yields our claim. □

### 3.3.3 Mass-fragmentations

Processes with values in the space $\mathcal{P}_\infty$ of partitions of $\mathbb{N}$ are certainly interesting objects in their own right; however, the focus of this text is rather on mass-partitions, which are probably more natural notions to consider. In particular, recall that our initial motivation is to gain insight on processes of fragmentation of a mass in which dislocations can occur immediately, as the latter could not be handled by the techniques used in Chapter 1. We shall now strive to transfer the approach developed for the study of self-similar fragmentations with values in $\mathcal{P}_\infty$, to the more concrete setting of mass-partitions.

Throughout this section, we will be dealing with a $\mathcal{P}_\infty$-valued self-similar fragmentation $\Pi$ with index of self-similarity $\alpha$, erosion coefficient $\mathtt{c}$ and dislocation measure $\nu$. Recall that, by assumption, the random partitions $\Pi(t)$ possess asymptotic frequencies for all $t \geq 0$ a.s.; the purpose of this section is to investigate the process of ranked asymptotic frequencies $|\Pi(\cdot)|^\downarrow$ which will be referred to as a (self-similar) *mass-fragmentation* later in the text. As it has been already observed, when $\mathtt{c} = 0$ and $\nu$ is finite, the latter is a self-similar fragmentation chain to which the results established in Chapter 1 apply. Roughly, our aim now is to show that these results can be extended to the general situation when the erosion coefficient can be positive and the dislocation measure infinite. For the sake of conciseness, we shall focus on some important issues, and refer to the literature for others.

From now on, it will be simpler to use the notation

$$X(t) = (X_1(t), \ldots) := |\Pi(t)|^\downarrow, \qquad t \geq 0,$$

for the mass-fragmentation process. We next present a couple of simple properties of the latter, which follow easily from our analysis of homogeneous fragmentations.

Plainly, the sigma-field generated by $X(t)$ is smaller than that generated by the exchangeable partition $\Pi(t)$ and, more precisely, we know from Kingman's Theorem 2.1 that the conditional distribution of $\Pi(t)$ given

$X(t) = \mathbf{s} \in \mathcal{P}_{\mathrm{m}}$ is $\varrho_{\mathbf{s}}$, that is of a paint-box based on $\mathbf{s}$. A slight variation of this argument yields the following stronger result.

**Lemma 3.15** *For every $t \geq 0$, the conditional distribution of $\Pi(t)$ given the mass-fragmentation process $(X(u), u \geq 0)$, is $\varrho_{\mathbf{s}}$, where $\mathbf{s} = X(t)$.*

*Proof* Recall that the process $\Pi = (\Pi(t), t \geq 0)$ is exchangeable. Clearly, asymptotic frequencies of blocks are invariant by the action of permutations (since for every permutation $\sigma$, we have $\sigma(n) = n$ for every $n$ sufficiently large, the blocks $B$ and $\sigma(B)$ have the same asymptotic frequency). It follows that for every integer $n$, the conditional distribution of $\Pi_{|[n]}(t)$ given $(X(u), u \geq 0)$ is exchangeable. Thus the conditional law of $\Pi(t)$ given $(X(u), u \geq 0)$ is exchangeable and, by Kingman's Theorem 2.1, it is that of a paint-box based on the conditional distribution of the ranked asymptotic frequencies $|\Pi(t)|^{\downarrow} = X(t)$. The latter is obviously measurable with respect to $(X(u), u \geq 0)$, which completes the proof. □

Our main goal in this section is to show that the mass-fragmentation process $X$ inherits the Markov property from the $\mathcal{P}_{\infty}$-valued process $\Pi$ and, more precisely, that its transitions fulfill the branching and self-similarity properties. In this direction, we start by defining a fragmentation operator for mass-partitions, which bears obvious similarities with Definition 3.1 (in particular, by a slight abuse, we shall use the same notation as there, which should not induce any confusion). Further, it is convenient to introduce the set

$$\tilde{\mathcal{P}}_{\mathrm{m}} = \left\{ \mathbf{x} = (x_1, \ldots) : x_i \geq 0 \text{ and } \sum_{i=1}^{\infty} x_i \leq 1 \right\}.$$

In other words, $\tilde{\mathcal{P}}_{\mathrm{m}}$ is the space of numerical sequences that yield a mass-partition when rearranged in decreasing order.

**Definition 3.5** *Let $\mathbf{x} = (x_1, \ldots) \in \tilde{\mathcal{P}}_{\mathrm{m}}$ and $\left(\mathbf{x}^{(i)} = (x_1^{(i)}, \ldots), i \in \mathbb{N}\right)$ be a sequence in $\tilde{\mathcal{P}}_{\mathrm{m}}$. We denote by $\mathrm{Frag}(\mathbf{x}, \mathbf{x}^{(\cdot)})$ and call the fragmentation of $\mathbf{x}$ by $\mathbf{x}^{(\cdot)}$ the mass-partition given by the decreasing rearrangement of the collection of real numbers $\left(x_i x_j^{(i)} : i, j \in \mathbb{N}\right)$.*

We stress that the mass-partitions $\mathbf{x}^{(i)}$ corresponding to indices $i$ for which $x_i = 0$ play no role in the fragmentation of $\mathbf{x}$ by $\mathbf{x}^{(\cdot)}$.

There is a simple connection between the fragmentation operator for exchangeable partitions and the fragmentation operator for mass-partitions. Recall Lemma 3.2.

**Lemma 3.16** *Let $\pi$ be a deterministic partition that possesses asymptotic frequencies and, $(\pi^{(i)}, i \in \mathbb{N})$ a sequence of exchangeable random partitions. We write $|\pi|$ (respectively, $|\pi^{(i)}|$) for the sequence of the asymptotic frequencies of the blocks of $\pi$ (respectively, of $\pi^{(i)}$ for every $i \in \mathbb{N}$). Then the random partition* $\mathrm{Frag}(\pi, \pi^{(\cdot)})$ *possesses asymptotic frequencies, and its decreasing rearrangement is given by* $\mathrm{Frag}(|\pi|, |\pi^{(\cdot)}|)$.

*Proof* Fix $i \in \mathbb{N}$ and consider the partition $\pi^{(i)}_{|\pi_i}$ of the $i$-th block $\pi_i$ of $\pi$ by the partition $\pi^{(i)}$. We know from Corollary 2.5 that each block $\pi_i \cap \pi^{(i)}_j$ possesses an asymptotic frequency given by $|\pi_i \cap \pi^{(i)}_j| = |\pi_i||\pi^{(i)}_j|$. Thus the ranked sequence of asymptotic frequencies of $\pi^{(i)}_{|\pi_i}$ is given by $|\pi_i||\pi^{(i)}|^{\downarrow}$, and our claim follows. □

Lemma 3.16 provides the key tool for checking that the Markov property of $\mathcal{P}_\infty$-valued self-similar fragmentations can be shifted to self-similar mass-fragmentations.

**Proposition 3.7** *The process $(X(t), t \geq 0)$ is Markovian and, more precisely Fellerian. Its semigroup can be described as follows. For every $t, t' \geq 0$, the conditional distribution of $X(t+t')$ given $X(t) = \mathbf{s} = (s_1, \ldots)$ is the law of* $\mathrm{Frag}(\mathbf{s}, \mathbf{s}^{(\cdot)})$*, where $\mathbf{s}^{(\cdot)} = (\mathbf{s}^{(1)}, \ldots)$ is a sequence of independent random mass-partitions and each $\mathbf{s}^{(i)}$ is distributed as $X(t' s_i^{\alpha})$ (with the convention that $X(t' s_i^{\alpha}) = \mathbf{0}$ if $s_i = 0$ and $\alpha < 0$).*

*Proof* Recall that $(\mathcal{F}_t)_{t \geq 0}$ stands for the natural filtration of the $\mathcal{P}_\infty$-valued fragmentation $\Pi$. We know that the conditional law of $\Pi(t+t')$ given $\mathcal{F}_t$ is that of $\mathrm{Frag}(\pi, \pi^{(\cdot)})$, where $\pi = \Pi(t)$ and $(\pi^{(i)}, i \in \mathbb{N})$ is a sequence of independent exchangeable partitions such that $\pi^{(i)}$ is distributed as $\Pi(|\pi_i|^{\alpha} t')$. We deduce from Lemma 3.16 that the conditional law of the ranked asymptotic frequencies $|\Pi(t+t')|^{\downarrow} = X(t+t')$ given $\mathcal{F}_t$ is that of $\mathrm{Frag}(\ell, \ell^{(\cdot)})$, where $\ell = |\pi|$ and $(\ell^{(i)}, i \in \mathbb{N})$ is a sequence of independent mass partitions such that $\ell^{(i)}$ is distributed as $|\Pi(\ell_i^{\alpha} t')|$.

Next, consider a permutation $\sigma$ of $\mathbb{N}$ such that $\ell_{\sigma(i)} = |\pi|_i^{\downarrow} = s_i$ for each $i \in \mathbb{N}$. Then write $\tilde{\ell}^{(i)} = \ell^{(\sigma(i))}$ and $\mathbf{s}^{(i)}$ for the decreasing rearrangement of $\tilde{\ell}^{(i)}$, so that $\mathbf{s}^{(\cdot)}$ is also a sequence of i.i.d. copies of $X(t') = |\Pi(t')|^{\downarrow}$. Plainly, there is the identity

$$\mathrm{Frag}(\ell, \ell^{(\cdot)}) = \mathrm{Frag}(\mathbf{s}, \tilde{\ell}^{(\cdot)}) = \mathrm{Frag}(\mathbf{s}, \mathbf{s}^{(\cdot)}).$$

This establishes the Markov property and the special form of the semigroup.

The Feller property follows. More precisely, on the one hand, it is easy to check that for every sequence $\mathbf{s}^{(\cdot)}$ of mass-partitions, the map $\mathbf{s} \to \mathrm{Frag}(\mathbf{s}, \mathbf{s}^{(\cdot)})$ is continuous. By dominated converge, this implies that if $\varphi : \mathcal{P}_{\mathrm{m}} \to \mathbb{R}$ is a continuous function, then the map $s \to \mathbb{E}\left(\varphi(\mathrm{Frag}(s, X^{(\cdot)}(t)))\right)$ is also continuous, where $X^{(\cdot)}(t)$ denotes a sequence of independent copies of $X^{(\cdot)}(t)$. On the other hand, we know that $\lim_{t\to 0} \Pi(t) = \mathbf{1}$ in probability, and we derive from continuity properties stated in Proposition 2.9 that $\lim_{t\to 0} X(t) = \mathbf{1}$ in probability. It follows readily that $\mathrm{Frag}(\mathbf{s}, X^{(\cdot)}(t))$ converges in probability to $\mathbf{s}$ as $t \to 0$, for every $\mathbf{s} \in \mathcal{P}_{\mathrm{m}}$. Thus, for any continuous function $\varphi : \mathcal{P}_{\mathrm{m}} \to \mathbb{R}$, we have

$$\lim_{t\to 0} \mathbb{E}\left(\varphi(\mathrm{Frag}(\mathbf{s}, X^{(\cdot)}(t))\right) = \varphi(\mathbf{s}),$$

which completes the proof. □

Proposition 3.7 shows that a self-similar mass-fragmentation enjoys both the branching and the scaling properties as stated in Proposition 1.2. Observe that its evolution is continuous and not discrete whenever the erosion coefficient is strictly positive or the dislocation measure infinite. In the converse direction, it can be proved that every $\mathcal{P}_{\mathrm{m}}$-valued process with càdlàg paths which fulfills these two properties, has the same distribution as the process of the ranked asymptotic frequencies of some self-similar $\mathcal{P}_{\infty}$-valued fragmentation; see [24] for details.

Our next goal is to get information on the infinitesimal generator of the self-similar mass-fragmentation. Following the classical approach of Stroock and Varadhan (see for example [99]), we address this problem from the point of view of martingales. That is, for a suitable continuous function $\mathbf{f} : \mathcal{P}_{\mathrm{m}} \to \mathbb{R}$, we aim at finding a continuous function $\mathbf{g} : \mathcal{P}_{\mathrm{m}} \to \mathbb{R}$ such that the process

$$\mathbf{f}(X(t)) - \int_0^t \mathbf{g}(X(r))dr, \qquad t \geq 0,$$

is a martingale. In this situation, one says that $\mathbf{f}$ belongs to the domain of the infinitesimal generator $\mathsf{G}$ of $X$ and that $\mathsf{G}\mathbf{f} = \mathbf{g}$. For the sake of simplicity, we shall only consider the case in which the index $\alpha$ of self-similarity is non-negative, and further focus on additive functionals; recall that in the case of fragmentation *chains*, the expression of the infinitesimal generator applied to additive functionals is given by (1.15).

**Proposition 3.8** *Suppose $\alpha \geq 0$. Let $f : [0, 1] \to \mathbb{R}$ be a function of class $\mathcal{C}^1$ with $f(0) = 0$, and define the map $\mathbf{f} : \mathcal{P}_{\mathrm{m}} \to \mathbb{R}$ by*

$$\mathbf{f}(\mathbf{s}) = \sum_{i=1}^{\infty} f(s_i), \qquad \mathbf{s} = (s_i, i \in \mathbb{N}) \in \mathcal{P}_{\mathrm{m}}.$$

*Then $\mathbf{f}$ belongs to the domain of the infinitesimal generator $\mathsf{G}$ of $X$, and*

$$\mathsf{G}\mathbf{f}(\mathbf{x}) = \sum_{i=1}^{\infty} x_i^{\alpha} \left( -\mathsf{c} x_i f'(x_i) + \int_{\mathcal{P}_{\mathrm{m}}} \nu(d\mathbf{s}) \left( \mathbf{f}(x_i \mathbf{s}) - f(x_i) \right) \right),$$

*where $\mathsf{c} \geq 0$ is the erosion coefficient and $\nu$ the rate of dislocations.*

*Proof* Consider a subordinator $\xi$ with killing rate $\mathsf{k}$, drift coefficient $\mathsf{d}$ and Lévy measure $\Lambda$. Let $g : [0, \infty] \to \mathbb{R}$ be a function with $g(\infty) = 0$, which is of class $\mathcal{C}^1$ on $[0, \infty[$, and which remains bounded as well as its derivative. It is easy to check (for example from the Lévy-Itô decomposition of subordinators and standard properties of Poisson random measures) that $g$ belongs to the domain of the infinitesimal generator $\Gamma$ of $\xi$. More precisely

$$\Gamma g(x) = -\mathsf{k} g(x) + \mathsf{d} g'(x) + \int_{]0,\infty[} (g(x+y) - g(x)) \Lambda(dy), \qquad x \in \mathbb{R}_+,$$

and $\Gamma g(\infty) = 0$. See for instance Section 31 in Sato [198] for a detailed proof. Specializing this when $g(x) = h(\mathrm{e}^{-x})$ for some function $h : [0, 1] \to \mathbb{R}$ of class $\mathcal{C}^1$ with $h(0) = h'(0) = 0$, we get that the process

$$h(\mathrm{e}^{-\xi_t}) - \int_0^t \Big( -\mathsf{k} h(\mathrm{e}^{-\xi_u}) - \mathsf{d} \mathrm{e}^{-\xi_u} h'(\mathrm{e}^{-\xi_u}) + \int_{]0,\infty[} (h(\mathrm{e}^{-\xi_u - y}) \\ - h(\mathrm{e}^{-\xi_u})) \Lambda(dy) \Big) du$$

is a martingale.

Recall that the time-substitution $T(\cdot)$ is defined by

$$T(t) := \inf \left\{ s \geq 0 : \int_0^s \exp(\alpha \xi_r) dr > t \right\}, \qquad t \geq 0,$$

so that $dT(t) = \exp(-\alpha\xi_{T(t)})dt$ whenever $T(t) < \infty$. As such a time-change preserves the local martingale property, setting $Z_t = \exp(-\xi_{T(t)})$ we get that

$$h(Z_t) - \int_0^t \Big( -\mathtt{k}h(Z_u) - \mathtt{d}Z_u h'(Z_u) + \int_{]0,\infty[} (h(Z_u \mathrm{e}^{-y}) - h(Z_u))\Lambda(dy) \Big) Z_u^\alpha du$$

is a local martingale. More precisely, this local martingale remains bounded on finite time-intervals, and thus is a true martingale.

Now, using Corollary 3.1, we specify the preceding for $Z_t = |\Pi_1(t)|$. Recall from Lemma 3.15 and Corollary 2.4 that for every $t \geq 0$, the conditional distribution of $|\Pi_1(t)|$ given $(X(u), u \geq 0)$ is that of a size-biased sample of $X(t) = (X_1(t), \ldots)$. Setting $f(x) = xh(x)$ and $\mathbf{f}(\mathbf{s}) = \sum_{i=1}^\infty f(s_i)$, we get that the projection of $h(Z_t)$ on the natural filtration of $X$ is $\mathbf{f}(X(t))$, and then that

$$\mathbf{f}(X(t)) - \int_0^t \left( \sum_{i=1}^\infty X_i^\alpha(u) Y_i(u) \right) du$$

is a martingale, where

$$Y_i(u) = -\mathtt{k}f(X_i(u)) + \mathtt{d}\left(f(X_i(u)) - X_i(u)f'(X_i(u))\right) + \int_{]0,\infty[} (\mathrm{e}^y f(X_i(u)\mathrm{e}^{-y}) - f(X_i(u)))\Lambda(dy)\,.$$

Using the expressions for $\mathtt{k}$, $\mathtt{d}$, and $\Lambda$ given in Theorem 3.2, we easily check the identity $Y(u) = \mathbf{Gf}(X(u))$, which completes the proof of our claim. □

We conclude the study of mass-fragmentations by considering their asymptotic behavior for large times, extending results which have been first proven in the setting of fragmentation chains in Chapter 1. For the sake of simplicity, we shall only consider here the homogeneous case, and further focus on the conservative case when the erosion coefficient is zero and no dust is created during sudden dislocations, that is

$$s_0 := 1 - \sum_{i=1}^\infty s_i = 0\,, \qquad \text{for } \nu \text{ almost every } \mathbf{s} \in \mathcal{P}_{\mathrm{m}}. \tag{3.12}$$

Recall then from Proposition 2.8(i) that the mass-partitions $X(t)$ are proper a.s.

It will be convenient to represent the sequence of asymptotic frequencies $|\Pi(t)|$ as a modified version of the empirical distribution,

$$\rho_t(dy) := \sum_{i=1}^\infty |\Pi_i(t)| \delta_{\frac{1}{t}\ln|\Pi_i(t)|}(dy)\,. \tag{3.13}$$

Note that $\rho_t$ can also be viewed as the conditional distribution of $t^{-1}\ln|\Pi_1(t)|$ given the mass-partition $X(t)$. We consider each $\rho_t$ as a random variable with values in the space of probability measures on $\mathbb{R}$, which is endowed with Prohorov's distance (so limits are taken in the sense of weak convergence of probability measures). The simplest results on the asymptotic behavior of homogeneous fragmentations are easy consequences of the classical law of large numbers and central limit theorem. In this direction, introduce the first and second right-derivatives of $\Phi$ at 0,

$$m := \Phi'(0+)\,, \quad \sigma^2 := -\Phi''(0+)\,;$$

so that in terms of the dislocation measure

$$m = -\int_{\mathcal{P}_{\mathrm{m}}}\left(\sum_{i=1}^{\infty} s_i \ln s_i\right)\nu(d\mathbf{s})\,, \quad \sigma^2 = \int_{\mathcal{P}_{\mathrm{m}}}\left(\sum_{i=1}^{\infty} s_i (\ln s_i)^2\right)\nu(d\mathbf{s})\,.$$

**Corollary 3.3** *Consider a homogeneous mass-fragmentation, and assume that the erosion coefficient is* $\mathtt{c}=0$ *and that (3.12) holds.*

(i) *Suppose that* $m<\infty$. *Then,*

$$\lim_{t\to\infty}\rho_t = \delta_{-m}\,,$$

*in probability.*

(ii) *Suppose that* $\sigma^2<\infty$, *and denote by* $\tilde\rho_t$ *the image of* $\rho_t$ *by the map* $x\to\sqrt{t}(x+m)/\sigma$ *and by* $\mathcal{N}(0,1)$ *the standard normal distribution, then*

$$\lim_{t\to\infty}\tilde\rho_t = \mathcal{N}(0,1)\,,$$

*in probability.*

*Proof* (i) The proof relies on first and second moment calculations. We start by making the following elementary observation. Let $\mathbf{s}\in\mathcal{P}_{\mathrm{m}}$ be a proper mass-partition, and $s_*$ and $s_\dagger$ two independent variables which are both size-biased sampled from $\mathbf{s}$. This means that $*$ and $\dagger$ are two i.i.d. integer-valued variables such that

$$\mathbb{P}(*=k)=\mathbb{P}(\dagger=k)=s_k\,, \qquad k\in\mathbb{N}\,.$$

For every continuous function $f:[0,1]\to\mathbb{R}$, we thus have

$$\sum_{i=1}^{\infty} s_i f(s_i) = \mathbb{E}\left(f(s_*)\right) \quad \text{and} \quad \left(\sum_{i=1}^{\infty} s_i f(s_i)\right)^2 = \mathbb{E}\left(f(s_*)f(s_\dagger)\right)\,.$$

When we apply these identities to an exchangeable random partition $\pi$ using Kingman's representation, the ranked asymptotic frequencies $|\pi|^{\downarrow} = s$ are in general random; and the size-biased samples $s_*$ and $s_\dagger$ should be only independent *conditionally* on **s**. In this direction, we may choose $s_* = |B_1|$ and $s_\dagger = |B_2|$, where $B_i$ is the block of $\pi$ containing $i$ (so in particular $B_1 = \pi_1$). In terms of $\Pi(t)$, we thus have the first and second moments identities

$$\mathbb{E}\left(\int_{]0,\infty[} f(y)\rho_t(dy)\right) = \mathbb{E}(f(t^{-1}\ln|B_1(t)|)),$$

$$\mathbb{E}\left(\left(\int_{]0,\infty[} f(y)\rho_t(dy)\right)^2\right) = \mathbb{E}(f(t^{-1}\ln|B_1(t)|))f(t^{-1}\ln|B_2(t)|)).$$

Now $\ln|B_1(t)|$ and $\ln|B_2(t)|$ are both distributed as $-\xi(t)$, where $\xi = (\xi(t), t \geq 0)$ is a subordinator with Laplace exponent $\Phi$. The hypothesis that $\Phi'(0+) = m < \infty$ ensures the (weak) law of large numbers for $\xi$, that is $t^{-1}\xi(t)$ converges in probability as $t \to \infty$ to $m$. We conclude that

$$\lim_{t\to\infty} \mathbb{E}\left(\int_{\mathbb{R}} f(y)\rho_t(dy)\right) = f(-m), \quad \lim_{t\to\infty} \mathbb{E}\left(\left(\int_{\mathbb{R}} f(y)\rho_t(dy)\right)^2\right) = f(-m)^2,$$

and therefore

$$\lim_{t\to\infty} \int_{\mathbb{R}} f(y)\rho_t(dy) = \int_{\mathbb{R}} f(y)\delta_{-m}(dy) \qquad \text{in } L^2(\mathbb{P}).$$

This readily yields our first claim.

(ii) The same argument as above yields the following identities for every bounded continuous function $f: \mathbb{R} \to \mathbb{R}$

$$\mathbb{E}\left(\int_{]0,\infty[} f(y)\tilde{\rho}_t(dy)\right) = \mathbb{E}\left(f\left(\frac{\ln|B_1(t)| + mt}{\sigma\sqrt{t}}\right)\right),$$

$$\mathbb{E}\left(\left(\int_{]0,\infty[} f(y)\tilde{\rho}_t(dy)\right)^2\right) = \mathbb{E}\left(f\left(\frac{\ln|B_1(t)| + mt}{\sigma\sqrt{t}}\right) f\left(\frac{\ln|B_2(t)| + mt}{\sigma\sqrt{t}}\right)\right).$$

Our assumption ensures the central limit theorem for the subordinator $\xi(t) = -\ln|B_1(t)|$, namely

$$\frac{\ln|B_1(t)| + mt}{\sigma\sqrt{t}} \text{ converges in law as } t \to \infty \text{ towards } N,$$

where $N$ has the $\mathcal{N}(0, 1)$-distribution, and of course the same holds when we replace $|B_1(t)|$ by $|B_2(t)|$. The key observation is that, although $|B_1(t)|$ and $|B_2(t)|$ are not independent, the fragmentation property implies that the variables $t^{-1/2}(\ln|B_1(t)| + mt)$ and $t^{-1/2}(\ln|B_2(t)| + mt)$ are asymptotically independent. Indeed, the first instant $t$ when the integers 1 and 2 belong to distinct blocks is a finite stopping time for which the strong Markov property thus applies. This readily yields the stated asymptotic independence.

Putting the pieces together, we see that

$$\lim_{t\to\infty} \mathbb{E}\left(\int_{\mathbb{R}} f(y)\tilde{\rho}_t(dy)\right) = \mathbb{E}(f(N)),$$

$$\lim_{t\to\infty} \mathbb{E}\left(\left(\int_{\mathbb{R}} f(y)\tilde{\rho}_t(dy)\right)^2\right) = \left(\mathbb{E}(f(N))\right)^2,$$

and we conclude that for every bounded continuous function $f$,

$$\lim_{t\to\infty} \int_{\mathbb{R}} f(y)\tilde{\rho}_t(dy) = \mathbb{E}(f(N)) \qquad \text{in } L^2(\mathbb{P}).$$

The convergence in probability $\tilde{\rho}_t$ to $\mathcal{N}(0,1)$ now follows from standard arguments. □

## 3.4 Comments

The idea of using Kingman's theory of random exchangeable partitions to investigate homogeneous fragmentation processes was suggested by Pitman [183] and developed in [34] from where most results of Sections 3.1 and 3.2 are taken. One can carry out a similar program using compositions (i.e. ordered partitions) in place of partitions, which yields in particular a nice representation of a general homogeneous fragmentation as a family of nested open subsets of the unit interval; see Basdevant [21]. Self-similar fragmentations with values in $\mathcal{P}_\infty$ were introduced (under slightly less restrictive requirements) in [36], where the reduction to the homogeneous case via suitable time-change, that is stated here in Theorem 3.3, was established. The key notion of stopping line was introduced for branching processes by Chauvin [70]; see also Jagers [128] for an extension to general branching processes. Self-similar mass-partitions have been studied by Berestycki [24], where a useful Poissonian construction, similar to that for $\mathcal{P}_\infty$-valued homogeneous fragmentations of Section 3.1.3, is proven.

Essentially all the results established in Chapter 1 for self-similar fragmentation chains can be extended to self-similar fragmentation processes, although for the sake of conciseness, only a few such extensions have been proven in the present chapter. We refer in particular to [38], [41], [47], [51], [119] and [120] for details. Roughly, the key ingredients for such extensions are the Poissonian construction (which, loosely speaking, provides an analog of the discrete genealogical structure of fragmentation chains), and information about the statistics of the tagged fragment. In this vein, the connection with self-similar Markov processes stated in Corollary 3.1 is crucial, and many results on self-similar Markov processes have natural applications to self-similar fragmentations. See for example [40] and [193] for recent developments. On the other hand, the fact that dislocations can occur immediately yields new sets of problems which were essentially trivial for fragmentation chains. For instance, we may consider small-time asymptotics (see [24]), estimates at a fixed time for the number of fragments of size greater than $\varepsilon$ as $\varepsilon \to 0$ (see [39] and [121]), . . .

It is interesting to stress that more generally, random exchangeable partitions are also remarkably well-suited for developing, in the setting of homogeneous fragmentations, the important technique of changes of probability measures for classical Galton-Watson processes and branching random walks initiated respectively by Lyons *et al.* [157] and Lyons [156]; see [51] for much more on this issue. We further refer to Basdevant [20] for an extension to time-inhomogeneous fragmentations. In a different direction, the representation of semifinite measures on partitions due to Kerov (Theorem 3 of Section 3 in Chapter 1 of [134]) may be related to Theorem 3.1; it would be interesting to make the connection clear.

The first explicit example of a self-similar fragmentation with immediate dislocations has been given by Aldous and Pitman [9] in connection with the standard additive coalescent; see the forthcoming Section 5.3.4. Aldous and Pitman's construction is based on a Poissonian logging of the Continuum Random Tree along its skeleton; see also [2] for an alternative approach based on the Poisson snake. A somewhat simpler construction of the same fragmentation process was then given in [31] using the Brownian excursion. In this vein, it is interesting to mention that the level sets of a Brownian excursion yield another natural (and even simpler) self-similar fragmentation process which is closely related to that of Aldous and Pitman; see Section 4 in [36]. These examples have been extended by Miermont [167, 168] to the stable random trees of Duquesne and Le Gall [86]; see also [1] for a further extension to more general Lévy trees. In this direction, Haas and Miermont [122] have pointed out that self-similar fragmentations with a negative index of self-similarity are closely related to a natural family of self-similar random trees.

Unfortunately, little is known about the exact distributions of self-similar fragmentations taken at a fixed time (except in a few special examples, see in particular the forthcoming Theorems 5.3 and 5.4). Theorem 3.2 and Corollary 3.1 specify the distribution of the tagged fragment, and thus the so-called structural distribution of $X(t)$ for every fixed $t \geq 0$; however, this is not sufficient to characterize the law of the random mass-partition $X(t)$. Miermont and Schweinsberg [169] have even shown that certain natural random mass-partitions, which have been presented in Section 2.2, cannot appear as the one-dimensional distribution of a self-similar fragmentation. An interesting open question in this direction would be for instance to obtain an explicit characterization of the EPPF of the random exchangeable partition $\Pi(1)$ in terms of the characteristics of the fragmentation.

# 4

# Exchangeable coalescents

Exchangeable coalescents form a natural family of Markov processes with values in the space of partitions of $\mathbb{N}$, in which blocks coagulate as time passes. We shall first present the celebrated coalescent of Kingman, which is often used as a model for the genealogy of large populations. Then we shall introduce a more general family of coalescents, called here exchangeable coalescents, in which coagulations may involve several blocks simultaneously. We shall also investigate the coagulation processes of mass-partitions associated with such exchangeable coalescents. The last section of this chapter is devoted to a representation of these exchangeable coalescents in terms of certain stochastic flows on the unit interval. Many ideas and techniques based on exchangeability, which were useful for investigating fragmentations, can also be fruitfully applied to coagulations.

## 4.1 Kingman's coalescent

Coalescence naturally arises when one studies the genealogy of populations; we first briefly explain why. Following Kingman [139], this will lead us to introduce a natural Markov process with values in the space $\mathcal{P}_\infty$ of partitions of $\mathbb{N}$.

### 4.1.1 Genealogy of populations in the Wright-Fisher model

Imagine at time $T > 0$ a population with size $n$ which can be identified with the set $[n] = \{1, \ldots, n\}$. Assume the population is haploid, meaning that each individual has exactly one parent at the previous generation, so we may follow its ancestral lineage backwards in time. Plainly, ancestral lineages coalesce, in the sense that when distinct individuals have the same ancestor at some generation $t$, they necessarily have the same ancestor at any generation $t' \leq t$. For every $t < T$, consider the partition, say $\pi(t) \in \mathcal{P}_n$, of the population into sub-families (i.e. blocks of $[n]$) having the same ancestor at the generation $T - t$. When $t$ increases, the partitions $\pi(t)$ get coarser and, more precisely, the partition $\pi(t+s)$ can be obtained from $\pi(t)$ by the coagulation of the sub-families such that their respective ancestors at the generation $T - t$ have the same ancestor at the generation $T - t - s$.

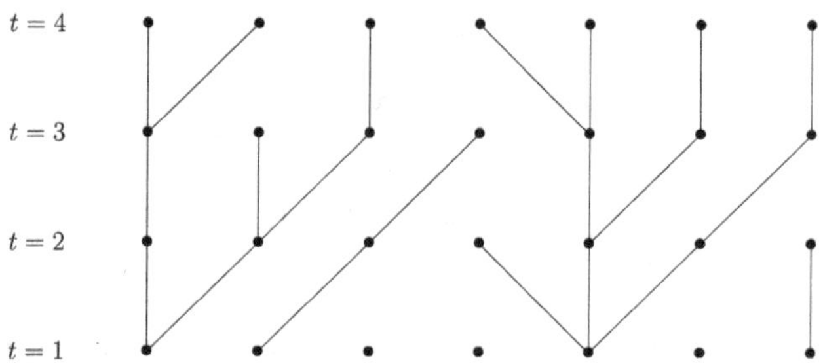


Coalescing ancestral lineages: population of fixed size 7 with 4 generations

To start the mathematical study of this phenomenon, we need a model for the evolution of populations. Here, we shall consider one of the simplest, which was introduced by Wright and Fisher around 1930; see [87]. In the Wright-Fisher model, time is discrete, the size of the population is fixed, the generations do not overlap and, finally, each individual at the generation $k+1$ picks its parent from the individuals at the generation $k$ according to the uniform probability, independently of the other individuals. In particular, the number of children $\xi$ of a typical individual has a binomial$(n, 1/n)$ distribution:

$$\mathbb{P}(\xi = k) = \binom{n}{k} n^{-k}(1 - 1/n)^{n-k}, \qquad k = 0, 1, \ldots, n.$$

Plainly, the probability that two distinct individuals at the same generation have the same parent at the preceding generation equals $1/n$. Since the generations are supposed to be independent, we see that the probability that the ancestral lines of these two individuals remain distinct during at least $k$ generations equals $(1 - 1/n)^k$. Hence, the time of coalescence of the ancestral lines, that is the age of the most recent common ancestor of two distinct individuals, has the geometric distribution with mean $n$. More generally, if we select $\ell \leq n$ distinct individuals at the same generation, then the probability that all have distinct parents at the preceding generation is the proportion of injections among the maps from $[n]$ to itself, that is $(1 - 1/n) \cdots (1 - (\ell - 1)/n)$. Thus the probability that the ancestral lines of these $\ell$ individuals do not coalesce before $k$ generations is $(1-1/n)^k \cdots (1-(\ell-1)/n)^k$.

This suggests a diffusion-approximation (cf. Ethier and Kurtz [99]) when the size $n$ of the population and the number of generations are large. Specifically, let us renormalize time in such a way that one time unit corresponds to $n$ generations, and let $n \to \infty$. Thus the probability that the ancestral lines of $\ell$ distinct individuals at the same generation remain distinct at least up to time $t$ (i.e. during at least $k = \lfloor nt \rfloor$ generations) converges when $n \to \infty$ to

$$\mathrm{e}^{-t(1+\cdots+(\ell-1))} = \exp(-t\ell(\ell-1)/2).$$

In other words, the time of the first coalescence for the ancestral lines of $\ell$ distinct individuals, converges in distribution to an exponential variable with parameter $\ell(\ell-1)/2$, that is the minimum of $\ell(\ell-1)/2$ independent standard exponential variables. Observe that there are precisely $\ell(\ell-1)/2$ pairs of ancestral lines that can be built from $\ell$ distinct individuals. These elementary observations have led Kingman to introduce a remarkable Markov process on the space of partitions of $\mathbb{N}$, which will be developed in the next section.

### 4.1.2 Construction of Kingman's coalescent

We start by introducing the simple notion of *coagulation* of pairs of blocks for partitions, referring to Section 2.3.1 for the basic notation. Let $\pi$ and $\pi'$ be two partitions of some set $E \subseteq \mathbb{N}$ (later in the text, we shall mostly be concerned with the cases $E = [n]$ and $E = \mathbb{N}$). We say that $\pi'$ can be obtained from $\pi$ by the coagulation of (exactly) two of its (non-empty) blocks if there exists $1 \le i < j$ such that $\pi_i, \pi_j \neq \emptyset$, $\pi'_i = \pi_i \cup \pi_j$, and for all $n' \neq i$, there is some $n \in \mathbb{N} \backslash \{i, j\}$ such that $\pi'_{n'} = \pi_n$. Recall the notation

$$\#\pi = \sup\{i : \pi_i \neq \emptyset\}$$

for the number of non-empty blocks of a partition $\pi$.

The discussion of the preceding section leads us to consider for each fixed $n > 0$ a Markov chain in continuous time with values in the space $\mathcal{P}_n$ of partitions of $[n]$, denoted by $\Pi^{[n]} = \left(\Pi^{[n]}(t), t \ge 0\right)$ and called the $n$-coalescent, which is governed by the following dynamics. The trivial partition $\mathbf{1}_{[n]} = ([n], \emptyset, \dots)$ is an absorbing state. If $\pi, \pi' \in \mathcal{P}_n$ are two partitions such that $\pi'$ can be obtained from $\pi$ by the coagulation of two of its blocks, then the jump rate from $\pi$ to $\pi'$ is one. All the other jump rates are equal to zero. This means that when the initial state of the chain is given by some partition $\pi$ with $\#\pi = k \ge 2$, $\Pi^{[n]}$ stays at $\pi$ for an exponential time with parameter $k(k-1)/2$ (i.e. the number of pairs of non-empty blocks), and then jumps at one of the $k(k-1)/2$ partitions which can be obtained from $\pi$ by the coagulation of two of its blocks, according to the uniform probability.

For the sake of simplicity, we shall assume throughout this section (except in the proof of the forthcoming Lemma 4.1) that $\Pi^{[n]}$ starts from the partition of $[n]$ into singletons, as the case of different initial configurations can easily be reduced to that one. We can then rephrase the description of the dynamics of $\Pi^{[n]}$ by saying that the process of the number of non-empty blocks,

$$\#\Pi^{[n]} = \left(\#\Pi^{[n]}(t), t \geq 0\right),$$

is a pure death process, that is a continuous-time Markov chain on $\mathbb{N}$ in which the only admissible steps are from $k$ to $k-1$; more precisely, the death rate at level $k \geq 1$ is $k(k-1)/2$. We call the sequence of the successive states visited by $\Pi^{[n]}$ the *state sequence*; specifically, for every $m = 1, \ldots, n$, we denote by $\kappa^{[n]}(m)$ the partition of $[n]$ with $m$ non-empty blocks that is obtained after $n-m-1$ coagulations in the $n$-coalescent process. In particular $\kappa^{[n]}(n)$ is the partition into singletons, and $\kappa^{[n]}(1)$ the trivial partition $\mathbf{1}_{[n]}$. The state sequence $(\kappa^{[n]}(n), \kappa^{[n]}(n-1), \kappa^{[n]}(1))$ is Markovian, that is

$$\mathbb{P}\left(\kappa^{[n]}(k) = \cdot \mid \kappa^{[n]}(n), \ldots, \kappa^{[n]}(k+1)\right) = \mathbb{P}\left(\kappa^{[n]}(k) = \cdot \mid \kappa^{[n]}(k+1)\right),$$

and is independent of the death process $\#\Pi^{[n]}$. Its transition probabilities from a partition $\pi$ with $\#\pi = k \geq 2$ is simply given by the uniform probability on the set of the $k(k-1)/2$ partitions which can be obtained from $\pi$ by the coagulation of two of its blocks. By definition, we have

$$\Pi^{[n]}(t) = \kappa^{[n]}(\#\Pi^{[n]}(t)), \qquad t \geq 0.$$

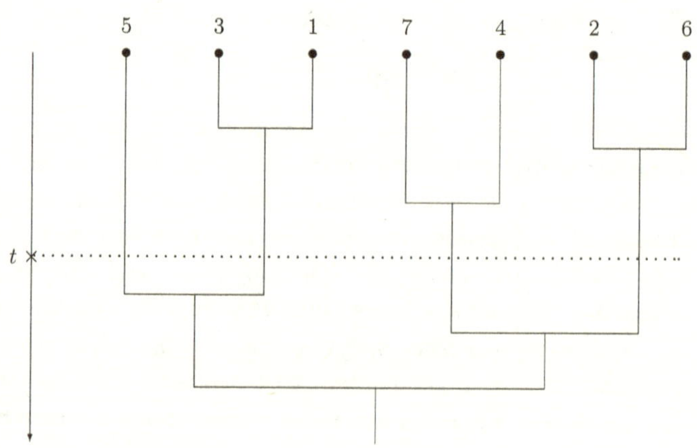


$k$-coalescent for $k = 7$ $\Pi^{[k]}(t) = (\{1,3\}, \{2,6\}, \{4,7\}, \{5\})$

The one-dimensional distributions of the state sequence $\kappa^{[n]}$ can be computed explicitly.

**Proposition 4.1** *Fix $n \in \mathbb{N}$ and let $(B_1, \ldots, B_k, \emptyset, \ldots)$ be some partition of $[n]$ with $B_k \neq \emptyset$. Then*

$$\mathbb{P}\left(\kappa^{[n]}(k) = (B_1, \ldots, B_k, \emptyset, \ldots)\right) = \frac{(n-k)!k!(k-1)!}{n!(n-1)!} \prod_{i=1}^{k} \#B_i! .$$

Observe that this formula shows that the random partition $\kappa^{[n]}(k)$ is exchangeable.

*Proof* The proof uses a backwards induction on the number of blocks $k$. The case when $k = n$ corresponds to the initial state of $\Pi^{[n]}$, that is the partition into singletons. So pick $2 \leq k \leq n$ and assume that the stated formula holds for all the partitions of $[n]$ with $k$ blocks or more. Observe that for every $1 \leq \ell < b$, there are $\frac{1}{2}\binom{b}{\ell}$ possible ways for splitting a block with size $b$ into two smaller blocks with sizes $\ell$ and $b - \ell$. We now use the identity

$$\mathbb{P}\left(\kappa^{[n]}(k-1) = (B_1, \ldots, B_{k-1}, \emptyset, \ldots)\right) = \frac{2}{k(k-1)} \sum \mathbb{P}(\kappa^{[n]}(k) = \pi) ,$$

where, in the right-hand side, the summation is taken over the family of partitions $\pi \in \mathcal{P}_n$ with $\#\pi = k$, such that the partition $(B_1, \ldots, B_{k-1}, \emptyset, \ldots)$ can result from $\pi$ after the coagulation of two blocks of the latter. This yields

$$\begin{aligned}
&\mathbb{P}\left(\kappa^{[n]}(k-1) = (B_1, \ldots, B_{k-1}, \emptyset, \ldots)\right) \\
&= \frac{2}{k(k-1)} \frac{(n-k)!k!(k-1)!}{n!(n-1)!} \\
&\quad \times \sum_{i=1}^{k-1} \sum_{\ell=1}^{\#B_i - 1} \frac{\#B_1! \cdots \#B_{k-1}!}{\#B_i!} \ell!(\#B_i - \ell)! \frac{1}{2}\binom{\#B_i}{\ell} \\
&= \frac{(n-k)!(k-1)!(k-2)!}{n!(n-1)!} \#B_1! \cdots \#B_{k-1}! \sum_{i=1}^{k-1} \sum_{\ell=1}^{\#B_i - 1} 1 .
\end{aligned}$$

Since

$$\sum_{i=1}^{k-1} \sum_{\ell=1}^{\#B_i - 1} 1 = \sum_{i=1}^{k-1} (\#B_i - 1) = n - (k-1) ,$$

this establishes the formula for partitions of $[n]$ with $k-1$ blocks. □

Proposition 4.1 enables us to compute the transition probabilities of the reversed state sequence $\kappa^{[n]}(1), \ldots, \kappa^{[n]}(n)$. Specifically, let $\xi = (B_1, \ldots, B_k, \emptyset, \ldots)$ be some partition of $[n]$ with $k \geq 2$ non-empty blocks, and $\gamma$ the partition of $[n]$ obtained from $\xi$ by the coagulation of the blocks $B_i$ and $B_j$ with $1 \leq i < j \leq k$. By Bayes' rule, we have

$$\mathbb{P}(\kappa^{[n]}(k) = \xi \mid \kappa^{[n]}(k-1) = \gamma) = \frac{2}{k(k-1)} \times \frac{\mathbb{P}(\kappa^{[n]}(k) = \xi)}{\mathbb{P}(\kappa^{[n]}(k-1) = \gamma)},$$

and after cancellations one gets

$$\mathbb{P}(\kappa^{[n]}(k) = \xi \mid \kappa^{[n]}(k-1) = \gamma) = \frac{2}{n-k+1} \times \frac{\#B_i!\#B_j!}{(\#B_i + \#B_j)!}.$$

It is interesting to observe that these transition probabilities can also be described as follows. First, since for every $m \in [\ell-1]$ there are $\frac{1}{2}\binom{\ell}{m}$ partitions of a block of size $\ell$ in two blocks of respective sizes $m$ and $\ell - m$, we have that conditionally on $\kappa^{[n]}(k-1)$, the probability that $\kappa^{[n]}(k)$ is obtained by splitting a given block with cardinal $\ell$ of $\kappa^{[n]}(k-1)$ equals

$$\frac{2}{n-k+1} \sum_{m=1}^{\ell-1} \frac{m!(\ell-m)!}{\ell!} \times \frac{1}{2}\binom{\ell}{m} = \frac{\ell-1}{n-k+1}.$$

Second, conditionally on this event, the probability that the split of the latter block yields two blocks, say $B$ and $B'$, is

$$\frac{2}{\ell-1} \times \frac{\#B!\#B'!}{\ell!}.$$

We now arrive at a fundamental property of $n$-coalescents, namely the following compatibility result.

**Lemma 4.1** *For every $n \geq 2$, the restriction $\Pi^{[n]}_{|[n-1]}$ of an $n$-coalescent to $[n-1]$ is an $(n-1)$-coalescent.*

*Proof* Fix $\gamma \in \mathcal{P}_{n-1}$ and pick an arbitrary $\pi \in \mathcal{P}_n$ such that $\gamma = \pi_{|[n-1]}$. We work with an $n$-coalescent $\Pi^{[n]}$ started from $\pi$.

Suppose first that the block of $\pi$ that contains $n$ is not reduced to the singleton $\{n\}$ and let $i < n$ be the smallest integer in this block. Plainly, for every $t \geq 0$, we may recover $\Pi^{[n]}(t)$ from its restriction to $[n-1]$ simply by adding $n$ to the block of $\Pi^{[n]}_{|[n-1]}(t)$ which contains $i$. This immediately implies that $\Pi^{[n]}_{|[n-1]}$ is a Markov chain with the same transitions as the $(n-1)$-coalescent.

Suppose now that $\{n\}$ is a block of $\pi$ and set $k = \#\pi$. Let $\tau$ denote the first jump time of $\Pi^{[n]}$, so $\tau$ has an exponential law with parameter $k(k-1)/2$, and is independent of $\pi' := \Pi^{[n]}(\tau)$. Consider the event $A$ that the block $\{n\}$ is not involved into the coagulation that occurs at time $\tau$, and $A'$ the complementary event. Clearly $\mathbb{P}(A) = 1 - 2/k$ and, conditionally on $A$, the restriction of $\pi'$ to $[n-1]$ is the uniform distribution on the set of $(k-1)(k-2)/2$ partitions obtained from $\pi_{|[n-1]}$ by the coagulation of two of its blocks.

On the event $A'$, let us denote by $\tau'$ the waiting time for the second jump of $\Pi^{[n]}$, so $\tau'$ has an exponential distribution with parameter $(k-1)(k-2)/2$. Moreover, by the first part of the proof, the restriction of $\pi'' := \Pi^{[n]}(\tau+\tau')$ to $[n-1]$ is independent of $\tau$ and $\tau'$, and is uniformly distributed on the set of the $(k-1)(k-2)/2$ partitions obtained from $\pi_{|[n-1]}$ by the coagulation of two of its blocks.

Now the process $\Pi^{[n]}_{|[n-1]}$ stays at $\pi_{|[n-1]}$ up to time $\tau + \mathbb{1}_{A'}\tau'$ and then jumps at $\pi'_{[n-1]}$ on the event $A$, and at $\pi''_{[n-1]}$ on the event $A'$. It is easily seen from the classical properties of independent exponential variables that $\tau + \mathbb{1}_{A'}\tau'$ has an exponential distribution with parameter $(k-1)(k-2)/2$ (this is essentially a consequence of the lack of memory). Moreover, this time is independent of the random partition of $[n-1]$ that equals $\pi'_{|[n-1]}$ on the event $A$ and $\pi''_{|[n-1]}$ on $A'$.

Putting the pieces together, we get that $\Pi^{[n]}_{|[n-1]}$ is a continuous time Markov chain with the following characteristics. When this chain starts from $\gamma \in \mathcal{P}_{n-1}$ with $\#\gamma = k \geq 2$, the next step of the chain occurs after an exponential time with parameter $k(k-1)/2$, and then the chain jumps at one of the $k(k-1)/2$ partitions obtained from $\gamma$ by the coagulation of two of its blocks, independently of the waiting time and according to the uniform distribution. In other words, $\Pi^{[n]}_{|[n-1]}$ is an $(n-1)$-coalescent. □

The compatibility stated in Lemma 4.1 can be combined with Kolmogorov's extension theorem. Specifically, this enables us to construct simultaneously for all $n \in \mathbb{N}$, a family of processes $\left(\Pi^{[n]}(t), t \geq 0\right)$ such that each $\Pi^{[n]}$ is an $n$-coalescent, and for every $t \geq 0$, $\Pi^{[n]}(t)$ coincides with the restriction of $\Pi^{[n+1]}(t)$ to $[n]$. In particular, the sequence $\Pi^{[1]}(t), \Pi^{[2]}(t), \ldots$ is compatible, and by Lemma 2.5, this yields the following definition.

**Definition 4.1** *There exists a unique (in law) process denoted by* $\Pi^{\mathrm{K}} = (\Pi^{\mathrm{K}}(t), t \geq 0)$*, with values in* $\mathcal{P}_\infty$ *and such that for every* $n \in \mathbb{N}$*, the process induced by the restriction to* $[n]$*,* $\Pi^{\mathrm{K}}_{|[n]} = (\Pi^{\mathrm{K}}_{|[n]}(t), t \geq 0)$*, is an n-coalescent. The process* $\Pi^{\mathrm{K}}$ *is called* **Kingman's coalescent**.

We point out that each restriction $\Pi^{\mathrm{K}}_{|[n]}$ has càdlàg (i.e. right-continuous with left-limits) paths with values in $\mathcal{P}_n$, and by the definition of the metric on $\mathcal{P}_\infty$ (cf. Lemma 2.6), this implies that $\Pi^{\mathrm{K}}$ has càdlàg paths valued in $\mathcal{P}_\infty$, a.s. Note that it is implicitly assumed that $\Pi^{\mathrm{K}}(0)$ is the partition of $\mathbb{N}$ into singletons. We also stress that

$$\text{Kingman's coalescent is an exchangeable process,} \tag{4.1}$$

in the sense that for every permutation $\varsigma$ of $\mathbb{N}$, $\left(\varsigma\left(\Pi^{\mathrm{K}}(t)\right), t \geq 0\right)$ has the same distributions as $\Pi^{\mathrm{K}}$. Indeed, it should be plain from the description of its jump rates that the $n$-coalescent is an exchangeable process. By compatibility, this implies that the processes $\Pi^{\mathrm{K}}$ and $\varsigma\left(\Pi^{\mathrm{K}}\right)$ have the same finite-dimensional distributions, and since both processes have càdlàg paths a.s., they thus have the same law (see for example Theorem 14.5 in [57] or Section VI.1 in [127]).

We now present a simple description of the evolution of Kingman's coalescent.

**Theorem 4.1** (i) *Kingman's coalescent comes down from infinity, that is for every $t > 0$, the number $\#\Pi^{\mathrm{K}}(t)$ of the non-empty blocks of $\Pi^{\mathrm{K}}(t)$ is finite a.s. More precisely, $(\#\Pi^{\mathrm{K}}(t), t > 0)$ is a pure death process with death rate $k(k-1)/2$ at level $k$.*

(ii) *For every $n \in \mathbb{N}$, let $\mathtt{K}(n)$ denote the state of $\Pi^{\mathrm{K}}$ when $\#\Pi^{\mathrm{K}} = n$. Then the state sequence $(\ldots, \mathtt{K}(n+1), \mathtt{K}(n), \ldots, \mathtt{K}(1))$ is Markovian and independent of the death process $\#\Pi^{\mathrm{K}}$.*

(iii) *The conditional distribution of $\mathtt{K}(n)$ given $\mathtt{K}(n+1)$ is the uniform probability on the set of the $n(n+1)/2$ partitions of $\mathbb{N}$ which can be obtained from $\mathtt{K}(n+1)$ by the coagulation of two of its blocks.*

*Proof* Clearly, for an arbitrary partition $\pi \in \mathcal{P}_\infty$, the number of non-empty blocks of $\pi$ can be obtained from its restrictions as the increasing limit

$$\#\pi = \lim_{n\to\infty} \#\pi_{|[n]} \,. \tag{4.2}$$

So let us fix $k \in \mathbb{N}$ and compute the probability that $\Pi^{\mathrm{K}}_{|[n]}(t)$ has at least $k$ non-empty blocks. By the description of $\#\Pi^{\mathrm{K}}_{|[n]}$ as a death process, we have

$$\mathbb{P}\left(\#\Pi^{\mathrm{K}}_{|[n]}(t) \geq k\right) = \mathbb{P}\left(\sum_{j=k}^{n} \frac{2}{j(j-1)}\mathbf{e}_j > t\right) \leq \mathbb{P}\left(\sum_{j=k}^{\infty} \frac{2}{j(j-1)}\mathbf{e}_j > t\right),$$

where $\mathbf{e}_1, \ldots$ is a sequence of i.i.d. standard exponential variables. Since

$$\lim_{k\to\infty} \sum_{j=k}^{\infty} \frac{2}{j(j-1)}\mathbf{e}_j = 0$$

with probability one, we conclude that $\mathbb{P}\left(\#\Pi^{\mathrm{K}}_{|[n]}(t) \geq k\right)$ tends to 0 as $k \to \infty$, that is Kingman's coalescent comes down from infinity.

The description of $\#\Pi^{\mathrm{K}}(\cdot)$ as a death process follows now from that for $\#\Pi^{\mathrm{K}}_{|[n]}(\cdot)$ and (4.2). Moreover if we denote by $\mathtt{K}^{[n]}(\cdot)$ the state sequence of the $n$-coalescent $\Pi^{\mathrm{K}}_{|[n]}$, then for every fixed integer $k$, (4.2) yields the identity

$$\mathtt{K}^{[n]}(j) = \mathtt{K}_{|[n]}(j)\,, \qquad j = 1, \ldots, k\,,$$

provided that $n$ is sufficiently large. The stated properties for the state sequence $\mathtt{K}$ are now readily derived from those of $\mathtt{K}^{[n]}$. □

We stress that Theorem 4.1(i) entirely characterizes the law of $\#\Pi^{\mathrm{K}}$, as there is a unique entrance law from $\infty$ for such a pure death process. Indeed, its one-dimensional distributions are necessarily given by

$$\mathbb{P}\left(\#\Pi^{\mathrm{K}}(t) \leq k\right) = \mathbb{P}\left(\sum_{j=k+1}^{\infty} \frac{2}{j(j-1)}\mathbf{e}_j \leq t\right), \qquad t > 0, k \in \mathbb{N},$$

where $\mathbf{e}_1, \ldots$ is a sequence of i.i.d. standard exponential variables.

We now complete this section with the observation that the time of total coalescent, that is the first instant $t$ such that $\Pi^{\mathrm{K}}(t)$ is the trivial partition $\mathbf{1}_{[n]}$ (which can also be viewed as the age of the most recent common ancestor if we think of Kingman's coalescent as a model for the genealogy of large populations), is distributed as

$$\sum_{j=2}^{\infty} \frac{2}{j(j-1)}\mathbf{e}_j\,.$$

In particular, the expectation of this variable equals 2.

### 4.1.3 Interval representation of Kingman's coalescent

Theorem 4.1 describes precisely the transitions of Kingman's coalescent; however, this is not entirely satisfactory as the one-dimensional distributions are not specified. The purpose of this section is to present an explicit construction of Kingman's coalescent based on a natural family of partitions of the unit interval, and which can be viewed as a multidimensional version of the paint-box construction (in this direction, recall from (4.1) that $\Pi^{\mathrm{K}}(\cdot)$ is an exchangeable process). The starting point is provided by Corollary 2.1(ii) which states a simple stability property of the uniform distributions on the simplexes $\Delta_n$ in connection with uniform coagulation.

To that end, let us introduce $U_1, \ldots$ and $U'_1, \ldots$, two independent sequences of i.i.d. uniform variables on $[0, 1]$. Set $\vartheta(1) = ]0, 1[$ and for each $n \in \mathbb{N}$, consider the random interval-partition

$$\vartheta(n+1) = ]0, 1[ \setminus \{U'_1, \ldots, U'_n\} .$$

It is then seen from the proof of Corollary 2.1(ii), that the sequence

$$(\ldots, |\vartheta(n+1)|^{\downarrow}, |\vartheta(n)|^{\downarrow}, \ldots, |\vartheta(1)|^{\downarrow})$$

is Markovian, and more precisely, the conditional distribution of the mass-partition $|\vartheta(n)|^{\downarrow}$ given $|\vartheta(n+1)|^{\downarrow}$ is the uniform distribution on the set of the $(n+1)n/2$ mass-partitions that can be obtained from $|\vartheta(n+1)|^{\downarrow}$ by merging two of its non-zero terms and then rearranging in decreasing order. This property can be shifted to paint-boxes as follows.

We denote by $\mathtt{K}'(n)$ the paint-box constructed from the random interval-partition $\vartheta(n)$ and the $U_i$, so that two indices $i$ and $j$ belong to the same block of $\mathtt{K}'(n)$ if and only if the variables $U_i$ and $U_j$ belong to the same interval component of $\vartheta(n)$. It follows readily from above that $(\ldots, \mathtt{K}'(n+1), \mathtt{K}'(n), \ldots, \mathtt{K}'(1))$ is a Markov chain governed by the same transitions as the state sequence $\mathtt{K}(\cdot)$ of Kingman's coalescent. However, we do not know yet whether $\mathtt{K}(n)$ and $\mathtt{K}'(n)$ have the same law.

Finally, recall Theorem 4.1 and introduce a death process $(D(t), t > 0)$ distributed as $(\#\Pi^{\mathrm{K}}(t), t > 0)$, that is with death rate $k(k-1)/2$ at level $k \in \mathbb{N}$. We assume that $D(\cdot)$ is independent of the preceding quantities, set $\Pi'(t) = \mathtt{K}'(D(t))$ for every $t > 0$ and let $\Pi'(0)$ be the partition of $\mathbb{N}$ into singletons. We are now able to state the following.

**Proposition 4.2** *The process* $(\Pi'(t), t \geq 0)$ *constructed above is a version of Kingman's coalescent.*

*Proof* By construction, the process $(\Pi'(t), t > 0)$ is Markovian and has the same transition probabilities as Kingman's coalescent; we have to check that it has the same one-dimensional distributions as $\Pi^{\mathrm{K}}$.

Let us introduce for every $n \in \mathbb{N}$, the first passage time of the death process $D$ at level $n$,

$$T(n) = \inf \{t > 0 : D(t) = n\} .$$

Plainly,

$$\lim_{n\to\infty} T(n) = 0 \qquad \text{a.s.} \tag{4.3}$$

Observe also that $(D(T(n)+t), t \geq 0)$ is a death process started from $n$, with death rate $j(j-1)/2$ at level $j$, and which is of course independent of the sequence $\mathrm{K}'(\cdot)$.

Then define for each $t \geq 0$ a $\mathcal{P}_n$-valued random variable $\Gamma^{[n]}(t)$, by declaring that $i, j = 1, \ldots, n$ belong to the same block of $\Gamma^{[n]}(t)$ if and only if the blocks $\mathrm{K}'_i(n) = \Pi'_i(T(n))$ and $\mathrm{K}'_j(n) = \Pi'_j(T(n))$ are part of the same block of the partition $\Pi'(T(n)+t)$. By inspection of the jump rates of the $n$-coalescent $\Gamma^{[n]}(\cdot)$ that are seen from the dynamics of $\Pi'(\cdot)$, we get that $(\Gamma^{[n]}(t), t \geq 0)$ is an $n$-coalescent.

Next, fix $k \in \mathbb{N}$. On the one hand, it is immediate that whenever $n$ is sufficiently large, for every $i \leq k$, the restriction of the $i$-th block of $\mathrm{K}'(n)$ to $[k]$, $\mathrm{K}'_i(n) \cap [k]$, is reduced to the singleton $\{i\}$. The very construction of $\Gamma^{[n]}$ implies that

$$\Gamma^{[n]}_{|[k]}(t) = \Pi'_{|[k]}(T(n)+t), \qquad t \geq 0,$$

provided that $n$ is large enough. On the other hand, by the compatibility property stated in Lemma 4.1, the restriction of $\Gamma^{[n]}$ to $[k]$ is a $k$-coalescent. Taking the limit as $n \to \infty$ and using (4.3), we conclude that for each $k \in \mathbb{N}$, $\Pi'_{|[k]}(\cdot)$ is a $k$-coalescent, which completes the proof. □

Note that by the exchangeability property (4.1), the state sequence $\mathrm{K}(\cdot)$ is also formed by exchangeable random partitions, and hence each possesses asymptotic frequencies. As an immediate consequence of Proposition 4.2 and Lemma 2.2(ii), the distribution of the latter is given as follows.

**Corollary 4.1** *For every integer $n \geq 2$, $|\mathrm{K}(n)|^{\downarrow}$ has the law of the decreasing rearrangement of a variable that is uniformly distributed on the simplex $\Delta_{n-1}$.*

In this vein, it is interesting to observe from Corollary 2.1(i) that time-reversal transforms the process of the ranked asymptotic frequencies of the state sequence in Kingman's coalescent into a sequence $(|\mathrm{K}(1)|^{\downarrow}, |\mathrm{K}(2)|^{\downarrow}, \ldots)$, which can be viewed as a fragmentation sequence.

## 4.2 Simultaneous and multiple coagulations

The fundamental point in the construction of Kingman's coalescent is the use of the space $\mathcal{P}_\infty$ of partitions of $\mathbb{N}$ as the natural state space. Restricting partitions to a finite set is a powerful discretization technique which enables the application of the elementary theory of continuous time Markov chains. In particular, this is the key to circumventing the major difficulty that infinitely many coagulations occur immediately after the initial time; just as in the preceding chapter, it enabled us to construct fragmentation processes in which dislocations occur immediately. In this section, we shall develop this idea further by introducing a general class of coalescent processes where coagulations of infinitely many blocks may occur and different merging may take place simultaneously. These coalescents have been considered first by Möhle and Sagitov [170] and Schweinsberg [199], extending earlier works by Pitman [183] and Sagitov [197]. They bear natural relations with a model for the genealogy of large populations which generalizes the Wright-Fisher model; see the comments in the forthcoming Section 4.4

### 4.2.1 Coagulation of partitions

We start by introducing the basic notion of coagulation of partitions which has a central role in this chapter.

**Definition 4.2** *Let $B \subseteq \mathbb{N}$ and $k' \in \overline{\mathbb{N}}$. A pair of partitions $(\pi, \pi') \in \mathcal{P}_B \times \mathcal{P}_{k'}$ is called* **admissible** *if the number of non-empty blocks of $\pi$ is $\#\pi \leq k'$. For every admissible pair of partitions $(\pi, \pi')$, we call* **coagulation of $\pi$ by $\pi'$** *and write $\mathrm{Coag}(\pi, \pi')$ for the partition $\pi'' = (\pi''_j, j \in \mathbb{N})$ of $B$ given by*

$$\pi''_j := \bigcup_{i \in \pi'_j} \pi_i, \qquad j \in \mathbb{N},$$

*where $(\pi_i, i \in \mathbb{N})$ and $(\pi'_j, j \in \mathbb{N})$ denote the sequence of the blocks of $\pi$ and $\pi'$, respectively.*

For instance, for $B = [10]$, if

$$\pi = (\{1, 6, 7\}, \{2, 4, 5\}, \{3, 8\}, \{9, 10\}) \quad \text{and} \quad \pi' = (\{1, 3\}, \{2, 4\}),$$

then $\mathrm{Coag}(\pi, \pi')$ results from the merging of the first and third and, respectively, the second and fourth blocks of $\pi$. We get

$$\mathrm{Coag}(\pi, \pi') = (\{1, 3, 6, 7, 8\}, \{2, 4, 5, 9, 10\}).$$

Plainly, the partition $\pi'' = \mathrm{Coag}(\pi, \pi')$ is *coarser* than $\pi$, in the sense that each block of the latter is contained in some block of the former. We also stress that the blocks $\pi''_j$ above are labeled according to the increasing order of their least elements, in agreement with our convention.

The partition into singletons will have a special role in this section. It is convenient to introduce the notation

$$\mathbf{0}_{[\infty]} := (\{1\}, \{2\}, \ldots).$$

Indeed, $\mathbf{0}_{[\infty]}$ serves as a neutral element for the coagulation operator, in the sense that

$$\mathrm{Coag}(\pi', \mathbf{0}_{[\infty]}) = \mathrm{Coag}(\mathbf{0}_{[\infty]}, \pi') = \pi'.$$

More generally, for every block $B \subseteq \mathbb{N}$, we denote by $\mathbf{0}_B$ the partition of $B$ into singletons, that is $\mathbf{0}_B$ is the restriction of $\mathbf{0}_{[\infty]}$ to $B$. Henceforth, we shall also use the notation $\mathbf{0}$ for the mass-partition in $\mathcal{P}_{\mathrm{m}}$ which is identical to 0; we hope this slight abuse of notation should be helpful rather than confusing.

It is important to observe the following easy fact.

**Lemma 4.2** *Coagulation is a Lipschitz-continuous function from $\mathcal{P}_\infty \times \mathcal{P}_\infty$ to $\mathcal{P}_\infty$, and associative in the sense that*

$$\mathrm{Coag}\,(\pi, \mathrm{Coag}\,(\pi', \pi'')) = \mathrm{Coag}\,(\mathrm{Coag}\,(\pi, \pi'), \pi'')$$

*whenever $(\pi, \pi')$ and $(\pi', \pi'')$ are admissible pairs.*

*Proof* Indeed, the fact that the labels of the blocks of a partition are assigned according to the order of their least element yields that for every $n \in \mathbb{N}$ and $\pi, \pi' \in \mathcal{P}_\infty$

$$\mathrm{Coag}(\pi, \pi')_{|[n]} = \mathrm{Coag}(\pi_{|[n]}, \pi') = \mathrm{Coag}(\pi_{|[n]}, \pi'_{|[n]}). \tag{4.4}$$

These identities and the very definition of the topology on $\mathcal{P}_\infty$ imply our first claim. The second is immediately checked. □

Another useful observation is that the coagulation operator preserves exchangeability in the following sense.

**Lemma 4.3** *Let $\pi$ and $\pi'$ be two independent exchangeable random partitions. Then the random partition* $\mathrm{Coag}(\pi, \pi')$ *is also exchangeable.*

*Proof* Let $\sigma$ be some permutation of $\mathbb{N}$, and $\sigma(\pi'')$ the image of $\pi'' = \mathrm{Coag}(\pi, \pi')$ by the action of $\sigma$. So the blocks of $\sigma(\pi'')$ are the images of those of $\pi''$ by the inverse permutation $\sigma^{-1}$, that is they are given by

$$\sigma^{-1}(\pi''_j) = \bigcup_{i \in \pi'_j} \sigma^{-1}(\pi_i), \qquad j \in \mathbb{N},$$

where $(\pi_i, i \in \mathbb{N})$ and $(\pi'_j, j \in \mathbb{N})$ denote the sequence of the blocks of $\pi$ and $\pi'$, respectively. The $\sigma^{-1}(\pi_i)$ form the family of the blocks of $\sigma(\pi) := \tilde{\pi}$, which is distributed as $\pi$. We stress that in general, these blocks are not labeled according to the increasing order of their least elements, so one should not believe that $\sigma(\pi'') = \mathrm{Coag}(\sigma(\pi), \pi')$.

Nonetheless, let $\sigma'$ be the permutation of $\mathbb{N}$ such that the $i$-th block of $\sigma(\pi)$ is $\sigma^{-1}(\pi_{\sigma'(i)})$ (if the number of non-empty blocks of $\pi$ is finite, we decide that $\sigma'(n) = n$ for every $n > \#\pi$, which then specifies $\sigma'$ uniquely). By construction, we now have $\sigma(\pi'') = \mathrm{Coag}(\sigma(\pi), \sigma'(\pi'))$.

Finally, observe that $\sigma'$ is independent of $\pi'$, and it follows from the exchangeability of $\pi'$ that the pair $(\sigma(\pi), \sigma'(\pi'))$ has the same law as $(\pi, \pi')$. This shows that $\sigma(\pi'')$ has the same law as $\pi''$. □

We also point out that more generally, the same argument (with heavier notation) shows that if $\pi^{(1)}, \ldots, \pi^{(n)}$ are independent exchangeable random partitions of $\mathbb{N}$, and we define by induction $\gamma^{(1)} := \pi^{(1)}$ and $\gamma^{(i+1)} := \mathrm{Coag}(\gamma^{(i)}, \pi^{(i+1)})$ for $i = 1, \ldots, n-1$, then the $n$-tuple $(\gamma^{(1)}, \ldots, \gamma^{(n)})$ is jointly exchangeable, in the sense that its distribution is invariant by the action of permutations of $\mathbb{N}$ on the space $(\mathcal{P}_\infty)^n$.

Finally, it is natural to compare the coagulation operator with the fragmentation operator introduced in Section 3.1.1. First, given a partition $\pi$ and a sequence of partitions $\pi^{(\cdot)}$, it is easy to construct a partition $\pi'$ (depending on $\pi$ and $\pi^{(\cdot)}$) such that

$$\mathrm{Coag}\left(\mathrm{Frag}(\pi, \pi^{(\cdot)}), \pi'\right) = \pi ,$$

and in the converse direction, if $(\pi, \pi')$ is a pair of admissible partitions, then one can construct a sequence $(\pi^{(i)}, i \in \mathbb{N})$ of partitions of $\mathbb{N}$ (depending on $\pi$ and $\pi'$) such that

$$\mathrm{Frag}\left(\mathrm{Coag}(\pi, \pi'), \pi^{(\cdot)}\right) = \pi .$$

So, in some weak sense, the operations of coagulation and fragmentation can be viewed as inverses of each other. We also point out that, at least at the level of mathematical formalism, coagulation is a simpler notion, essentially because its definition only involves a pair of compatible partitions, whereas an infinite sequence of partitions is needed to define a fragmentation.

### 4.2.2 Exchangeable coalescents and coagulation rates

We now turn our attention to Markov processes with values in the space $\mathcal{P}_\infty$ of partitions of $\mathbb{N}$. In this direction, the notion of coagulation leads naturally to the following definition.

**Definition 4.3** *Fix* $n \in \overline{\mathbb{N}}$, *and let* $\Pi = (\Pi(t), t \geq 0)$ *be a Markov process with values in* $\mathcal{P}_n$ *which is continuous in probability.*

(i) $\Pi$ *is called an* **exchangeable coalescent** *if its semigroup can be described as follows. For every* $t, t' \geq 0$, *the conditional distribution of* $\Pi(t+t')$ *given* $\Pi(t) = \pi$ *is the law of* $\mathrm{Coag}(\pi, \pi')$, *where* $\pi'$ *is some exchangeable random partition (whose law only depends on* $t'$*).*

(ii) *We call* $\Pi$ *a* **standard** *exchangeable coalescent if* $\Pi$ *is an exchangeable coalescent started from* $\mathbf{0}_{[n]}$, *the partition of* $[n]$ *into singletons.*

**Remark.** Plainly, Kingman's coalescent $\Pi^{\mathrm{K}}$ is a remarkable example of a standard exchangeable coalescent.

Later in this text, we shall be mostly interested in $\mathcal{P}_\infty$-valued processes. We make a few simple observations related to this definition which will be important in this setting.

First, the elementary property (4.4) of the coagulation operator implies that if $\Pi$ is an exchangeable coalescent with values in $\mathcal{P}_\infty$, then for every $n \in \mathbb{N}$, the restriction $\Pi_{|[n]}$ of $\Pi$ to $[n]$ is also an exchangeable coalescent.[1] More precisely, for every $t, t' \geq 0$, the conditional distribution of $\Pi_{|[n]}(t+t')$ given $\Pi_{|[n]}(t) = \pi$ is the law of $\mathrm{Coag}(\pi, \Pi(t')) = \mathrm{Coag}(\pi, \Pi_{|[n]}(t'))$. Conversely, it follows from the Compatibility Lemma 2.5 that if $\Pi$ is a process with values in $\mathcal{P}_\infty$ such that its restriction $\Pi_{|[n]}$ is an exchangeable coalescent for each $n \in \mathbb{N}$, then $\Pi$ itself is an exchangeable coalescent.

Second, we may work with a version of $\Pi$ with regular paths. Indeed, since $\mathcal{P}_n$ is a finite space, we may consider for each $n$ a version of the Markov chain $\Pi_{|[n]}$ with càdlàg paths, so that, by the very definition of the distance on $\mathcal{P}_\infty$, $\Pi$ then also has càdlàg paths a.s. From now on, we shall always consider such regular version.

Third, we point out that the exchangeable random partition $\pi'$ in (i) which is used to specify the semigroup, has obviously the law of $\Pi(t')$ when $\Pi$ is standard (simply take $t = 0$ in the definition).

[1] Just as for exchangeable fragmentations, the compatibility property (4.4) is crucial. Indeed, the restriction map $\pi \to \pi_{[n]}$ is not injective, and therefore, the restriction to $[n]$ of a Markov process with values in $\mathcal{P}_\infty$ may well not be Markovian.

Finally, there is no loss of generality in focussing on standard exchangeable coalescents. Indeed, thanks to the associativity of the coagulation operator, if $\pi \in \mathcal{P}_\infty$ and $\Pi$ is a standard exchangeable coalescent, then the process $(\mathrm{Coag}(\pi, \Pi(t)), t \geq 0)$ is an exchangeable coalescent started from $\pi$, with the same semigroup as $\Pi$. In particular, the semigroup of an exchangeable coalescent is given in terms of its standard version $\Pi$ by

$$\pi \to \mathbb{E}\left(\varphi(\mathrm{Coag}(\pi, \Pi(t)))\right), \qquad \pi \in \mathcal{P}_\infty, \tag{4.5}$$

where $\varphi : \mathcal{P}_\infty \to \mathbb{R}$ stands for a generic measurable bounded function. Recall that the space of partitions $\mathcal{P}_\infty$ is metrizable and compact.

**Proposition 4.3** *The semigroup of an exchangeable coalescent enjoys the Feller property. This means that for every continuous function $\varphi : \mathcal{P}_\infty \to \mathbb{R}$, the map (4.5) is continuous for each $t \geq 0$, and*

$$\lim_{t\to 0} \mathbb{E}\left(\varphi(\mathrm{Coag}(\pi, \Pi(t))\right) = \varphi(\pi), \qquad \pi \in \mathcal{P}_\infty.$$

*Proof* This follows immediately from Lemma 4.2. □

The Feller property provides another argument for the existence of a càdlàg version of the process. It also ensures that the (completed) natural filtration of $\Pi$ is right-continuous and enables us to extend the Markov property to stopping times. See for example Section III.2 in Revuz and Yor [192] for details.

Let us now investigate the transitions of an exchangeable coalescent, using the simple structure of their restrictions. Indeed, exchangeable coalescents with values in a finite space like $\mathcal{P}_n$ are Markov chains. Thus, the distribution of the restricted chain $\Pi_{|[n]}$ can be characterized by its jump rates. In this direction, denote the jump rate of the standard chain $\Pi_{|[n]}$ from $\mathbf{0}_{[n]}$ to $\pi$ by

$$q_\pi := \lim_{t\to 0+} \frac{1}{t} \mathbb{P}\left(\Pi_{|[n]}(t) = \pi\right), \qquad \pi \in \mathcal{P}_n \backslash \{\mathbf{0}_{[n]}\}.$$

It is immediately seen from this definition that the jump rates inherit an exchangeability property from $\Pi$, in the sense that for every permutation $\sigma$ of $[n]$, there is the identity

$$q_\pi = q_{\sigma(\pi)}. \tag{4.6}$$

Note that for every $n' \geq n$, $q_\pi$ also gives the jump rate of $\Pi_{|[n']}$ from $\pi'$ to $\mathrm{Coag}(\pi', \pi)$, where $\pi' \in \mathcal{P}_{n'}$ with $\#\pi' = n$, and that any jump which is not of this type must have a rate equal to zero. Hence, the family $(q_\pi, \pi \in$

$\mathcal{P}_n \backslash \{\mathbf{0}_{[n]}\}$ and $n \in \mathbb{N}$) characterizes the transitions of all the restricted chains $\Pi_{|[n]}$, and thus of the exchangeable coalescent $\Pi$.

A first basic result about these jump rates is that they can be represented by a single measure on $\mathcal{P}_\infty$. In this direction, recall the notation

$$\mathcal{P}_{n',\pi} = \left\{\pi' \in \mathcal{P}_{n'} : \pi'_{|[n]} = \pi\right\},$$

where $\pi \in \mathcal{P}_n$ and $n' \in \{n, n+1, \ldots, \infty\}$.

**Proposition 4.4** *Let $(q_\pi : \pi \in \mathcal{P}_n \backslash \{\mathbf{0}_{[n]}\}$ and $n \in \mathbb{N})$ be the family of jump rates of some exchangeable coalescent $\Pi$. There exists a unique measure $\mu$ on $\mathcal{P}_\infty$, called the* **coagulation rate** *of $\Pi$, such that $\mu(\{\mathbf{0}_{[\infty]}\}) = 0$ and*

$$\mu\left(\mathcal{P}_{\infty,\pi}\right) = q_\pi$$

*for every $n \in \mathbb{N}$ and every partition $\pi \in \mathcal{P}_n \backslash \{\mathbf{0}_{[n]}\}$.*

*Proof* The proof of the existence of the coagulation rate $\mu$ follows exactly the same argument as that for the proof of the existence of the splitting rate in the setting of exchangeable fragmentations; see Proposition 3.2. The quantity $\mu\left(\{\pi \in \mathcal{P}_\infty : \pi_{|[n]} \neq \mathbf{0}_{[n]}\}\right)$ represents the total jump rate (i.e. the parameter of the holding time) from $\mathbf{0}_{[n]}$ for the restricted chain $\Pi_{|[n]}$, and hence must be finite. □

### 4.2.3 Poissonian construction

The purpose of this section is to present an explicit construction of exchangeable coalescents from their coagulation rates. To start with, let us observe that the coagulation rate $\mu$ of an exchangeable coalescent is a measure $\mu$ on $\mathcal{P}_\infty$ which fulfills the following properties:

$$\mu(\{\mathbf{0}_{[\infty]}\}) = 0 \quad \text{and} \quad \mu(\pi \in \mathcal{P}_\infty : \pi_{|[n]} \neq \mathbf{0}_{[n]}) < \infty \quad \text{for every } n \in \mathbb{N}, \tag{4.7}$$

and

$$\mu \text{ is exchangeable, that is invariant by the action of permutations.} \tag{4.8}$$

Indeed, (4.7) is plain from Proposition 4.4, whereas (4.8) derives from (4.6) and Proposition 4.4. Conversely, it is easy to show that conditions (4.7) and (4.8) are sufficient for a measure to be the coagulation rate of some exchangeable coalescent.

**Lemma 4.4** *Let $\mu$ be a measure on $\mathcal{P}_\infty$ which fulfills (4.7) and (4.8). There exists an exchangeable coalescent with coagulation rate $\mu$.*

*Proof* We merely provide a sketch of the proof as precise construction will be provided in Proposition 4.5 below. For every $n \in \mathbb{N}$, we can use the family

$$q_{\pi} = \mu\left(\mathcal{P}_{\infty,\pi'}\right), \qquad \pi' \in \mathcal{P}_k \backslash \{\mathbf{0}_{[k]}\} \text{ and } k \leq n$$

as the jump rates of some Markov process $\Pi^{[n]}$ with values in $\mathcal{P}_n$. More precisely, for every $\pi \in \mathcal{P}_n$ with $\#\pi = k$ non-empty blocks and every $\pi' \in \mathcal{P}_k \backslash \{\mathbf{0}_{[k]}\}$, the jump rate of $\Pi^{[n]}$ from $\pi$ to $\mathrm{Coag}(\pi, \pi')$ is $q_{\pi'}$, and all the other jump rates are 0. By additivity of $\mu$ and (3.2), we see that (3.3) holds. It is now immediate that the restriction of $\Pi^{[n]}$ to $[k]$ is again Markovian, with the same jump rates as $\Pi^{[k]}$. By the same argument based on Kolmogorov's extension theorem that we used just before Definition 4.1, we can construct a unique (in distribution) Markov process $\Pi = (\Pi(t), t \geq 0)$ with values in $\mathcal{P}_\infty$ and started from $\mathbf{0}_{[\infty]}$, such that $\Pi_{|[n]}$ is a version of $\Pi^{[n]}$ for every $n \in \mathbb{N}$.

The exchangeability assumption (4.8) implies that $\Pi$ is an exchangeable coalescent. Indeed, by the very construction of Markov chains with assigned jump rates, it implies that if $\tau$ stands for the first jump time of $\Pi_{|[n]}$, then the random partition $\Pi_{|[n]}(\tau)$ is exchangeable. By iteration and Lemma 4.3, we deduce that for every $t \geq 0$, $\Pi_{|[n]}(t)$ is also an exchangeable random partition, and we conclude that $\Pi$ is an exchangeable coalescent with coagulation rate $\mu$. □

We now present a more explicit construction which does not require the appeal to Kolmogorov's extension Theorem, and sheds light on the Poissonian structure of exchangeable coalescent. Specifically, let $\mu$ be some measure on $\mathcal{P}_\infty$ which fulfills (4.7) and (4.8). Introduce a Poisson random measure $\mathtt{M}$ on $[0, \infty[ \times \mathcal{P}_\infty$ with intensity $dt \otimes \mu(d\pi)$, and for each $n \in \mathbb{N}$, let $\mathtt{M}_n$ be the image of $\mathtt{M}$ by the map $(t, \pi) \to (t, \pi_{|[n]})$. So $\mathtt{M}_n$ is a Poisson measure on $[0, \infty[ \times \mathcal{P}_n$ with intensity $dt \times \mu_n(d\pi))$, where $\mu_n$ denotes the measure on $\mathcal{P}_n$ obtained as the image of $\mu$ by the restriction map $\pi \to \pi_{|[n]}$. In particular, for every $\pi \in \mathcal{P}_n \backslash \{\mathbf{0}_{[n]}\}$, the process $(\mathtt{M}_n([0, t] \times \{\pi\}), t \geq 0)$ that measures the fiber based on $\pi$ as time passes, is Poisson with intensity $q_\pi = \mu\left(\mathcal{P}_{\infty,\pi}\right)$ and to different partitions in $\mathcal{P}_n \backslash \{\mathbf{0}_{[n]}\}$ correspond independent processes.

We denote by $\left\{(t_i, \pi^{(i)}), i \in \mathbb{N}\right\}$ the family of the atoms of $\mathtt{M}_n$ on $[0, \infty[ \times \left(\mathcal{P}_n \backslash \{\mathbf{0}_{[n]}\}\right)$, ranked in increasing order of their first coordinate. We set $\Pi^{[n]}(t) = \mathbf{0}_{[n]}$ for $t \in [0, t_1[$, and then define recursively

$$\Pi^{[n]}(t) = \mathrm{Coag}(\Pi^{[n]}(t_i-), \pi^{(n)}(t_i)), \qquad \text{for every } t \in [t_i, t_{i+1}[\,.$$

**Proposition 4.5** *In the notation above, for every $t \geq 0$, the sequence of random partitions $\left(\Pi^{[n]}(t), n \in \mathbb{N}\right)$ is compatible. If we denote by $\Pi(t)$ the unique partition of $\mathcal{P}_\infty$ such that $\Pi_{|[n]}(t) = \Pi^{[n]}(t)$ for every $n \in \mathbb{N}$, then the process $\Pi = (\Pi(t), t \geq 0)$ is a standard exchangeable coalescent with coagulation rate $\mu$.*

*Proof* Fix $n \geq 2$ and write $(t_1, \pi^{(1)})$ for the first atom of $\mathtt{M}_n$ on $[0, \infty[\times \left(\mathcal{P}_n \backslash \{\mathbf{0}_{[n]}\}\right)$. Plainly, $\Pi^{[n-1]}(t) = \Pi^{[n]}_{|[n-1]}(t)$ for every $t \in [0, t_1[$.

Consider first the case when $\pi^{(1)}_{|[n-1]} \neq \mathbf{0}_{[n-1]}$. Then $(t_1, \pi^{(1)}_{|[n-1]})$ is the first atom of $\mathtt{M}_{n-1}$ on $[0, \infty[\times \left(\mathcal{P}_{n-1} \backslash \{\mathbf{0}_{[n-1]}\}\right)$, and it follows from (4.4) that $\Pi^{[n-1]}(t) = \Pi^{[n]}_{|[n-1]}(t)$ for every $t \in [t_1, t_2[$. Next, consider the case $\pi^{(1)}_{|[n-1]} = \mathbf{0}_{[n-1]}$. Then $\mathtt{M}_{n-1}$ has no atoms on $[0, t_2[\times \left(\mathcal{P}_{n-1} \backslash \{\mathbf{0}_{[n-1]}\}\right)$, and it follows again from (4.4) that $\Pi^{[n-1]}(t) = \Pi^{[n]}_{|[n-1]}(t) = \mathbf{0}_{[n-1]}$ for every $t \in [0, t_2[$. By iteration, this shows that the restriction of $\Pi^{[n]}$ to $[n-1]$ coincides with $\Pi^{[n-1]}$.

It is immediate from the Poissonian construction that each $\Pi^{[n]}$ is a continuous time Markov chain on $\mathcal{P}_n$. More precisely, for every $\pi \in \mathcal{P}_n$ with $\#\pi = k$ non-empty blocks and every $\pi' \in \mathcal{P}_k \backslash \{\mathbf{0}_{[k]}\}$, the jump rate of $\Pi^{[n]}$ from $\pi$ to $\mathrm{Coag}(\pi, \pi')$ is given by the intensity of the Poisson process $\mathtt{M}_n([0, \cdot] \times \{\pi'\})$, and we know that the latter equals $q_{\pi'} = \mu\left(\mathcal{P}_{\infty, \pi'}\right)$. Plainly, all the other jump rates are 0, and by an application of Lemma 4.3, we have thus that $\Pi^{[n]}$ is an exchangeable coalescent.

It should now be plain that the process $\Pi$ which is specified by its restrictions $\Pi_{|[n]} = \Pi^{[n]}$, is a standard exchangeable coalescent with coagulation rate $\mu$. □

### 4.2.4 Characterization of coagulation rates

Our next goal is to characterize explicitly the coagulation rates of exchangeable coalescents, that is the measures on $\mathcal{P}_\infty$ which fulfill (4.7) and (4.8). In this direction, we first give two fundamental examples.

First, for every pair $(i, j) \in \mathbb{N}^2$ with $i < j$, we write $\mathtt{K}(i, j)$ for the partition of $\mathbb{N}$ whose blocks consist of the pair $\{i, j\}$ and the singletons $\{k\}$ for $k \neq i, j$. If $\delta_\pi$ stands for the Dirac point mass at $\pi \in \mathcal{P}_\infty$, then the measure

$$\mu^{\mathtt{K}} := \sum_{1 \leq i < j < \infty} \delta_{\mathtt{K}(i,j)}$$

fulfills conditions (4.7) and (4.8). It is immediately seen from the Poissonian construction that $\mu^{\mathtt{K}}$ is the coagulation rate of Kingman's coalescent.

To construct the second example, recall that $\varrho_{\mathbf{s}}$ denotes the distribution of a paint-box based on $\mathbf{s} \in \mathcal{P}_{\mathrm{m}}$. Next, we consider mixtures of paint-boxes. Recall from Proposition 2.9 that the map $s \to \varrho_{\mathbf{s}}$ is continuous, and that we denote by $\mathbf{0} = (0, \ldots)$ the mass-partition identical to 0. Consider a sigma-finite measure $\nu$ on $\mathcal{P}_{\mathrm{m}}$ such that

$$\nu(\{\mathbf{0}\}) = 0 \quad \text{and} \quad \int_{\mathcal{P}_{\mathrm{m}}} \left( \sum_{i=1}^{\infty} s_i^2 \right) \nu(d\mathbf{s}) < \infty \,, \tag{4.9}$$

and define a measure on $\mathcal{P}_{\infty}$ by

$$\varrho_{\nu}(d\pi) = \int_{\mathbf{s} \in \mathcal{P}_{\mathrm{m}}} \varrho_{\mathbf{s}}(d\pi)\nu(d\mathbf{s}) \,.$$

**Lemma 4.5** *The measure $\varrho_{\nu}$ is the coagulation rate of some exchangeable coalescent.*

*Proof* Each $\varrho_{\mathbf{s}}$ is an exchangeable probability measure on $\mathcal{P}_{\infty}$. Exchangeability is preserved by mixing, so $\varrho_{\nu}$ is an exchangeable measure on $\mathcal{P}_{\infty}$. For all $\mathbf{s} \in \mathcal{P}_{\mathrm{m}} \backslash \{\mathbf{0}\}$, the measures $\varrho_{\mathbf{s}}$ assign zero mass to the partition into singletons $\mathbf{0}_{[\infty]}$, so (4.8) holds for the mixture $\varrho_{\nu}$.

Next, for $\mathbf{s} \in \mathcal{P}_{\mathrm{m}}$, we see from the paint-box construction that

$$\varrho_{\mathbf{s}}(\pi_{|[2]} \neq \mathbf{0}_{[2]}) = \sum_{i=1}^{\infty} s_i^2 \,.$$

Thus (4.9) ensures that condition (4.7) holds for $n = 2$ and $\mu = \varrho_{\nu}$. By exchangeability of $\varrho_{\nu}$, we conclude that in fact (4.7) holds for any $n \in \mathbb{N}$. This establishes our statement, thanks to Lemma 4.4. □

We now state the main result of this section, which claims that every coagulation rate can be obtained as a linear combination of the two preceding examples.

**Theorem 4.2** *Consider a measure $\mu$ on $\mathcal{P}_{\infty}$ which fulfills (4.7) and (4.8). Then there exists a unique $\mathrm{c} \geq 0$ and a unique measure $\nu$ on $\mathcal{P}_{\mathrm{m}}$ that fulfills (4.9), such that*

$$\mu = \mathrm{c}\mu^{\mathrm{K}} + \varrho_{\nu}.$$

*Specifically, the following holds:*

(i) *For $\mu$-almost every $\pi \in \mathcal{P}_{\infty}$, $\pi$ possesses asymptotic frequencies $|\pi|$.*
(ii) *The measure on $\mathcal{P}_{\mathrm{m}}$ given by*

$$\nu(d\mathbf{s}) = \mathbb{1}_{\{s \neq \mathbf{0}\}} \mu(|\pi|^{\downarrow} \in d\mathbf{s})$$

*fulfills (4.9), and there is the identity*

$$\mathbb{1}_{\{|\pi|^{\downarrow}\neq \mathbf{0}\}}\mu(d\pi) = \varrho_{\nu}(d\pi)\,, \qquad \pi \in \mathcal{P}_{\infty}.$$

(iii) *There exists a real number* $\mathtt{c} \geq 0$ *such that*

$$\mathbb{1}_{\{|\pi|^{\downarrow}=\mathbf{0}\}}\mu(d\pi) = \mathtt{c}\mu^{\mathrm{K}}(d\pi)\,.$$

Later in the text, we shall refer to $\mathtt{c}$ as the *coefficient of binary coagulation* and to $\nu$ as the *measure of multiple coagulations* of the exchangeable coalescent $\Pi$. The description of the coagulation rate of an exchangeable coalescent in terms of $\mathtt{c}$ and $\nu$ is a close relative to that of the splitting rate in Theorem 3.2.

*Proof* The proof is a slight modification of that of Theorem 3.2; essentially the only significant difference is that the role of the partition $\mathbf{0}_{\mathbb{N}}$ into singletons (which is the neutral element for the coagulation operator) here replaces that of the trivial partition $\mathbf{1}_{\mathbb{N}}$ (which can be viewed as the neutral element for the fragmentation operator) in the setting of homogeneous fragmentations. We shall nonetheless provide details for the sake of completeness.

(i) For every $n \in \mathbb{N}$, introduce the measure

$$\mu_n(d\pi) = \mathbb{1}_{\{\pi_{|[n]}\neq \mathbf{0}_{[n]}\}}\mu(d\pi)\,, \qquad \pi \in \mathcal{P}_{\infty}\,.$$

Then $\mu_n$ is a finite measure on $\mathcal{P}_{\infty}$ which is invariant by the action of permutations that coincide with the identity on $[n]$. Let $\vec{\mu}_n$ be the image of $\mu_n$ by the $n$-shift on partitions, namely the map $\pi \to \vec{\pi}$ defined by

$$i \overset{\vec{\pi}}{\sim} j \iff i+n \overset{\pi}{\sim} j+n\,, \qquad i, j \in \mathbb{N}.$$

Then $\vec{\mu}_n$ is a finite exchangeable measure on $\mathcal{P}_{\infty}$, and by Kingman's Theorem 2.1, $\vec{\mu}_n$ can be expressed as a mixture of paint-boxes:

$$\vec{\mu}_n(d\pi) = \int_{\mathcal{P}_{\mathrm{m}}} \varrho_{\mathrm{s}}(d\pi)\vec{\mu}_n(|\pi|^{\downarrow} \in ds)\,. \tag{4.10}$$

As shift does not affect asymptotic frequencies, $\mu_n$-almost every partition has asymptotic frequencies. Since $\mu$ can be obtained as the increasing limit of the $\mu_n$, this establishes the first claim.

(ii) Let us write $\{i \sim j\}$ for the event that two integers $i$ and $j$ belong to the same block of $\pi$. By (4.10), we have for every $\mathbf{s} \in \mathcal{P}_{\mathrm{m}}$ that

$$\mu_n(n+1 \sim n+2 \mid |\pi|^{\downarrow} = s) = \sum_{k=1}^{\infty} s_k^2\,.$$

Hence, if we denote by $\nu_n$ the restriction to $\mathcal{P}_m\backslash\{\mathbf{0}\}$ of the image measure of $\mu_n$ by the map $\pi \to |\pi|^{\downarrow}$, then

$$\mu_n(n+1 \sim n+2) \geq \int_{\mathcal{P}_m} \left(\sum_{i=1}^{\infty} s_i^2\right) \nu_n(d\mathbf{s}).$$

On the one hand, the finite measure $\nu_n$ increases as $n \uparrow \infty$ to the measure $\nu$ defined in the statement, so

$$\lim_{n\to\infty} \int_{\mathcal{P}_m} \left(\sum_{i=1}^{\infty} s_i^2\right) \nu_n(d\mathbf{s}) = \int_{\mathcal{P}_m} \left(\sum_{i=1}^{\infty} s_i^2\right) \nu(d\mathbf{s}).$$

On the other hand, by the exchangeability of $\mu$,

$$\mu_n(n+1 \sim n+2) \leq \mu(n+1 \sim n+2) = \mu(1 \sim 2) < \infty,$$

which shows that $\nu$ fulfills (4.9).

Finally, fix $k \in \mathbb{N}$ and pick any $\pi^{[k]} \in \mathcal{P}_k\backslash\{\mathbf{0}_{[k]}\}$. We have by monotone convergence

$$\mu(\pi_{|[k]} = \pi^{[k]}, |\pi|^{\downarrow} \neq \mathbf{0}) = \lim_{n\to\infty} \mu(\pi_{|[k]} = \pi^{[k]}, |\pi|^{\downarrow} \neq \mathbf{0}, \pi_{|\{k+1,\dots,k+n\}} \neq \mathbf{0}_{\{k+1,\dots,k+n\}}).$$

In the notation introduced in (i), we see from an obvious permutation that

$$\mu(\pi_{|[k]} = \pi^{[k]}, |\pi|^{\downarrow} \neq \mathbf{0}, \pi_{|\{k+1,\dots,k+n\}} \neq \mathbf{0}_{\{k+1,\dots,k+n\}}) = \vec{\mu}_n(\pi_{|[k]} = \pi^{[k]}, |\pi|^{\downarrow} \neq \mathbf{0}).$$

Applying (4.10) and then letting $n$ tend to $\infty$, we conclude that

$$\mu(\pi_{|[k]} = \pi^{[k]}, |\pi|^{\downarrow} \neq \mathbf{0}) = \int_{\mathcal{P}_m} \varrho_{\mathbf{s}}(\pi_{|[k]} = \pi^{[k]}) \nu(d\mathbf{s}).$$

As $k$ is arbitrary, this establishes (ii).

(iii) Consider $\tilde{\mu}$, the restriction of $\mu$ to the event $\{1 \sim 2, |\pi|^{\downarrow} = \mathbf{0}\}$, which has finite mass. Its image by the 2-shift as defined in (i) for $n = 2$, is an exchangeable finite measure on $\mathcal{P}_\infty$ for which almost every partition has asymptotic frequencies $\mathbf{0}$. Thus $\tilde{\mu}$ is proportional to the Dirac mass at $\mathbf{0}_{\mathbb{N}}$, the partition of $\mathbb{N}$ into singletons. By exchangeability of $\mu$ and the finiteness of $\tilde{\mu}$, the set of partitions $\pi$ for which there is some $n \geq 3$ in the same block as 1 and 2 must have zero $\tilde{\mu}$ measure. Thus $\tilde{\mu} = \mathtt{c}\mu^{\mathrm{K}}(1, 2)$ for some $\mathtt{c} \geq 0$, and by exchangeability, we conclude that $\mathbb{1}_{\{|\pi|^{\downarrow}=\mathbf{0}\}}\mu(d\pi) = \mathtt{c}\mu^{\mathrm{K}}(d\pi)$. □

**Remark.** One can compute the jump rates $q_\pi$ of the restricted chains $\Pi_{|[n]}$ explicitly in terms of the rate of binary coagulation $\mathtt{c}$ and the measure of multiple coagulation $\nu$; however, the formulas than can be obtained are rather involved in this general setting (see Schweinsberg [199]). Nonetheless, in the special case of simple coalescents discussed in the forthcoming Section 4.4, expressions for the jump rates become more tractable, see (4.12) below.

## 4.3 Exchangeable mass-coalescents

In this section, we consider the process of the ranked asymptotic frequencies associated with a standard exchangeable coalescent $\Pi$.

### 4.3.1 Markov property

We start by introducing the operator of coagulation of mass-partitions by a partition of $\mathbb{N}$. Dust plays a special role in this definition, as microscopic particles of dust can coagulate and form macroscopic masses. In other words, the mass of dust of the coagulation of some mass-partition **s** may be strictly smaller than the initial mass of dust of **s**. Recall that $\tilde{\mathcal{P}}_{\mathrm{m}}$ stands for the space of numerical sequences whose decreasing rearrangement is a mass-partition.

**Definition 4.4** *Let* $\mathbf{s} \in \tilde{\mathcal{P}}_{\mathrm{m}}$ *and* $\pi \in \mathcal{P}_\infty$ *a partition of* $\mathbb{N}$ *which possesses asymptotic frequencies. We write* $\mathrm{Coag}(s, \pi)$ *for the mass-partition obtained by the decreasing rearrangement of the terms of the sequence*

$$s_0|\pi_i| + \sum_{j\in\pi_i} s_j\,, \qquad i \in \mathbb{N}\,,$$

*where* $s_0 = 1 - \sum_{j=1}^{\infty} s_j$ *denotes the mass of dust for* **s**.

The following elementary lemma makes the connection with Definition 4.2.

**Lemma 4.6** *Let* $\pi \in \mathcal{P}_\infty$ *be a deterministic partition which possesses asymptotic frequencies, and such that the blocks of* $\pi$ *with zero asymptotic frequency are either empty or singletons. Write*

$$S = \bigcup_{i\in\mathbb{N}:\,|\pi_i|=0} \pi_i\,,$$

*for the set of singletons of* $\pi$. *So* $S$ *has an asymptotic frequency and* $|S| = 1 - \sum_{i=1}^{\infty} |\pi_i|$. *Let* $\pi'$ *be a random exchangeable partition. Then the random partition* $\pi'' = \mathrm{Coag}(\pi, \pi')$ *possesses asymptotic frequencies, and more precisely, there is the identity*

$$|\pi''|^{\downarrow} = \mathrm{Coag}(|\pi|, \pi')\,.$$

*Proof* Thanks to Kingman's Theorem 2.1, we may suppose that $\pi'$ is given by a paint-box based on some interval-partition $\vartheta \in \mathcal{P}_{\mathrm{I}}$ and an i.i.d. sequence of uniform variables, $U_1, \ldots$ Let $I$ be some interval component of $\vartheta$, and focus on the block $B' := \{i \in \mathbb{N} : U_i \in I\}$ of $\pi'$ and on the corresponding block of $\pi''$:

$$B'' := \bigcup_{i \in B'} \pi_i .$$

The points in $S$ are the singletons of $\pi$; we denote by $\sigma : S \to \mathbb{N}$ the injective map such that for $j \in S$, $\sigma(j)$ is the index of the block $\{j\}$ in $\pi$, that is $\{j\} = \pi_{\sigma(j)}$. We also introduce $J := \{j \in \mathbb{N}, |\pi_j| > 0\}$.

For every integer $n$ we have

$$\frac{1}{n}\#(B'' \cap [n]) = \frac{1}{n} \sum_{j \in S \cap [n]} \mathbb{1}_{\{U_{\sigma(j)} \in I\}} + \sum_{j \in J \cap B'} \frac{1}{n}\#(\pi_j \cap [n]) .$$

Since $(U_{\sigma(j)}, j \in S)$ form an i.i.d. sequence of uniform variables, we deduce from the law of large numbers and the assumption that $S$ has an asymptotic frequency that when $n$ tends to $\infty$, the first term in the sum in the right-handle side converges to $|S||I| = |S||B'|$. Since each block $\pi_j$ $(j \in J)$ has an asymptotic frequency, it follows from Fatou's lemma that

$$\liminf_{n\to\infty} \frac{1}{n}\#(B'' \cap [n]) \geq |S||B'| + \sum_{j \in J \cap B'} |\pi_j| = |S||B'| + \sum_{j \in B'} |\pi_j| .$$

Next, fix $\eta > 0$, consider the finite set $J(\eta) := \{j \in \mathbb{N}, |\pi_j| > \eta\}$, and introduce

$$B_\eta := \bigcup_{j \in J \setminus J(\eta)} \pi_j ,$$

so that $B_\eta$ is the complementary set of $S \cup \left(\bigcup_{j \in J(\eta)} \pi_j\right)$. It follows from our assumptions that $B_\eta$ has an asymptotic frequency which tends to 0 as $\eta \to 0$. On the other hand, there is the obvious inclusion

$$B'' \subseteq \{j \in S : U_{\sigma(j)} \in I\} \cup B_\eta \cup \left( \bigcup_{j \in B' \cap J(\eta)} \pi_j \right) ,$$

from which we deduce the lower-bound

$$\limsup_{n\to\infty} \frac{1}{n}\#(B'' \cap [n]) \leq |S||B'| + |B_\eta| + \sum_{j \in J(\eta) \cap B'} |\pi_j| .$$

Letting $\eta$ tend to 0, we conclude that the block $B''$ has an asymptotic frequency given by $|B''| = |S||B'| + \sum_{j \in B'} |\pi_j|$. □

The random partition $\Pi(t)$ given by the state of some exchangeable coalescent $\Pi$ evaluated at time $t \geq 0$ is exchangeable, and we denote by $|\Pi|^{\downarrow} = \left(|\Pi(t)|^{\downarrow}, t \geq 0\right)$ the process of the ranked asymptotic frequencies of $\Pi$. Here is a first basic property.

**Proposition 4.6** *The process of the ranked asymptotic frequencies $|\Pi|^{\downarrow}$ is Markovian and, more precisely, for every $t, t' \geq 0$, the conditional distribution of $|\Pi|^{\downarrow}(t+t')$ given $|\Pi|^{\downarrow}(t) = s$ is that of* $\text{Coag}(s, \Pi(t'))$. *Moreover, this semigroup fulfills the Feller property.*

*Proof* Let $(\mathcal{F}_t)_{t\geq 0}$ denote the natural filtration of $\Pi$, and $\Phi : \mathcal{P}_{\mathrm{m}} \to \mathbb{R}$ some continuous function. The Markov property of $\Pi$ implies that for every $t, t' \geq 0$, on the event $\Pi(t) = \pi$,

$$\mathbb{E}\left(\Phi(|\Pi(t+t')|^{\downarrow}) \mid \mathcal{F}_t\right) = \mathbb{E}\left(\Phi(|\text{Coag}(\pi, \Pi(t'))|^{\downarrow})\right).$$

We know from Proposition 2.8 that $\pi$ fulfills the requirements of Lemma 4.6 with probability one, and as $\Pi(t')$ is exchangeable, this implies that $|\Pi|^{\downarrow}$ is Markovian and has the semigroup given in the statement.

The Feller property follows readily from the continuity of the coagulation operator (cf. Lemma 4.2) and Proposition 2.9. □

### 4.3.2 Dust in exchangeable mass-coalescents

Recall from Theorem 4.2 that the coagulation rate $\mu$ can be expressed in terms of a coefficient $\mathsf{c} \geq 0$ of binary coagulation, and of a measure $\nu$ on $\mathcal{P}_{\mathrm{m}}$ that fulfills (4.9) and specifies the rate of multiple coagulations. We conclude this section by considering the question of whether the asymptotic frequencies of $\Pi(t)$ are proper or improper, and in the latter case, we characterize the distribution of the mass of dust in terms of $\mathsf{c}$ and $\nu$. Recall that $(\mathcal{F}_t)_{t\geq 0}$ stand for the natural filtration of $\Pi$.

**Proposition 4.7** *The random partition $\Pi(t)$ has improper asymptotic frequencies with positive probability if and only if*

$$\mathsf{c} = 0 \quad \text{and} \quad \int_{\mathcal{P}_{\mathrm{m}}} (1 - s_0)\nu(ds) < \infty, \qquad \text{where } s_0 = 1 - \sum_{i=1}^{\infty} s_i.$$

*In this case, if we denote by*

$$\mathrm{D}(t) = 1 - \sum_{i=1}^{\infty} |\Pi_i(t)|, \qquad t \geq 0,$$

*the mass of dust at time t, then the law of* $\mathtt{D}(t)$ *is specified by its entire moments*

$$\mathbb{E}\left(\mathtt{D}(t)^k\right) = \exp\left(-t\int_{\mathcal{P}_{\mathrm{m}}}(1-s_0^k)\nu(ds)\right), \qquad k \in \mathbb{N}.$$

*Moreover, the multiplicative increments of the process* $(\mathtt{D}(t), t \geq 0)$ *are independent and stationary, in the sense that for every* $t, t' \geq 0$*, the conditional distribution of* $\mathtt{D}(t+t')$ *given* $\mathcal{F}_t$ *is that of* $\mathtt{D}(t)\tilde{\mathtt{D}}(t')$*, where* $\tilde{\mathtt{D}}(t')$ *is independent of* $\mathcal{F}_t$ *and has the same law as* $\mathtt{D}(t')$*. This characterizes the finite-dimensional distributions of the process* $(\mathtt{D}(t), t \geq 0)$.

**Remark.** Another way of formulating Proposition 4.7 is to consider the right-continuous version $\xi = (\xi(t), t \geq 0)$ of the process $(-\ln \mathtt{D}(t), t \geq 0)$. Then $\xi$ is a subordinator, with Laplace exponent

$$\Phi(q) = \int_{\mathcal{P}_{\mathrm{m}}}(1-s_0^q)\nu(ds), \qquad q > 0.$$

This formula for $\Phi$ is of the Lévy-Khintchine type. In particular, the drift coefficient is zero, the Lévy measure is the image of the measure of multiple coagulations $\nu$ by the map $s \to -\ln s_0$, and the killing rate given by $\mathtt{k} = \nu(s_0 = 0)$. We refer to Section 3.3.1 for the terminology. It may be interesting to compare this result with Theorem 3.2 for exchangeable fragmentation.

*Proof* Recall that an exchangeable partition has proper asymptotic frequencies if and only if none of its blocks is a singleton; see Proposition 2.8. By exchangeability, we thus see that $\Pi(t)$ has proper asymptotic frequencies a.s. if and only if $\mathbb{P}(\Pi_1(t) = \{1\}) = 0$.

By the Poissonian construction of Section 4.2.3, the event $\Pi_1(t) = \{1\}$ occurs if and only if all the atoms $(r, \pi)$ of the Poisson random measure $\mathtt{M}$ with $r \leq t$ fulfill $\pi_1 = \{1\}$. Because $\mathtt{M}$ has intensity $dt \otimes \mu(d\pi)$, the latter event has probability $\exp(-at)$, where $a = \mu(\{\pi \in \mathcal{P}_\infty : \pi_1 \neq \{1\}\})$. An easy calculation based on the expression $\mu = \mathtt{c}\mu^{\mathrm{K}} + \varrho_\nu$ for the coagulation rate $\mu$ (cf. Theorem 4.2) shows that

$$a = \infty \mathbb{1}_{\{\mathtt{c}>0\}} + \int_{\mathcal{P}_{\mathrm{m}}}(1-s_0)\nu(ds),$$

which establishes our first claim.

From now on, we suppose that $\mathtt{c} = 0$ and $\int(1-s_0)\nu(ds) < \infty$. In order to compute the entire moments of $\mathtt{D}(t)$, we observe from the paint-box construction that for every $\mathbf{s} \in \mathcal{P}_m$ and $k \in \mathbb{N}$, we have

$$\varrho_{\mathbf{s}}(\{\pi \in \mathcal{P} : 1, \ldots, k \text{ are singletons of } \pi\}) = s_0^k,$$

where $\varrho_{\mathbf{s}}$ denotes the law of an exchangeable random partition with ranked asymptotic frequencies given by $\mathbf{s}$. The same argument as above shows that the event that $1, \ldots, k$ are singletons of $\Pi(t)$ occurs if and only if $1, \ldots, k$ are singletons of $\pi$ for all partitions $\pi$ such that $(r, \pi)$ is an atom of the Poisson random measure $\mathtt{M}$ for some $r \leq t$. By immediate Poissonian calculations, we get

$$\begin{aligned}
\mathbb{E}\left(\mathtt{D}(t)^k\right) &= \mathbb{P}(1, \ldots, k \text{ are singletons of } \Pi(t)) \\
&= \exp\left(-t\mu(\{\pi \in \mathcal{P} : \exists i \leq k \text{ which is not a singleton of } \pi\}\right)) \\
&= \exp\left(-t\int_{\mathcal{P}_m} (1 - s_0^k)\nu(d\mathbf{s})\right),
\end{aligned}$$

where the last equality stems from the expression $\mu = \varrho_\nu$ for the coagulation rate. Of course, $\mathtt{D}(t)$ takes its values in $[0, 1]$ and thus is determined by its entire moments.

Finally, we check that the multiplicative increments of $\mathtt{D}(\cdot)$ are independent and stationary. In this direction, consider $\mathbf{s} \in \mathcal{P}_m$ and $\pi$ a partition which possesses asymptotic frequencies. We write $\mathbf{s}' = \text{Coag}(\mathbf{s}, \pi)$, so that by Definition 4.4, the mass of dust $s_0' = 1 - \sum_{i=1}^{\infty} s_i'$ of $s'$ is given by

$$s_0' = 1 - \sum_{i=1}^{\infty}\left(s_0|\pi_i| + \sum_{j\in\pi_i} s_j\right) = 1 - \sum_{j=1}^{\infty} s_j - s_0\sum_{i=1}^{\infty}|\pi_i| = s_0\left(1 - \sum_{i=1}^{\infty}|\pi_i|\right).$$

Combining this observation with Proposition 4.6 completes the proof. □

## 4.4 Simple coalescents and flows of bridges

In this section, we will be interested in an important and natural sub-family of exchangeable coalescents, which was introduced independently by Pitman [183] and Sagitov [197], such that when a coagulation occurs, all the blocks involved in the coagulation merge as a single block.

**Definition 4.5** (i) *Call a partition $\pi \in \mathcal{P}_\infty$* **simple** *if and only if at most one of its blocks is neither empty nor reduced to a singleton.*
(ii) *A mass-partition $\mathbf{s} \in \mathcal{P}_m$ is called* **simple** *if and only if it is given by $\mathbf{s} = (x, 0, \ldots)$ for some $0 \leq x \leq 1$.*

(iii) *An exchangeable coalescent* $\Pi$ *is called* **simple** *if and only if its coagulation rate* $\mu$ *is supported by simple partitions.*

We may use the representation given in Theorem 4.2 of the coagulation rate of an exchangeable coalescent $\Pi$ in the form $\mu = \mathtt{c}\mu^{\mathrm{K}} + \varrho_\nu$ as the sum of a binary coagulation rate and a multiple coagulation rate with rate $\nu$, where $\mathtt{c} \geq 0$ and $\nu$ fulfills (4.9). It should be clear that $\Pi$ is simple if and only if $\nu$ is supported by simple mass-partitions.

Later in this text, it will be convenient to use the canonical projection $(x, 0, \ldots) \to x$ and identify the sub-space of simple mass-partitions with the unit interval, and then $\nu$ as a measure on $[0, 1]$ with

$$\nu(\{0\}) = 0 \quad \text{and} \quad \int_{[0,1]} x^2 \nu(dx) < \infty \,. \tag{4.11}$$

An alternative formulation is that $\Pi$ is a simple if and only if for every $n \in \mathbb{N}$ and every $\pi \in \mathcal{P}_n \backslash \{\mathbf{0}_{[n]}\}$, the jump rate $q_\pi$ of $\Pi_{|[n]}$ from $\mathbf{0}_{[n]}$ to $\pi$ equals zero, except when $\pi$ is simple. In that case, we may compute the jump rates explicitly in terms of the rates $\mathtt{c} \geq 0$ and $\nu$. Indeed, for every $2 \leq k \leq n$, if $\pi \in \mathcal{P}_n \backslash \{\mathbf{0}_{[n]}\}$ is simple and has one block with $k$ elements, then

$$q_\pi := q_{n,k} = \mathtt{c}\,\mathbb{1}_{\{k=2\}} + \int_{]0,1]} x^k (1-x)^{n-k} \nu(dx) \,. \tag{4.12}$$

We mention that Pitman [183] uses the finite measure $\Lambda(dx) = \mathtt{c}\delta_0 + x^2\nu(dx)$ on $[0, 1]$ to characterize the coagulation rates rather than $\mathtt{c}$ and $\nu$, and in this setting (4.12) becomes

$$q_\pi := q_{n,k} = \int_{[0,1]} x^{k-2} (1-x)^{n-k} \Lambda(dx) \,.$$

The purpose of this section is to show that simple coalescents can be encoded by certain stochastic flows. An interesting coalescent process, introduced by Bolthausen and Sznitman [61], arises naturally in this setting. The presentation follows closely [44].

### 4.4.1 Compositions of simple bridges

In this section, it is convenient to think of a mass-partition $\mathbf{s} \in \mathcal{P}_{\mathrm{m}}$ as the ranked sequence of the masses assigned to atoms by some random probability measure. More precisely, we work on the unit interval $[0, 1]$, throw the atoms at random according to the uniform distribution, independently of each other, and let the possible dust be uniformly spread. In other words, let us introduce a sequence $V_1, \ldots$ of i.i.d. uniformly distributed variables on $[0, 1]$ and consider the random probability measure

$$db_{\mathbf{s}}(x) = s_0 dx + \sum_{i=1}^{\infty} s_i \delta_{V_i}(dx), \qquad x \in [0, 1],$$

where $\delta_a$ stands for the Dirac point mass at $a$. The distribution function $b_{\mathbf{s}}$ is a right-continuous increasing function on $[0, 1]$ with $b_{\mathbf{s}}(0) = 0$ and $b_{\mathbf{s}}(1) = 1$; the ranked sequence of jump sizes of $b_{\mathbf{s}}$ is given by $\mathbf{s}$, and the jump locations by the i.i.d. uniform variables.

We record this in the following definition.

**Definition 4.6** *Let $\mathbf{s} \in \mathcal{P}_{\mathrm{m}}$ be fixed, and consider a sequence $V_1, \ldots$ of i.i.d. uniformly distributed variables on $[0, 1]$. We call* **s-bridge** *any process distributed as*

$$b_{\mathbf{s}}(x) := s_0 x + \sum_{i=1}^{\infty} s_i 1\!\!1_{\{V_i \le x\}}, \qquad x \in [0, 1],$$

*where $s_0 = 1 - \sum_{i=1}^{\infty} s_i$. We refer to $\mathbf{s}$ as the jump-sizes of $b_{\mathbf{s}}$. Finally, a process distributed as a mixture of $\mathbf{s}$-bridge (i.e. when one randomizes the mass-partition) is just called a bridge.*

**Example** It can immediately be checked that if $\varsigma$ is a subordinator on the time-interval $[0, 1]$, then both the normalized process $\varsigma(\cdot)/\varsigma(1)$, and the conditioned process $\varsigma(\cdot)$ given $\varsigma(1) = 1$ (cf. Sections 2.2.2–4), are bridges in the sense of Definition 4.6.

Plainly, $b_{\mathbf{s}}(0) = 0$, $b_{\mathbf{s}}(1) = 1$, and the random measure $db_{\mathbf{s}}(x)$ is exchangeable, in the sense that its image by any injective map $[0, 1] \to [0, 1]$ that preserves the Lebesgue measure has the same distribution as $db_{\mathbf{s}}(x)$. It follows in particular that the bridge $b_{\mathbf{s}}$ has exchangeable increments, that is for every $n \in \mathbb{N}$, the law of the $n$-tuple

$$(b_{\mathbf{s}}(1/n), b_{\mathbf{s}}(2/n) - b_{\mathbf{s}}(1/n), \ldots, b_{\mathbf{s}}(1) - b_{\mathbf{s}}((n-1)/n))$$

is invariant under permutation. We refer to Kallenberg [133] for fundamental properties of bridges with exchangeable increments.

The space of probability distributions on $[0, 1]$ is endowed with the topology of weak convergence of measures, which is equivalent, for example, to the topology induced by the $L^2(dx)$-distance for the distribution functions, and this space is compact. The following useful result of continuity in distribution completes Proposition 2.9.

**Lemma 4.7** *Consider for each $n \in \overline{\mathbb{N}}$ a random variable $S^{(n)}$ with values in $\mathcal{P}_m$. Let $b^{(n)}$ be a bridge with jump-sizes $S^{(n)}$. The following conditions are equivalent:*

(i) *When $n \to \infty$, $S^{(n)}$ converges in distribution to $S^{(\infty)}$.*
(ii) *When $n \to \infty$, $b^{(n)}$ converges in distribution to $b^{(\infty)}$.*

*Proof* Suppose (i). By the same argument as in the proof of Proposition 2.9 based on Skorokhod's representation theorem, we may and will assume that $\lim_{\to\infty} S^{(n)} = S^{(\infty)}$ a.s., which enables us to focus on the case when $S^{(n)} = \mathbf{s}^{(n)}$ is a deterministic mass-partition for all $n \in \overline{\mathbb{N}}$.

Recall that $V_1, \ldots$ is a sequence of i.i.d. uniform variables; we take for every $n \in \overline{\mathbb{N}}$

$$b^{(n)}(x) := s_0^{(n)} x + \sum_{i=1}^{\infty} s_i^{(n)} \mathbb{1}_{\{V_i \leq x\}}, \qquad x \in [0, 1].$$

Write $\mathcal{R}_x$ for the sigma-field generated by $\left(V_i \mathbb{1}_{\{V_i \geq x\}}, i \in \mathbb{N}\right)$, so $(\mathcal{R}_x)_{0 \leq x \leq 1}$ is a reversed filtration, and for every $0 \leq x \leq 1$ and $n \in \overline{\mathbb{N}}$, the variable $b^{(n)}(x-)$ is $\mathcal{R}_x$-measurable.

It is easily seen that for each $i \in \mathbb{N}$, the processes

$$x \to x^{-1} \mathbb{1}_{\{V_i < x\}}$$

are $(\mathcal{R}_x)_{0 \leq x \leq 1}$-reversed martingales. By linearity, we deduce from the expression above of the bridges $b^{(n)}$ that the processes

$$\left(x^{-1} b^{(n)}(x-), 0 < x \leq 1\right)$$

are $(\mathcal{R}_x)_{0 \leq x \leq 1}$-reversed martingales with bounded variation. Observe that the reversed martingale $x \to x^{-1}(b^{(\infty)}(x-) - b^{(n)}(x-)$ equals 0 for $x = 1$ and has a jump of size $V_i^{-1}(s_i^{(\infty)} - s_i^{(n)})$ at $V_i$. It follows that for every $x \in ]0, 1]$

$$\mathbb{E}(|b^{(\infty)}(x) - b^{(n)}(x)|^2) \leq x^{-2}(1-x) \sum_{i=1}^{\infty} (s_i^{(\infty)} - s_i^{(n)})^2.$$

Since $\lim_{n\to\infty}(s_i^{(\infty)} - s_i^{(n)}) = 0$ and $|s_i^{(\infty)} - s_i^{(n)}| \leq 1/i$, it follows by dominated convergence that $b^{(n)}(x)$ converges in $L^2(\mathbb{P})$ to $b^{(\infty)}(x)$, and thus (ii) holds.

Now suppose (ii). By compactness of the space of mass-partitions, from any sub-sequence of $(S^{(n)}, n \in \mathbb{N})$ we can extract a sub-sequence that converges in law, say with limit distributed as some random mass-partition $\tilde{S}$. By the first part of the proof, the bridge $b^{(\infty)}$ is distributed as a bridge with jump-sizes $\tilde{S}$, and it follows that $\tilde{S}$ has the same law as $S^{(\infty)}$. Thus $S^{(n)}$ converges in law to $S^{(\infty)}$ as $n \to \infty$. □

We will use the notation

$$b_{\mathbf{s}}^{-1}(r) := \inf \{v \in [0,1] : b_{\mathbf{s}}(v) > r\} , \qquad r \in ]0,1[ ,$$

for the right-continuous inverse of a bridge $b_{\mathbf{s}}$.

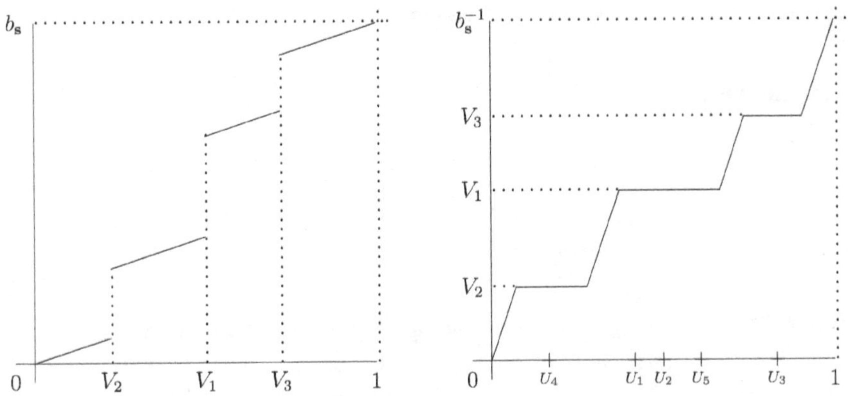


s-bridge $b_{\mathbf{s}}$, its inverse $b_{\mathbf{s}}^{-1}$ and the associated paint-box $\pi = (\{1,2,5\}, \{3\}, \{4\})$

Observe that when $U$ is an independent uniform variable, the conditional law of $b_{\mathbf{s}}^{-1}(U)$ given $b_{\mathbf{s}}$ is $db_{\mathbf{s}}$. Kingman's paint-box construction can be conveniently rephrased using the inverse of bridges. Specifically, introduce the usual sequence $U_1, \ldots$ of i.i.d. uniform variables, which we assume to be independent of the bridge $b_{\mathbf{s}}$, and consider the equivalence relation

$$i \overset{\pi}{\sim} j \iff b_{\mathbf{s}}^{-1}(U_i) = b_{\mathbf{s}}^{-1}(U_j) .$$

As conditionally on the bridge $b_{\mathbf{s}}$, $b_{\mathbf{s}}^{-1}(U_1), \ldots$ is a sequence of i.i.d. variables with distribution $b_{\mathbf{s}}$, $\pi$ is clearly a paint-box with ranked asymptotic frequencies $\mathbf{s}$.

In the case of a simple mass-partition $\mathbf{s} = (x, 0, \ldots)$, it is convenient to use the notation $b_x = b_{(x,0,\ldots)}$, so that

$$b_x(r) = (1-x)r + x\mathbb{1}_{\{V \le r\}} , \qquad r \in [0,1] , \tag{4.13}$$

where $V$ is a uniform random variable on $[0, 1]$. We then call $b_x$ a *simple bridge*. Plainly, the paint-box $\pi$ constructed above is simple a.s. if and only if the bridge $b$ is simple.

In this section, we shall establish a few elementary properties of simple bridges which are easy to check and will suffices for our purposes; we mention however, that they remain valid in fact for arbitrary bridges (cf. [44]). Suppose now that $x < 1$; the continuous inverse of $b_x$ is thus given by

$$b_x^{-1}(r) = \begin{cases} r/(1-x) & \text{if } r \in [0, (1-x)V], \\ V & \text{if } r \in ](1-x)V, x+(1-x)V[, \\ (r-x)/(1-x) & \text{if } r \in [(1-x)V+x, 1]. \end{cases} \tag{4.14}$$

We make the following key observation.

**Lemma 4.8** *Define for every $i \in \mathbb{N}$*

$$U_i' := b_x^{-1}(U_j), \qquad j \in \pi_i.$$

*Then $U_1', \ldots$ form a sequence of i.i.d. uniform variables which is independent of the exchangeable partition $\pi$.*

*Proof* The proof relies on the following elementary observation. Fix $n \in \mathbb{N}$, pick $k \in \{0, \ldots, n\}$, and let $n_1 < \cdots < n_k$ be $k$ distinct integers in $[n]$. Work conditionally on $V = v \in ]0, 1[$. The probability of the event that $U_{n_j} \notin ](1-x)v, x+(1-x)v[$ for every $j = 1, \ldots, k$, that is the singletons $\{n_1\}, \ldots, \{n_k\}$ are blocks of $\pi$, equals $(1-x)^k$. Moreover, conditionally on this event, the variables $U_{n_1}, \ldots, U_{n_k}$ are $k$ independent variables which are uniformly distributed on $[0, (1-x)v] \cup [(1-x)v+x, 1]$. It follows that $b_x^{-1}(U_{n_1}), \ldots, b_x^{-1}(U_{n_k})$ are $k$ independent uniform variables.

Suppose first that $k \le n-2$ and consider the simple partition $\gamma \in \mathcal{P}_n$ such that $\{n_1\}, \ldots, \{n_k\}$ are exactly the singleton-blocks of $\gamma$. It follows immediately from the above observation that

$$\mathbb{P}\left(\pi_{|[n]} = \gamma, U_1' \in du_1, \ldots, U_{k+1}' \in du_{k+1}\right) = x^{n-k}(1-x)^k du_1 \cdots du_{k+1}.$$

Finally, consider the event that $\pi_{|[n]} = \mathbf{0}_{[n]}$ is the partition into singletons, which means that there is at most one index $i \in [n]$ such that $U_i \in ](1-x)V, x+(1-x)V[$. Apply the elementary observation above in the cases $k = n-1$ and $k = n$. We see that

$$\begin{aligned} &\mathbb{P}\left(\pi_{|[n]} = \mathbf{0}_{[n]}, U_1' \in du_1, \ldots, U_n' \in du_n\right) \\ &\quad = (n-1)x(1-x)^{n-1} du_1 \cdots du_n + (1-x)^n du_1 \cdots du_n, \end{aligned}$$

which completes the proof. □

Lemma 4.8 suggests the use the i.i.d. uniform variables $U'_1, \ldots$ as the basis for an independent paint-box. More precisely, let $x' \in ]0,1[$, and consider an independent simple bridge $b_{x'}$, that is

$$b_{x'}(r) = (1-x')r + x'\mathbb{1}_{\{V' \leq r\}}, \qquad r \in [0,1],$$

where $V'$ is a uniform variable which is independent of $V$. We write $\pi'$ for the paint-box defined from the inverse bridge $b_{x'}^{-1}$ and the variables $U'_i$ as above, so that

$$i \overset{\pi'}{\sim} j \iff b_{x'}^{-1}(U'_i) = b_{x'}^{-1}(U'_j).$$

Note from Lemma 4.8 that the random partitions $\pi$ and $\pi'$ are independent.

**Corollary 4.2** *Consider the composition of bridges $b^\circ := b_x \circ b_{x'}$ and its continuous inverse $b^{\circ -1} = b_{x'}^{-1} \circ b_x^{-1}$.*

(i) *The random partition $\pi^\circ$ defined by*

$$i \overset{\pi^\circ}{\sim} j \iff b^{\circ -1}(U_i) = b^{\circ -1}(U_j),$$

*coincides with the coagulation* $\mathrm{Coag}(\pi, \pi')$ *of $\pi$ by $\pi'$.*

(ii) *If we define*

$$U_i^\circ = b^{\circ -1}(U_j), \qquad j \in \pi_i^\circ,$$

*then $U_1^\circ, \ldots$ is a sequence of i.i.d. variables which is independent of $\pi$, $\pi'$, and a fortiori of $\pi^\circ$.*

(iii) *$b^\circ$ is a bridge in the sense of Definition 4.6. More precisely, $b^\circ$ is an $\mathbf{s}$-bridge, where $\mathbf{s}$ stands for the ranked sequence of the asymptotic frequencies of $\pi^\circ$.*

*Proof* (i) By definition, $i$ and $j$ belong to the same block of $\pi^\circ$ if and only if

$$b_{x'}^{-1} \circ b_x^{-1}(U_i) = b_{x'}^{-1} \circ b_x^{-1}(U_j).$$

Let $k$ and $\ell$ be the respective indices of the blocks of $\pi$ which contain $i$ and $j$, that is $i \in \pi_k$ and $j \in \pi_\ell$. Then $b_x^{-1}(U_i) = U'_k$ and $b_x^{-1}(U_j) = U'_\ell$, and we see that $i$ and $j$ belong to the same block of $\pi^\circ$ if and only if $b_{x'}^{-1}(U'_k) = b_{x'}^{-1}(U'_\ell)$, that is if and only if $k$ and $\ell$ belong to the same block of $\pi'$. This shows that $\pi^\circ = \mathrm{Coag}(\pi, \pi')$.

(ii) Recall from Lemma 4.8 that $\pi$ is independent of the variables $U'_i$ and $b_{x'}$. On the other hand, observe that for every $i \in \mathbb{N}$, there is the identity

$$U_i^\circ = b_{x'}^{-1}(U'_j), \qquad j \in \pi'_i.$$

Another application of Lemma 4.8 now shows that the $U_i^\circ$ form an i.i.d. sequence of uniform variables which is independent of $\pi$ and $\pi'$.

(iii) The jump locations of $b^\circ$ are necessarily of the type $U_i^\circ$ for some $i \in \mathbb{N}$. More precisely, $U_i^\circ$ is the location of a jump of $b^\circ$ if and only if the $i$-th block $\pi_i^\circ$ is not reduced to a singleton and, in this case, the size of the jump coincides with the asymptotic frequency $|\pi_i^\circ|$ of $\pi_i^\circ$. As we know from (ii) that the variables $U_i^\circ$ are independent of $\pi^\circ$, this shows that $b^\circ$ is an $\mathbf{s}$-bridge with $\mathbf{s} = |\pi^\circ|^\downarrow$. □

We stress that the argument used for proving Corollary 4.2 can be iterated to deal with compositions of finitely many independent simple bridges. More precisely, if $b_{x_1}, \dots, b_{x_n}$ are independent simple bridges, and if we set $b := b_{x_1} \circ b_{x_2} \circ \cdots \circ b_{x_n}$, then Corollary 4.2 still holds when we replace $b_x$ by $b$. We shall often use this straightforward extension of Corollary 4.2 later in the text.

### 4.4.2 Flows of bridges and coagulation

We now apply the observations of the preceding section to dwell on a natural representation of simple exchangeable coalescents in terms of flows of bridges. We first consider the elementary case when the coefficient of binary coagulation is $\mathtt{c} = 0$ and the measure of multiple coagulations $\nu$ is a finite measure on $]0, 1[$. The Poissonian construction of Section 4.2.3 leads us to introduce a Poisson random measure on $\mathbb{R} \times ]0, 1[$ with intensity $dt \otimes \nu(dx)$. Next, conditionally on the Poisson measure, we associate to each atom $(t, x)$ a bridge, denoted by $b^{(t)}$, which is distributed as $b_x$ and such that to different atoms correspond independent bridges. The assumption that $\nu$ is finite ensures that for every $t < t'$, there are only finitely many atoms on $]t, t'] \times ]0, 1[$. More precisely, we write $(t_1, x_1), \dots, (t_N, x_N)$ for these atoms, where $N$ is a Poisson variable with intensity $(t' - t)\nu(]0, 1[)$ and

$$t < t_1 < \cdots < t_N \le t' .$$

This enables us to define a random function $B_{t,t'} : [0, 1] \to [0, 1]$ by

$$B_{t,t'} = b^{(t_1)} \circ \cdots \circ b^{(t_N)} .$$

By conditioning on the number $N$ of atoms of the Poisson measure on $]t, t'] \times ]0, 1]$ and by applying the extension of Corollary 4.2, we see that $B_{t,t'}$ is a bridge in the sense of Definition 4.6.

Let $U_1, \ldots$ be a sequence of i.i.d. uniform variables which is independent of the family $(B_{t,t'}, -\infty < t \le t' < \infty)$. For every $t \ge 0$, define an exchangeable random partition $\Pi(t)$ by

$$i \overset{\Pi(t)}{\sim} j \iff B_{0,t}^{-1}(U_i) = B_{0,t}^{-1}(U_j).$$

**Lemma 4.9** *The process $(\Pi(t), t \ge 0)$ constructed above is a standard simple exchangeable coalescent with coagulation rate $\mu = \varrho_\nu$.*

*Proof* This follows immediately from the Poissonian construction of an exchangeable coalescent (cf. Proposition 4.5) and Corollary 4.2. □

Lemma 4.9 provides a nice representation of simple exchangeable coalescents in terms of the composition of simple bridges, but this representation only concerns the rather elementary case when the coagulation rate is finite. We shall now see that this restriction can be removed using an approximation scheme. Specifically, let $\mathtt{c} \ge 0$ and $\nu$ be an arbitrary measure on $\mathcal{P}_{\mathrm{m}}$ which fulfills (4.9). Suppose that $\nu$ is supported by simple mass-partitions, so it can be identified with a measure on $[0, 1]$ which fulfills (4.11). We may then find a sequence $(\nu^{(n)}, n \in \mathbb{N})$ of finite measures on $[0, 1]$ such that

$$x^2\nu^{(n)}(dx) \text{ converges weakly as } n \to \infty \text{ to } \mathtt{c}\delta_0 + x^2\nu(dx). \tag{4.15}$$

For each $n \in \mathbb{N}$, let us denote by

$$B^{(n)} = (B_{t,t'}^{(n)}, -\infty < t \le t' < \infty)$$

a family of bridges constructed as above from the finite measure $\nu^{(n)}$. Recall also that we consider bridges as random variables with values in the space $L^2([0, 1], dr)$.

**Theorem 4.3** *In the notation above, the following holds:*

(i) *As $n \to \infty$, $(B_{t,t'}^{(n)}, -\infty < t \le t' < \infty)$ converges in the sense of weak convergence of finite-dimensional distributions to, say, $(B_{t,t'}, -\infty < t \le t' < \infty)$, which fulfills the following properties:*

  (i.a) *For every $t \le t' \le t''$, $B_{t,t''} = B_{t,t'} \circ B_{t',t''}$ a.s.*
  (i.b) *$B_{t,t'}$ is a bridge and its law only depends on $t' - t$.*
  (i.c) *If $t_1' < t_2' < \cdots < t_n'$, $B_{t_1',t_2'}, B_{t_2',t_3'}, \ldots, B_{t_{n-1}',t_n'}$ are independent.*
  (i.d) *$B_{0,0} = \mathrm{Id}$ and $\lim_{t\to 0+} B_{0,t}(r) = r$ in $L^2(dr \otimes d\mathbb{P})$.*

(ii) *For every $t \geq 0$, denote the right-continuous inverse of $B_{0,t}$ by*

$$B^{-1}_{t,t'}(r) := \inf\left\{v \in [0,1] : B_{t,t'}(v) > r\right\}, \qquad r \in ]0,1[\,.$$

*Let $U_1, \ldots$ be a sequence of i.i.d. uniform variables which is independent of $(B_{0,t}, t \geq 0)$, and define for every $t \geq 0$ the random partition $\Pi(t)$ by*

$$i \overset{\Pi(t)}{\sim} j \iff B^{-1}_{0,t}(U_i) = B^{-1}_{0,t}(U_j)\,.$$

*Then $(\Pi(t), t \geq 0)$ is a standard simple exchangeable coalescent with coagulation rate $\mu = \mathtt{c}\mu^{\mathrm{K}} + \varrho_\nu$.*

As a reference to the properties (i.a–d), we say that the family $(B_{t,t'}, -\infty < t \leq t' < \infty)$ is a *flow of bridges* on $[0,1]$. More generally, one can show that there is a bijective correspondence between on the one hand the laws of flows of bridges, and on the other hand the laws of standard exchangeable coalescents; see Bertoin and Le Gall [44].

*Proof* Write $q^{(n)}_\pi$ for the jump rates of the exchangeable coalescent $\Pi^{(n)}$ with coagulation rate $\mu^{(n)} = \mu_{\nu^{(n)}}$, so that by (4.12)

$$q^{(n)}_\pi = \int_{]0,1]} x^\ell (1-x)^{k-\ell} \nu^{(n)}(dx)\,,$$

if $\pi \in \mathcal{P}_k \backslash \{\mathbf{0}_{[k]}\}$ is simple and has one block with $\ell \in \{2, \ldots, k-1\}$ elements. The assumption (4.15) ensures that

$$\lim_{n\to\infty} q^{(n)}_\pi = q_\pi := \mathtt{c}\,\mathbb{1}_{\{\ell=2\}} + \int_{]0,1]} x^\ell(1-x)^{k-\ell}\nu(dx)\,.$$

This shows that for every $k \in \mathbb{N}$, the sequence of restricted chains $\Pi^{(n)}_{|[k]}$ converges in the sense of finite-dimensional distributions to $\Pi_{|[k]}$, where $\Pi$ is an exchangeable coalescent with coagulation rate $\mu = \mathtt{c}\mu^{\mathrm{K}} + \varrho_\nu$. By a straightforward compatibility argument, this implies that the sequence of random exchangeable partitions $\Pi^{(n)}(\cdot)$ converges in law in the sense of finite-dimensional distributions as $n \to \infty$ towards $\Pi(\cdot)$.

We may now combine Proposition 2.9 and Lemma 4.7, and deduce that for each $-\infty < t \leq t' < \infty$, $B^{(n)}_{t,t'}$ converges in law to some bridge. On the other hand, the family $(B^{(n)}_{t,t'}, -\infty < t \leq t' < \infty)$ is a flow of bridges and, thanks to (i.a) and (i.c), the one-dimensional convergence in distribution that we just showed readily extends to convergence of finite-dimensional distributions. Kolmogorov's Extension Theorem ensures the existence of a limiting family $(B_{t,t'}, -\infty < t \leq t' < \infty)$; moreover, that properties (i.a–c) hold for the latter is now obvious, as well as (ii) (again by Proposition 2.9 and Lemma 4.7).

So we only need to verify (i.d). In this direction, recall that the exchangeable coalescent $\Pi$ enjoys the Feller property, in particular it is continuous in probability. Thus (i.d) follows also from Proposition 2.9 and Lemma 4.7. □

We will now point out that under additional assumptions, one can reinforce Theorem 4.3 into an almost sure convergence result. More precisely, suppose throughout the rest of this section that the coefficient of binary coagulations is $\mathtt{c} = 0$ and that the measure of multiple coagulations fulfills

$$\int_{]0,1[} x\nu(dx) < \infty . \tag{4.16}$$

Just as when $\nu$ is finite, consider a Poisson random measure $\mathtt{M}$ on $\mathbb{R}\times ]0,1[$ with intensity $dt \otimes \nu(dx)$, and conditionally on $\mathtt{M}$, associate to each atom $(t, x)$ a bridge $b^{(t)}$ distributed as $b_x$, in such a way that to different atoms correspond independent bridges.

For every $n \in \mathbb{N}$, the restriction of the Poisson measure to $\mathbb{R}\times ]1/n, 1[$ is a Poisson measure with intensity $dt \otimes \nu^{(n)}(dx)$, where $\nu^{(n)}(dx) = \mathbb{1}_{\{x>1/n\}}\nu(dx)$ is a finite measure. We can thus construct an elementary flow of bridges $B^{(n)}$ from the latter, and since $x^2\nu_n(dx)$ converges weakly to $x^2\nu(dx)$, Theorem 4.3 shows that $B^{(n)}$ converges weakly as $n \to \infty$ towards the flow of bridges corresponding to the simple exchangeable coalescent with coagulation rate $\mu = \varrho_\nu$. We now show that in fact an almost sure convergence holds.

**Proposition 4.8** *Under the preceding assumptions, for every $t < t'$, the sequence of inverse bridges $B_{t,t'}^{(n)-1}$ converges uniformly on $[0, 1]$ with probability one.*

*Proof* The key of the proof relies on the observation that for every $x \in ]0, 1[$, the inverse $b_x^{-1}$ of the simple bridge $b_x$ is Hölder continuous with parameter $1/(1-x)$, and that also $||b_x^{-1} - \mathrm{Id}||_\infty \leq x/(1-x)$. Our goal is to show that $(B_{t,t'}^{(n)-1}, n \in \mathbb{N})$ is a Cauchy sequence in $\mathcal{C}([0, 1])$. In this direction, fix $n \in \mathbb{N}$, let $v_1 < \cdots < v_k$ be the sequence of times in $]t, t']$ on which the Poisson measure has an atom $(v, x)$ with $x > 1/n$. We also agree that $v_0 = t$ and $v_{k+1} = t'$, and write $b^{(j)}$ for the simple bridge corresponding to $v_j$.

Next take an arbitrary $n' > n$, and define

$$\beta^{(0)} = B_{v_0,v_1-}^{(n')}\,,\ \beta^{(1)} = B_{v_1,v_2-}^{(n')}\,,\ \ldots\,,\ \beta^{(k)} = B_{v_k,v_{k+1}-}^{(n')}\,,$$

so that

$$B_{t,t'}^{(n')} = \beta^{(0)} \circ b^{(1)} \circ \beta^{(1)} \circ \cdots \circ b^{(k)} \circ \beta^{(k)} .$$

Denote by $c^{(i)}$ the inverse of $b^{(i)}$ and by $\gamma^{(i)}$ that of $\beta^{(i)}$, so we have to evaluate

$$||\gamma^{(k)} \circ c^{(k)} \circ \gamma^{(k-1)} \circ c^{(k-1)} \circ \cdots \circ \gamma^{(1)} \circ c^{(1)} \circ \gamma^{(0)} - c^{(k)} \circ c^{(k-1)} \circ \cdots \circ c^{(1)}||_\infty .$$

The basic estimates at the beginning of this proof yield by an easy iteration the following upper bound for the preceding quantity

$$2\left(y_k + \frac{1}{1-x_k}\left(y_{k-1} + \frac{1}{1-x_{k-1}}\left(\cdots + \frac{1}{1-x_1}y_0\right)\right)\right) \leq 2\left(\sum_{i=0}^{k} y_i\right)\prod_{i=1}^{k}\frac{1}{1-x_i},$$

where $x_i$ denotes the second coordinate of the atom that occurs at $v_i$ and

$$y_i := \int_{]v_i,v_{i+1}[\times]1/n',1/n]} x\mathtt{M}(dx) .$$

Now define $C := \prod 1/(1-x)$ where the product is taken over the atoms $(v, x)$ of $\mathtt{M}$ such that $t < v \leq t'$. On the one hand, the assumption (4.16) and the first-moment formula of Lemma 2.3 for Poisson measures imply that $\sum x < \infty$ a.s., and hence $C < \infty$ a.s. On the other hand, we have

$$\sum_{i=0}^{k} y_i = \int_{]t,t']\times]1/n',1/n]} x\mathtt{M}(dx) ,$$

and it follows again from (4.16) and the first moment formula that this quantity tends to 0 a.s. as $n, n' \to \infty$. Our claim follows. □

### 4.4.3 The dual flow and a population model

In this section, we shall see that the flow of bridges which is used in Theorem 4.3 to construct a simple exchangeable coalescent, can be interpreted in terms of a natural population model. Throughout this section, $(B_{t,t'}, -\infty < t \leq t' < \infty)$ denotes a flow of bridges associated to some simple exchangeable coalescent $\Pi$; we also implicitly assume that the coagulation rate $\mu \neq 0$ to avoid the useless discussion of a trivial case.

As we explained in Section 4.1.1, coagulations arise when one studies the genealogy of populations, and for this purpose, one has to work as time goes *backwards*. Therefore, a population model based on a flow of bridges should rather be defined via the *dual* flow, namely

$$\hat{B}_{t,t'} := B_{-t',-t} , \qquad -\infty < t \leq t' < \infty .$$

Plainly, $\hat{B}_{t,t'}$ is a bridge whose law only depends on $t' - t$, and which converges in probability to Id when $t' - t \to 0$. Furthermore the bridges

$\hat{B}_{t_1,t_2}, \ldots, \hat{B}_{t_{n-1},t_n}$ are independent for every $t_1 < \cdots < t_n$, and the following cocycle property holds for every $t < t' < t''$:

$$\hat{B}_{t',t''} \circ \hat{B}_{t,t'} = \hat{B}_{t,t''} .$$

We write $\rho_t$ for the probability measure with distribution function $\hat{B}_{0,t}$, that is

$$\rho_t(dy) = d\hat{B}_{0,t}(y) , \qquad 0 \leq y \leq 1.$$

We immediately derive from the cocycle property of the dual flow that $(\rho_t, t \geq 0)$ is a Markov process with values in the space $\mathcal{M}_1$ of the probability measures on $[0, 1]$. Recall that the latter is a compact metric space when endowed with Prohorov's distance. Further, it follows readily from the fact that $B_{0,t}$ has no fixed discontinuities, that $(\rho_t, t \geq 0)$ is in fact a Feller process, and in particular it possesses a càdlàg modification. From now on, we implicitly deal with this càdlàg version. Note also that $\rho_0(dy) = dy$ is the Lebesgue measure on $[0, 1]$.

The process $(\rho_t, t \geq 0)$ has an interesting interpretation as a population model which has been considered first by Donnelly and Kurtz [82, 83], and that we now present. We may think of $\rho_t(dr)$ as the size of the progeny at time $t$ of the fraction $dr$ of the initial population. Consider for simplicity the case of when the coagulation rate is finite, that is $\mu = \varrho_\nu$ where $\nu$ is a finite measure on $]0, 1[$. Recall the discrete Poissonian construction of the flow of bridges that was presented at the beginning of Section 4.4.2. We see that the process $(\rho_t, t \geq 0)$ (or equivalently, $(\hat{B}_{0,t}, t \geq 0)$) is a continuous time Markov chain, and that the jump times of this chain are given by a Poisson process with intensity $\nu(]0, 1[)$. More precisely, if $t_n$ is the instant of the $n$-th jump, then

$$\hat{B}_{0,t_n} = b_X \circ \hat{B}_{0,t_{n-1}} ,$$

where $b_X$ is a simple bridge which is independent of $B_{0,t_{n-1}}$, such that its jump size $X$ is a random variable with distribution $\nu(\cdot)/\nu(]0, 1[)$, and its jump location $U$ an independent uniform variable. This means that

$$\rho_{t_n} = (1-X)\rho_{t_{n-1}} + X\delta_Y ,$$

where conditionally on $\rho_{t_{n-1}}$, $X$ and $Y := \hat{B}^{-1}_{0,t_{n-1}}(U)$ are independent random variables, with $Y$ distributed according to $\rho_{t_{n-1}}$. In terms of the evolution of the population, this means that an individual picked at random in the population at time $t_{n-1}$ generates a proportion $X$ of the population at time $t_n$. The rest of the population at time $t_{n-1}$ is reduced by a factor $(1-X)$ so that, at time $t_n$, the total size of the population is still 1. This description bears obvious similarities with that for the evolution of the Moran and the Fleming-Viot processes; see for example Chapter 1 of Etheridge [98].

We can now interpret the coalescent in terms of the genealogy of this population model. More precisely, fix some time $T > 0$, and consider the population at time $T$, which is identified as $[0, 1]$. Pick a sequence of individuals uniformly at random, that is consider a sequence $U_1, \ldots$ of i.i.d. uniform variables which is independent of the flow $(B_{t,t'}, 0 \leq t \leq t' \leq T)$. Two individuals $i$ and $j$ have the same ancestor $r \in ]0, 1[$ at the generation $T-t$ if and only if $U_i$ and $U_j$ both belong to the interval $]\hat{B}_{T-t,T}(r-), \hat{B}_{T-t,T}(r)[$. In other words, for each $t \in [0, T]$, we may consider the partition $\Pi(t)$ of $\mathbb{N}$ defined by

$$i \overset{\Pi(t)}{\sim} j \Longleftrightarrow \hat{B}^{-1}_{T-t,T}(U'_i) = \hat{B}^{-1}_{T-t,T}(U'_j);$$

where the blocks of the partition consist of the families of individuals which have the same ancestor at the generation $T-t$. Lemma 4.9 shows that $(\Pi(t), 0 \leq t \leq T)$ is a simple exchangeable coalescent with coagulation rate $\mu = \varrho_\nu$. Further, we can use Theorem 4.3 to extend this to situations where $\nu$ is an infinite measure on $]0, 1]$ with $\int_{]0,1]} x^2 \nu(dx) < \infty$.

Let us now introduce some natural definition in this framework. To start with, we observe that Dirac point masses are absorbing states for the population model $\rho = (\rho_t, t \geq 0)$, in the sense that if $\rho_t = \delta_z$ for some $z \in [0, 1]$ (which means that the entire population at time $t$ has the same ancestor $z$ in the initial population), then the same holds for every time $t' \geq t$. This leads us to define the *fixation time*

$$\zeta := \inf\{t \geq 0 : \rho_t = \delta_z \text{ for some } z \in [0, 1]\}.$$

One says that *fixation* occurs when the fixation time $\zeta$ is finite. The following useful bound is due to Schweinsberg [200].

**Proposition 4.9** *Set for every integer $n \geq 2$*

$$\varphi(n) := \frac{n(n-1)}{2}\mathrm{c} + \int_{]0,1]} ((1-x)^n - 1 + nx)\nu(dx).$$

*Then the expectation of the fixation time is bounded by*

$$\mathbb{E}(\zeta) \leq \sum_{n=2}^{\infty} 1/\varphi(n).$$

*As a consequence, fixation occurs with probability one provided that the series in the right-hand side converges (this holds in particular when the coefficient c of binary coagulation is not zero).*

More precisely, Schweinsberg [200] has proved that, as soon as the measure $\nu$ has no atom at 1, the condition of convergence of the series in Proposition 4.9 is also necessary for fixation. In the same vein, this condition is necessary and sufficient for the coalescent $\Pi$ to come down from infinity, in the sense that $\#\Pi(t) < \infty$ a.s. for every $t > 0$, where $\#\pi$ stands for the number of non-empty blocks of a partition $\pi$. In terms of the population model, this means that for any $t > 0$, we can find a finite number of individuals in the initial population which generate the entire population at time $t$.

The proof of Proposition 4.9 relies on the following technical lemma. Recall the notation (4.12).

**Lemma 4.10** *The function $\varphi$ increases, and for every $n \geq 2$, there is the identity*

$$\varphi(n) = \sum_{k=2}^{n} (k-1) \binom{n}{k} q_{n,k} \,.$$

*Proof* For every $0 < x \leq 1$, function $b \to bx - 1 + (1-x)^b$ increases on $b \geq 2$, which implies the first claim. The second follows easily from the definition of $q_{n,k}$ and the binomial formula. □

We can now tackle the proof of Proposition 4.9.

*Proof* Let $n \in \mathbb{N}$ denote a fixed integer, and consider for every $t \geq 0$ the number of non-empty blocks $\#\Pi_{|[n]}(t)$. It should be plain from the dynamics of the restricted chain $\Pi_{|[n]}$ that the process $\#\Pi_{|[n]}$ is a Markov chain with values in $[n]$ with non-increasing paths, which is absorbed at 1. More precisely, for every $\ell = 2, \ldots, n$ and $k = 2, \ldots, \ell$, when the coalescent chain $\Pi_{|[n]}$ has $\ell$ blocks and a coagulation involving $k$ of its blocks occurs, then $\#\Pi_{|[n]}$ decreases by $k-1$. Hence the jump rate $r_{\ell,\ell-k+1}$ of $\#\Pi_{|[n]}$ from $\ell$ to $\ell - k + 1$ is given in terms of the jump rates (4.12) of the coalescent by

$$r_{\ell,\ell-k+1} = \binom{\ell}{k} q_{\ell,k} \,.$$

In other words, the infinitesimal generator $\mathsf{G}^{[n]}$ of $\#\Pi_{|[n]}$ is specified by

$$\mathsf{G}^{[n]} f(\ell) = \sum_{k=2}^{\ell} \binom{\ell}{k} q_{\ell,k} \left( f(\ell - k + 1) - f(\ell) \right) , \qquad \ell \in [n] \,.$$

Now assume that the series $\sum_{b=2}^{\infty} 1/\varphi(b)$ converges (since otherwise there is nothing to prove), and define

$$f(\ell) := \sum_{k=\ell+1}^{\infty} 1/\varphi(k), \qquad \ell \geq 1.$$

Recall Lemma 4.10. Since $1/\varphi$ decreases, we have for $2 \leq k \leq \ell$

$$f(\ell - k + 1) - f(\ell) \geq (k-1)/\varphi(\ell)$$

and therefore

$$\mathsf{G}^{[n]} f(\ell) \geq \frac{1}{\varphi(\ell)} \sum_{k=2}^{\ell} (k-1) \binom{\ell}{k} q_{\ell,k} = 1.$$

The process

$$f(\#\Pi_{|[n]}(t)) - \int_0^t \mathsf{G}^{[n]} f(\#\Pi_{|[n]}(s)) ds, \qquad t \geq 0$$

is a martingale, and an application of the optional sampling theorem at the absorption time

$$\zeta_n := \inf\left\{t \geq 0 : \#\Pi_{|[n]}(t)) = 1\right\},$$

yields the bound

$$\mathbb{E}(\zeta_n) \leq \mathbb{E}\left(\int_0^{\zeta_n} \mathsf{G}^{[n]} f(\#\Pi_{|[n]}(s)) ds\right) = f(1) - f(n).$$

Plainly, the sequence $(\zeta_n, n \in \mathbb{N})$ increases and $\lim_{n\to\infty} \zeta_n := \zeta_\infty$ is the time of entire coalescence for $\Pi$. Further, the latter obviously has the same distribution as the fixation time $\zeta$ of the population model, which proves that $\mathbb{E}(\zeta) \leq f(1)$. □

We finally turn our interest to the one-point motion of the dual flow, namely

$$\hat{B}_{0,t}(y) = \rho_t([0, y])$$

for a fixed $y \in [0, 1]$. Again from the cocycle property, we see that this process is a Markov process, and that its semigroup $Q_t$ does not depend on $y$: For every $x \in [0, 1]$, $Q_t(x, \cdot)$ is simply the law of $B_{0,t}(x)$. Moreover, this semigroup is again Feller thanks to the absence of fixed discontinuities for $B_{0,t}$. It is easy to verify that $\rho_t(\{y\}) = 0$ for every $t \geq 0$, a.s. (if $\varepsilon > 0$ and $T_\varepsilon = \inf\{t \geq 0 : \rho_t(\{y\}) \geq \varepsilon\}$, apply the strong Markov property of $\rho$ at time $T_\varepsilon$ to see that on the event $\{T_\varepsilon < \infty\}$ there will exist rational values of $t$ for which $\rho_t(\{y\}) > 0$, which does not occur with probability one). From the right-continuity of paths of $\rho_t$, we now deduce that the paths of $\hat{B}_{0,t}(y)$ are also right-continuous a.s. (they will indeed be càdlàg by the Feller property).

**Proposition 4.10** *The process* $(\hat{B}_{0,t}(y), t \geq 0)$ *is a martingale which converges a.s. to* 0 *or* 1. *More precisely,*

$$\mathbb{P}\left(\lim_{t\to\infty} \hat{B}_{0,t}(y) = 1\right) = y .$$

*Proof* The martingale property stems from the fact that for any bridge $B$, one has $\mathbb{E}(B(y)) = y$. Now the martingale $\hat{B}_{0,t}(y)$ is bounded and thus converges a.s., and we have to check that the only possible limit points are 0 and 1.

Plainly, the probability that 1 and 2 belong to distinct blocks of the partition $\Pi(t)$ equals $\exp(-tq_{\{1,2\}})$, where $q_{\{1,2\}}$ is the jump rate of $\Pi_{|[2]}$ from the partition into singleton. Expressing this probability in terms of jump-sizes $S(t)$ of $B_{0,t}$, we get

$$\mathbb{E}\left(\sum_{i=1}^{\infty}(S_i(t))^2\right) = 1 - \exp(-tq_{\{1,2\}}) ,$$

a quantity which converges to 1 when $t \to \infty$. Hence the ranked sequence of jump-sizes of $B_{0,t}$ converge in probability to $(1, 0, \dots)$, and thus, by Lemma 4.7, $B_{0,t}$ (or, equivalently, $B_{-t,0}$) converges in distribution to the bridge that has a unique jump with size one. The rest of the proof is now straightforward. □

Proposition 4.10 means that as time tends to infinity, the offspring at time $t$ of a fraction of size $y$ of the initial population tends to one with probability $y$ and to 0 with probability $1 - y$. When fixation occurs a.s., there exists a (random) real number $\mathbf{e} \in [0, 1]$ such that $\rho_t(dy) = \delta_{\mathbf{e}}(dy)$ whenever $t$ is sufficiently large; we call $\mathbf{e}$ the *primitive Eve* of the population. The primitive Eve can be identified as the critical point

$$\mathbf{e} = \inf\left\{y \in [0, 1] : \lim_{t\to\infty} \hat{B}_{0,t}(y) = 1\right\} = \sup\left\{y \in [0, 1] : \lim_{t\to\infty} \hat{B}_{0,t}(y) = 0\right\} .$$

Note that Proposition 4.10 shows that the primitive Eve is uniformly distributed on [0, 1].

### 4.4.4 The Bolthausen-Sznitman coalescent

In this section, we will investigate a remarkable example of flow of bridges based on stable subordinators. Specifically, for every $\alpha \in ]0, 1]$, let $\varsigma_\alpha = (\varsigma_\alpha(t), t \geq 0)$ be a standard subordinator with index $\alpha$ (see Section 2.2.5).

Recall that this means that $\varsigma_\alpha$ is an increasing process with independent and stationary increments, and its one-dimensional distributions are characterized via their Laplace transforms by

$$\mathbb{E}\left(\exp(-q\varsigma_\alpha(t))\right) = \exp(-tq^\alpha), \qquad q \geq 0.$$

The fundamental property of stable subordinators that we will use in this section is the so-called *subordination scheme*. Specifically, for every $\beta \in ]0, 1]$, if $\varsigma'_\beta$ denotes a standard stable subordinator with index $\beta$ which is independent of $\varsigma_\alpha$, then it is immediately checked that the compound process $\varsigma_\alpha \circ \varsigma'_\beta$ is a standard stable subordinator with index $\alpha\beta$.

In fact, we shall use a variant of the subordination scheme for bridges. More precisely, $\varsigma_\alpha$ clearly enjoys the scaling property: for every $a > 0$,

$$(a\varsigma_\alpha(a^{-\alpha}t), t \geq 0) \text{ has the same law as } \varsigma_\alpha .$$

This implies that the distribution of the process

$$b_\alpha(r) := \varsigma_\alpha(rt)/\varsigma_\alpha(t), \qquad r \in [0, 1], \tag{4.17}$$

does not depend on $t$; and following Definition 2.6, we call PD$(\alpha, 0)$-bridge any process distributed as $b_\alpha$. More precisely, the ranked sequence of the jump sizes of $b_\alpha$ is a random mass-partition with the PD$(\alpha, 0)$-distribution, and the locations of these jumps, that is the locations of the jumps of $\varsigma_\alpha(t\cdot)$, form a sequence of i.i.d. uniform variables which is independent of the sequence of the jump sizes. Thus $b_\alpha$ is a bridge in the sense of Definition 4.6. We now state the version of the subordination scheme that will be useful in this section.

**Lemma 4.11** *Fix* $\alpha, \beta \in ]0, 1]$, *let* $b_\alpha$ *be a* PD$(\alpha, 0)$*-bridge and* $b'_\beta$ *a* PD$(\beta, 0)$*-bridge which is independent of* $b_\alpha$. *Then the compound process* $b_\alpha \circ b'_\beta$ *is a* PD$(\alpha\beta, 0)$*-bridge.*

*Proof* Let $\varsigma_\alpha$ and $\varsigma'_\beta$ be two independent standard stable subordinators with indices $\alpha$ and $\beta$. Then the process $b_\alpha$ defined by (4.17) for $t = \varsigma'_\beta(1)$ is a PD$(\alpha, 0)$-bridge which is independent of $\varsigma'_\beta$, and a fortiori of the PD$(\beta, 0)$-bridge $b'_\beta(r) := \varsigma'_\beta(r)/\varsigma'_\beta(1)$. The claim now follows from the fact that $\varsigma_\alpha \circ \varsigma'_\beta$ is a standard stable subordinator with index $\alpha\beta$. □

Recall Definition 2.6 and Lemma 4.3. Lemma 4.11 readily yields the following.

**Corollary 4.3** *Fix $0 < \alpha, \beta < 1$. Let $\pi^{(\alpha)}$ be a* PD$(\alpha, 0)$*-partition and $\pi^{(\beta)}$ an independent* PD$(\beta, 0)$*-partition Then the exchangeable partition* Coag $(\pi^{(\alpha)}, \pi^{(\beta)})$ *is a* PD$(\alpha\beta, 0)$*-partition.*

*Proof* Let $b_\alpha$ be a PD$(\alpha, 0)$-bridge. Write $S^{(\alpha)} = (S_1^{(\alpha)}, \ldots)$ for the random mass-partition with PD$(\alpha, 0)$-law given by the ranked sequence of the jump-sizes of $b_\alpha$, and for every $i \in \mathbb{N}$, let $U_i^{(\alpha)}$ be the location of the jump with size $S_i^{(\alpha)}$. Recall that $U_1^{(\alpha)}, \ldots$ form a sequence of i.i.d. uniform variables which is independent of $S^{(\alpha)}$.

Denote the inverse bridge by

$$b_\alpha^{-1}(r) := \inf\{s \in [0, 1] : b_\alpha(s) > r\}, \qquad 0 \le r < 1,$$

and let $U_1, \ldots$ be another sequence of i.i.d. uniform variables which is independent of $b_\alpha$. We introduce the PD$(\alpha, 0)$-partition $\pi^{(\alpha)}$ as the paint-box based on $b_\alpha^{-1}$ and the $U_i$, which is defined by

$$i \overset{\pi^{(\alpha)}}{\sim} j \iff b_\alpha^{-1}(U_i) = b_\alpha^{-1}(U_j).$$

For every $i \in \mathbb{N}$, define also $U_i' = b_\alpha^{-1}(U_j)$ whenever $U_j \in \pi_i^{(\alpha)}$.

Just as in Lemma 4.8, we now claim that $U_1', \ldots$ is a sequence of i.i.d. uniform variables which is independent of $\pi^{(\alpha)}$. Indeed, if we write $\vartheta_\alpha = [0, 1] \backslash \{b_\alpha(r), r \in [0, 1]\}^{\mathrm{cl}}$ for the complementary of the closed range of $b_\alpha$, then by definition $\pi^{(\alpha)}$ coincides with the paint-box based on $\vartheta_\alpha$ and the $U_i$. The ranked sequence of the lengths of the interval components of $\vartheta_\alpha$ is $S^{(\alpha)}$, and thus it follows from Lemma 2.7 that $\pi^{(\alpha)}$ is independent of $(U_i^{(\alpha)}, i \in \mathbb{N})$. Next, recall that $S^{(\alpha)}$ is proper and $S_i^{(\alpha)} > 0$ for every $i \in \mathbb{N}$ a.s., so all the blocks of $\pi^{(\alpha)}$ have strictly positive asymptotic frequencies. This enables us to define, for every $i \in \mathbb{N}$, $\sigma(i)$ by $|\pi_i^{(\alpha)}| = S_{\hat{\sigma}(i)}^{(\alpha)}$, so that $\sigma : \mathbb{N} \to \mathbb{N}$ is a random permutation which is independent of $(U_i^{(\alpha)}, i \in \mathbb{N})$. It follows that $(U_i' = U_{\hat{\sigma}(i)}^{(\alpha)}, i \in \mathbb{N})$ is a sequence of i.i.d. uniform variables which is independent of $\pi^{(\alpha)}$.

The rest of the proof is now straightforward. Let $b_\beta$ be an independent PD$(\beta, 0)$-bridge and $b_\beta^{-1}$ its inverse. We define the paint-box $\pi^{(\beta)}$ by

$$i \overset{\pi^{(\beta)}}{\sim} j \iff b_\beta^{-1}(U_i') = b_\beta^{-1}(U_j');$$

we know from above that $\pi^{(\alpha)}$ and $\pi^{(\beta)}$ are independent. We can repeat the argument of Corollary 4.2(i) and get that Coag$(\pi^{(\alpha)}, \pi^{(\beta)}) := \pi^{(\alpha\beta)}$ is given by the paint-box

$$i \overset{\pi^{(\alpha\beta)}}{\sim} j \iff b_\beta^{-1} \circ b_\alpha^{-1}(U_i) = b_\beta^{-1} \circ b_\alpha^{-1}(U_j).$$

We conclude the proof by observing that $b_\beta^{-1} \circ b_\alpha^{-1}$ is the inverse of the bridge $b_\alpha \circ b_\beta$, which is a PD$(\alpha\beta, 0)$-bridge by Lemma 4.11. □

For every $t \geq 0$, let $P_t^{\mathrm{BS}}$ be the operator on the space of continuous function $\Phi : \mathcal{P}_\infty \to \mathbb{R}$ defined by

$$P_t^{\mathrm{BS}}\Phi(\pi) := \mathbb{E}\left(\Phi\left(\mathrm{Coag}(\pi, \pi^{(\mathrm{e}^{-t})})\right)\right), \qquad \pi \in \mathcal{P}_\infty,$$

where $\pi^{(\mathrm{e}^{-t})}$ stands for a PD$(\mathrm{e}^{-t}, 0)$-partition. Corollary 4.3 combined with the associativity property of the coagulation operator (see Lemma 4.2) shows that the family of operators $(P_t^{\mathrm{BS}}, t \geq 0)$ is a Markovian semigroup. More precisely, it gives the transition probabilities of some exchangeable coalescent process $\Pi^{\mathrm{BS}}(\cdot)$. This semigroup was introduced by Bolthausen and Sznitman [61], which explains the superscript BS in the notation.

**Proposition 4.11** *The exchangeable coalescent $\Pi^{\mathrm{BS}}(\cdot)$ based on the Bolthausen-Sznitman semigroup $(P_t^{\mathrm{BS}}, t \geq 0)$ is simple. Its coagulation rate is given by*

$$\mu^{\mathrm{BS}} = \int_0^1 \varrho_x x^{-2} dx\,,$$

*where $\varrho_x$ is the law of a random exchangeable partition with ranked asymptotic frequencies $(x, 0, \dots)$. In other words, the coefficient of binary coagulation is $\mathtt{c} = 0$ and the rate of multiple coagulations is given by the measure $\nu(dx) = x^{-2}dx$, $x \in ]0, 1[$.*

*Proof* Fix $n \geq 2$, and consider some partition $\pi^{[n]} \in \mathcal{P}_n$ with $\pi^{[n]} \neq \mathbf{0}_{[n]}$. We have to evaluate the jump rate

$$q_{\pi^{[n]}} = \lim_{t\to 0} \frac{1}{t}\mathbb{P}\left(\Pi^{\mathrm{BS}}_{|[n]}(t) = \pi^{[n]}\right).$$

In this direction, suppose that the number of non-empty blocks of $\pi^{[n]}$ is $\#\pi^{[n]} = k$ and that these $k$ blocks have respective sizes $n_1, \dots, n_k$ where $n_1 + \cdots + n_k = n$. Recall from Theorem 2.3(ii) the EPPF for PD$(\alpha, 0)$-partitions; this yields

$$\frac{1}{t}\mathbb{P}\left(\Pi^{\mathrm{BS}}_{|[n]}(t) = \pi^{[n]}\right) = t^{-1}\mathrm{e}^{-t(k-1)}\frac{(k-1)!}{(n-1)!}\prod_{i=1}^{k}(1-\mathrm{e}^{-t})_{n_i-1\uparrow},$$

with $(a)_{\ell\uparrow} = a(a+1)\cdots(a+\ell-1)$ for $\ell \in \mathbb{N}$ and $(a)_0 = 1$. Now we have $(1-\mathrm{e}^{-t})_{\ell\uparrow} = O(t)$ as $t \to 0+$ whenever $\ell \geq 1$, and thus

$$\prod_{i=1}^{k}(1-\mathrm{e}^{-t})_{n_i-1\uparrow} = o(t)\,, \qquad t \to 0\,,$$

except when $\pi^{[n]}$ is simple. In the latter case, if $\pi^{[n]}$ has a block with size $\ell \in \{2,\ldots,n\}$, then we have $k = n-\ell+1$, and we find

$$\begin{aligned} q_\pi &= \lim_{t\to 0}\frac{1}{t}\mathbb{P}\left(\Pi^{\mathrm{BS}}_{|[n]}(t) = \pi^{[n]}\right) \\ &= \frac{\Gamma(n-\ell+1)\Gamma(\ell-1)}{\Gamma(n)} \\ &= \int_0^1 x^{\ell-2}(1-x)^{n-\ell}dx\,. \end{aligned}$$

The comparison with (4.12) completes the proof. □

Pitman [183] discovered a remarkable duality between fragmentation and coagulation operators based on Poisson-Dirichlet variables, which extends considerably Corollary 4.3. To state this result, it is convenient to introduce for every $\alpha \in ]0,1[$ and $\theta > -\alpha$ the following notation. For every partition $\gamma \in \mathcal{P}_\infty$, we write $\mathrm{Coag}_{\alpha,\theta}(\gamma)$ for the distribution of $\mathrm{Coag}(\gamma,\pi)$ where $\pi$ is a PD$(\alpha,\theta)$-partition. Similarly, we write $\mathrm{Frag}_{\alpha,\theta}(\gamma)$ for the distribution of $\mathrm{Frag}(\gamma,\pi^{(\cdot)})$ where $\pi^{(\cdot)} = (\pi^{(1)},\ldots)$ is a sequence of independent PD$(\alpha,\theta)$-partitions. We may now state:

**Theorem 4.4** *Let $\Gamma,\Gamma'$ be two random partitions. For every $\alpha,\beta \in ]0,1[$ and $\theta > -\alpha\beta$, the following assertions are equivalent:*

(i) *$\Gamma$ is a* PD$(\alpha,\theta)$*-partition and conditionally on $\Gamma = \gamma$, $\Gamma'$ has the law* $\mathrm{Coag}_{\beta,\theta/\alpha}(\gamma)$.
(ii) *$\Gamma'$ is a* PD$(\alpha\beta,\theta)$*-partition and conditionally on $\Gamma' = \gamma'$, $\Gamma$ has the law* $\mathrm{Frag}_{\alpha,-\alpha\beta}(\gamma')$.

*Proof* (i) and (ii) provide two descriptions of the joint law of $(\Gamma,\Gamma')$, and we have to check that they coincide. In this direction, it suffices to show that their joint EPPF's are the same, that is for every $\gamma,\gamma' \in \mathcal{P}_\infty$ and $n \in \mathbb{N}$, the two descriptions yield the same value for the probabilities, say $p^{(\mathrm{i})}$ and $p^{(\mathrm{ii})}$, that $\Gamma_{|[n]} = \gamma_{|[n]}$ and $\Gamma'_{|[n]} = \gamma'_{|[n]}$. We focus on the case when $\gamma_{|[n]}$ is finer than $\gamma'_{|[n]}$, as otherwise this probability is obviously zero for both descriptions.

Suppose that $\gamma_{|[n]}$ has $K$ non-empty blocks with sizes $a_1, \ldots, a_K$ and that $\gamma'_{|[n]} = \text{Coag}(\gamma_{|[n]}, \eta)$ where $\eta \in \mathcal{P}_K$ has $k \leq K$ non-empty blocks with sizes $j_1, \ldots, j_k$. By the very definition of the coagulation operator, we have in the case (i)

$$p^{(\mathrm{i})} = \mathrm{p}_{\alpha,\theta}(a_1, \ldots, a_K)\mathrm{p}_{\beta,\theta/\alpha}(j_1, \ldots, j_k)$$

where $\mathrm{p}_{\alpha,\theta}$ stands for the EPPF of a $\mathrm{PD}_{\alpha,\theta}$-partition.

In this situation, $\gamma'_{|[n]}$ has $k$ non-empty blocks, say $B_1, \ldots, B_k$, with respective sizes $b_1, \ldots, b_k$. There exists a unique $k$-tuple $\eta^{(\cdot)} = (\eta^{(1)}, \ldots, \eta^{(k)})$ where $\eta^{(\ell)}$ is a partition of $B_\ell$ with $j_\ell$ non-empty blocks, such that $\gamma_{|[n]}$ coincides with the family of the blocks of $\eta^{(\ell)}$ for $\ell = 1, \ldots k$. Denote by $c_{\ell,1}, \ldots, c_{\ell,j_\ell}$ the lengths of the non-empty blocks of $\eta^{(\ell)}$, and observe that $(c_{\ell,i}, i = 1, \ldots, j_\ell \text{ and } \ell = 1, \ldots, k)$ is a reordering of $(a_i, i = 1, \ldots, K)$.

By the very definition of the fragmentation operator and exchangeability, we have in the case (ii)

$$p^{(\mathrm{ii})} = \mathrm{p}_{\alpha\beta,\theta}(b_1, \ldots, b_k)\prod_{\ell=1}^{k} \mathrm{p}_{\alpha,-\alpha\beta}(c_{\ell,1}, \ldots, c_{\ell,j_\ell}).$$

Applying Pitman's sampling formula (cf. Theorem 2.3(ii)), we get the explicit formulas

$$p^{(\mathrm{i})} = \frac{(\theta/\alpha)_{K\uparrow}}{(\theta)_{n\uparrow}}\left(\prod_{i=1}^{K} -(-\alpha)_{a_i\uparrow}\right)\frac{(\theta/(\alpha\beta))_{k\uparrow}}{(\theta/\alpha)_{K\uparrow}}\prod_{\ell=1}^{k} -(-\beta)_{j_\ell\uparrow},$$

$$p^{(\mathrm{ii})} = \frac{(\theta/(\alpha\beta))_{k\uparrow}}{(\theta)_{n\uparrow}}\left(\prod_{\ell=1}^{k} -(-\alpha\beta)_{b_\ell\uparrow}\frac{(-\beta)_{j_\ell\uparrow}}{(-\alpha\beta)_{b_\ell\uparrow}}\prod_{i=1}^{j_\ell} -(-\alpha)_{c_{\ell,i}\uparrow}\right).$$

By obvious cancellations and the relations between the parameters noted above, we see that $p^{(\mathrm{i})} = p^{(\mathrm{ii})}$, which completes the proof. □

Theorem 4.4 has important consequences for the Bolthausen-Sznitman coalescent $\Pi^{\mathrm{BS}}$. Probably the most striking one is the following description of the time-reversed process as a time-inhomogeneous fragmentation.

**Corollary 4.4** *The reversed Bolthausen-Sznitman coalescent* $(\Pi^{\mathrm{BS}}(-\ln t), t \in ]0, 1])$ *is a time-inhomogeneous Markov process on* $\mathcal{P}_\infty$. *Its semigroup can be described as follows. For every* $0 < t \leq t' \leq 1$, *conditionally on* $\Pi^{\mathrm{BS}}(-\ln t) = \pi$, $\Pi^{\mathrm{BS}}(-\ln t')$ *is distributed as* $\mathrm{Frag}_{t',-t}(\pi)$.

*Proof* Recall that the semigroup $P_t^{\mathrm{BS}}$ of the Bolthausen-Sznitman coalescent can be expressed as the law of the random coagulation operator $\mathrm{Coag}_{e^{-t},0}(\cdot)$.

The claim follows immediately from Theorem 4.4 specified for $\alpha = t'$, $\beta = t/t'$, $\theta = 0$. □

We conclude this section by establishing an interesting identity in law between the process of the asymptotic frequencies of the first block in a Bolthausen-Sznitman coalescent, and the Dirichlet process with parameter 1 (cf. Definition 2.5).

**Corollary 4.5** *The process* $\left(|\Pi_1^{\mathrm{BS}}(t)|, t \geq 0\right)$ *has the same finite-dimensional distributions as* $(\gamma(1-\mathrm{e}^{-t}))/\gamma(1), t \geq 0)$ *where* $\gamma(\cdot)$ *is a standard gamma subordinator (i.e. with parameter* $(1, 1)$*).*

*Proof* On the one hand, we know that $\Pi^{\mathrm{BS}}(-\ln t)$ is a $\mathrm{PD}(t, 0)$ partition, so by the residual allocation model described in Proposition 2.7, $|\Pi_1^{\mathrm{BS}}(-\ln t)|$ is a beta$(1-t, t)$ variable. Taking logarithms, we deduce from the representation (2.3) of beta variables as a ratio of gamma variables that $\ln|\Pi_1^{\mathrm{BS}}(-\ln t)|$ can be expressed as $\ln \gamma(1-t) - \ln \gamma(1)$, where $\gamma(\cdot)$ is a standard gamma process, and the latter quantity is independent of $\ln \gamma(1)$.

On the other hand, it follows easily from the description of the reversed process $\Pi^{\mathrm{BS}}(-\ln t)$ as a time-inhomogeneous fragmentation process on $\mathcal{P}_\infty$ that the process

$$\ln|\Pi_1^{\mathrm{BS}}(-\ln t)|, \qquad 0 < t \leq 1,$$

has independent (non-stationary) increments; see Section 3.2.2. We readily derive from (2.3) that $(\ln \gamma(1) - \ln \gamma(1-t), 0 < t \leq 1)$ also has independent increments. As for processes with independent increments, one-dimensional distributions determine the multi-dimensional distributions, we conclude from the first part of the proof that the two processes have the same finite-dimensional distributions. □

## 4.5 Comments

Kingman coalescent is a very important model in population genetics; see in particular Durrett [87], Neuhauser [172], Nordborg [173], Tavaré [207], Wakeley [212], ... and references therein for alternative presentations and further developments. Perhaps the main conceptual breakthrough is that it provides a theoretical basis for studying the genealogy of a population backwards in time, whereas previous models (for example branching processes) only worked as time goes forward. For instance, Kingman's coalescent can be used to estimate the age of the most recent common ancestor of a population; we refer to [67, 135] and the references therein for some discussions about this question, somewhat controversial when applied to the human population.

Our presentation of Kingman's coalescent in connection with the elementary population model of Wright and Fisher should not be misleading: Kingman's coalescent appears more generally in the study of the genealogy of a fairly large class of population models, including for instance the model of Moran (see for example [87]) and certain branching processes (see [147]). Cannings [65, 66] has introduced an extension of the Wright-Fisher model for the evolution of haploid populations with a fixed size and non-overlapping generations, such that the number of children of individuals at the same generation is given by an exchangeable variable (see [202] for a natural example involving branching processes). Extending an earlier work of Sagitov [197], Möhle and Sagitov [170] and Schweinsberg [199] have shown that under suitable assumptions, taking limits in the genealogical process for Cannings' model yields precisely the class of exchangeable coalescents which is considered in this chapter. Exchangeable coalescents also appear in the limit of the genealogy of more sophisticated population models which incorporate selective mutations and recombination; see Durrett and Schweinsberg [89, 203]. The main results presented in Sections 4.2 and 4.3 are due to Pitman [183] and Schweinsberg [199], although the present approach differs from the original.

The work by Pitman [183] on coalescents with multiple collisions (called here simple coalescents) has been partly influenced by the introduction by Bolthausen and Sznitman [61] of the remarkable coalescent process which is presented in Section 4.4.4. The Bolthausen-Sznitman coalescent arises naturally in the study of certain random energy models in statistical mechanics (see Derrida [76], Ruelle [196], and the recent series of papers by Bovier and Kurkova [62]). Its interpretation in terms of stable subordination has been developed in [43] (see also [49] and [161]), and is closely related to an observation made by Neveu (unpublished) who stressed the connection between Derrida's GREM model and the genealogy of a remarkable continuous branching process (in turn the latter is essentially equivalent to stable subordination). In this vein, Birkner *et al.* [59] pointed out that more generally, the genealogy of branching processes with a stable branching mechanism can also be described in terms of certain simple exchangeable coalescents; see also [27, 28] for recent developments. In a different direction, we refer to Goldschmidt and Martin [116] for a clever construction of the Bolthausen-Sznitman coalescent based on simple recursive trees, which has several interesting consequences.

The link between simple exchangeable coalescents and stochastic flows in Sections 4.4.1–2 is essentially taken from [44] (the latter work presents a more general correspondence). We refer to [45, 46] for recent developments in this area, in particular for the representation of such flows as the solution of some stochastic differential equations, and applications to the asymptotic study of simple coalescents. The population model viewed as a generalized Fleming-Viot process in Section 4.4.3, has been introduced first by Donnelly and Kurtz [82, 83]. In this direction, the connections between the usual Fleming-Viot process and Kingman's coalescent are known, see [82] and Section 5.2 of the monograph by Etheridge [98].

The remarkable duality between coagulation and fragmentation operators based on certain Poisson-Dirichlet partitions, which is stated here in Theorem 4.4, is due to Pitman [183] (see also [42, 116] and the forthcoming Section 5.3.4 for further instances of such duality). It lies at the heart of the study of Ruelle's cascades by Basdevant [20]. This result, together with the obvious likenesses in the structures of exchangeable fragmentations and exchangeable coalescents, might suggest that these two classes of processes should be related by time-reversal. However, the situation is far less simple. For instance, consider an exchangeable coalescent $\Pi$ and an independent subordinator $\sigma$; it is easily checked that the compound process $\Pi \circ \sigma$ is again an exchangeable coalescent. Nonetheless the similar property fails in general when $\Pi$ is a fragmentation, as subordination does not preserve the branching property.

We have seen in Chapter 3 (respectively, Chapter 4) that random exchangeable partitions are remarkably well-suited for developing the theory of exchangeable fragmentation (respectively, coagulation) processes. It is interesting to stress that more generally, similar techniques can be used to investigate exchangeable fragmentation-coagulation processes, in which both fragmentation and coagulation can occur; see Berestycki [26]. In particular, the dynamics of an exchangeable fragmentation-coagulation process are entirely characterized by four parameters, namely a coefficient of erosion, a dislocation measure, a coefficient of binary coagulation, and a measure of multiple coagulations (compare with Theorems 3.1 and 4.2). Further, it is easy to check that such processes possess a unique stationary distribution; some of its properties can be determined explicitly in terms of the characteristics of the process. We refer to Berestycki [26] for a detailed analysis.

# 5
# Asymptotic regimes in stochastic coalescence

This chapter is concerned with systems in which particles coagulate pairwise and randomly as time passes. The key assumption is that the rate at which a pair of particles merges only depends on the two particles involved in the coagulation. Although defining rigorously the evolution is easy when there is only a finite number of particles, the extension to systems with an infinite number of particles and finite total mass requires some regularity assumptions on the rates of coagulation. In a different direction, we shall consider the hydrodynamic regime when the total mass of the system tends to infinity. For so-called sub-multiplicative kernels and under appropriate assumptions, the empirical distribution of particles in the system converges after a suitable renormalization to a deterministic measure depending on time. The latter solves a system of non-linear PDEs which is known as Smoluchowski's coagulation equation. The approach relies mainly on combinatorial arguments via a remarkable connection with certain random graphs models. Finally, we pay special attention to the so-called additive coalescent. We shall develop a combinatorial construction due to Pitman, which involves random trees and forests. This provides a key to understanding the asymptotic behavior of such coagulation dynamics when, roughly speaking, the process starts from dust, that is from a large number of small particles.

## 5.1 Stochastic coalescence

Call a symmetric function $\kappa : \mathbb{R}_+ \times \mathbb{R}_+ \to \mathbb{R}_+$ a *coagulation kernel*. Informally, imagine the Markovian evolution of a system of masses $(x_1, \dots)$ in which each pair of masses, say $\{x_i, x_j\}$, coagulates at a rate $\kappa(x_i, x_j)$, independently of the other pairs in the system. In this section, we shall first define rigorously the dynamics when the number of particles in the initial configuration is finite, and then, under regularity assumptions on $\kappa$, investigate limits when the number of particles tends to infinity while the total mass remains bounded. We refer to Section 1.1.1 for some elementary background on Markov chains which will be useful in the present framework.

### 5.1.1 Coalescent chains

Recall that $\mathcal{S}^{\downarrow}$ denotes the space of decreasing numerical sequences with limit 0. We use the subset of sequences with finitely many non-zero terms,

$$\mathcal{S}_f^{\downarrow} := \left\{ \mathbf{s} = (s_1, \dots) \in \mathcal{S}^{\downarrow} : s_n = 0 \text{ for } n \text{ sufficiently large} \right\} ,$$

as the space of configurations for finite particle systems. The presence of infinitely many zeros in a sequence $\mathbf{s} \in \mathcal{S}_f^{\downarrow}$ is mostly a matter of convenience, which will be explained later. As usual, we may think of terms of a sequence $\mathbf{s}$ as particles, ignoring the 0 terms.

For every $\mathbf{s} \in \mathcal{S}_f^{\downarrow}$ and $1 \le i < j$, we introduce the notation $\mathbf{s}^{i\oplus j}$ for the sequence in $\mathcal{S}_f^{\downarrow}$ obtained from $\mathbf{s}$ by merging its $i$-th and $j$-th terms (i.e. one removes $s_i$ and $s_j$ and adds $s_i + s_j$) and then ranking the resulting sequence in decreasing order. Note that $\mathbf{s}^{i\oplus j} = \mathbf{s}$ if and only if $s_j = 0$.

**Definition 5.1** *Let $\kappa$ be a coagulation kernel. A* **coalescent chain** *with values in $\mathcal{S}_f^{\downarrow}$ and coagulation kernel $\kappa$ is a continuous time Markov chain $X = (X(t), t \ge 0)$, with jump rates given by*

$$q(\mathbf{s}, \cdot) = \sum_{1 \le i < j, s_j > 0} \kappa(s_i, s_j) \delta_{\mathbf{s}^{i\oplus j}} ,$$

*where $\mathbf{s} = (s_1, \dots) \in \mathcal{S}_f^{\downarrow}$.*

The simplest example of coalescent chain is that with the constant coagulation kernel $\kappa(x, y) \equiv 1$. It corresponds to Kingman's coalescent, which was amply studied in Section 4.1. In the forthcoming Sections 5.2.1 and 5.3, we shall investigate two other important examples, namely the *multiplicative* and the *additive* coalescents which are associated to the kernels $\kappa(x, y) = xy$ and $\kappa(x, y) = x + y$, respectively.

We point out that a slightly different notion, which was introduced by Marcus [162] and Lushnikov [155], is often used in the physics literature; see the survey by Aldous [6] and the references therein. In this modified version, the rate at which a pair of particles $\{s_i, s_j\}$ merges is $\kappa(s_i, s_j)/m$ rather than $\kappa(s_i, s_j)$, where $m := \sum_{k=1}^{\infty} s_k$ is the total mass of the system. Of course, the elementary linear time-change $t \to t/m$ transforms a coalescent chain with coagulation kernel $\kappa$ in the sense of Definition 5.1 into that considered by Marcus and Lushnikov, and we shall henceforth focus on the former.

Let us give a dynamical description of a coalescent chain with coagulation kernel $\kappa$. A simple configuration, that is with a single non-zero term, $\mathbf{s} = (s_1, 0, \dots)$, is an absorbing state. In other words, the chain remains constant once there is a single particle left in the system. When one starts from $n \geq 2$ particles, say, $s_1, \dots, s_n > 0$, we may think that an exponential random variable $\mathbf{e}_{i,j}$ with parameter $\kappa(s_i, s_j)$ has been attached to each pair $\{i, j\}$ with $1 \leq i < j \leq n$. The first jump of the chain occurs at time $\min_{1\leq i<j\leq n} \mathbf{e}_{i,j}$, and if this minimum is reached for the pair, say, $\{k, \ell\}$, then at this instant the particles $s_k$ and $s_\ell$ merge. After the first coagulation, the system starts afresh, that is we introduce a new sequence of independent exponential times associated to each pair of particles, and so on. Clearly the number of particles decreases by 1 after each coagulation, and, provided that the coagulation kernel is never 0, after $n-1$ steps the system enters the simple configuration $(\sum_{i=1}^n s_i, 0, \dots)$, which is an absorbing state.

We record this series of remarks in the following statement.

**Proposition 5.1** *Let $X = (X(t), t \geq 0)$ be a coalescent chain in $\mathcal{S}_f^\downarrow$ with coagulation kernel $\kappa$ started from some finite configuration $\mathbf{s} = (s_1, \dots) \in \mathcal{S}_f^\downarrow$. The first coagulation time $T := \inf\{t \geq 0 : X(t) \neq X(0)\}$ has an exponential distribution with parameter*

$$\rho := \sum_{1\leq i<j: s_j>0} \kappa(s_i, s_j) .$$

*The configuration $X(T)$ after the first coagulation is independent of $T$, and its law is specified by*

$$\mathbb{E}(F(X(T)) \mid X(0) = \mathbf{s}) = \rho^{-1} \sum_{1\leq i<j: s_j>0} \kappa(s_i, s_j) F(\mathbf{s}^{i\oplus j}) ,$$

*where $F : \mathcal{S}_f^\downarrow \to \mathbb{R}$ stands for a generic measurable functional.*

From the analytic point of view, Definition 5.1 specifies the infinitesimal generator $\mathsf{G}$ of the coalescent chain. More precisely, for every functional $F : \mathcal{S}_f^\downarrow \to \mathbb{R}$ and $\mathbf{s} = (s_1, \dots) \in \mathcal{S}_f^\downarrow$, one has

$$\mathsf{G}F(\mathbf{s}) = \sum_{1\leq i<j} \big(F(\mathbf{s}^{i\oplus j}) - F(\mathbf{s})\big) \kappa(s_i, s_j) . \tag{5.1}$$

Note that the general term of the series in the right-hand side is 0 whenever $s_j = 0$, so we do not need to restrict the summation to indices $i, j$ such that $s_j > 0$. In the linear case when $\mathbf{f}(\mathbf{s}) = \sum_{i=1}^{\infty} f(s_i)$ for some function $f : [0, \infty[ \to \mathbb{R}$ with $f(0) = 0$, the formula reads

$$\mathtt{Gf}(\mathbf{s}) = \sum_{1 \le i < j} \left( f(s_i + s_j) - f(s_i) - f(s_j) \right) \kappa(s_i, s_j) . \qquad (5.2)$$

Observe that this quantity is always 0 for the identity function $f(x) = x$, which corresponds to the obvious fact that the total mass in the system, $\sum_{i=1}^{\infty} X_i(t)$, remains constant as time passes.

Then, following an idea of Fournier [104], we shall specify a classical Poissonian construction of interacting particle systems in the framework of coalescent chains. This yields an efficient coupling of coalescent chains started from different configurations which will be useful in the next section.

Set $\mathbb{T} = \{(i, j) \in \mathbb{N}^2 : i < j\}$ and consider a Poisson random measure $N$ on $\mathbb{T} \times \mathbb{R}_+ \times \mathbb{R}_+$ with intensity $\# \otimes du \otimes dt$, where $\#$ stands for the counting measure on $\mathbb{T}$. Informally, each atom of $N$ corresponds to a possible coagulation, the first coordinate specifies the pair of indices of the particles potentially involved, the second is used as a test to decide whether the coagulation is effective or not, and the third gives the instant when this possible coagulation may occur. More precisely, suppose that $((i, j), u, t)$ is an atom of $N$, and that immediately before time $t$, the configuration in the system is $\mathbf{y} = (y_1, \dots) \in \mathcal{S}_f^{\downarrow}$. Then the atoms $y_i$ and $y_j$ may coagulate at time $t$, and the coagulation does occur if and only if $u \le \kappa(y_i, y_j)$. In other words, the configuration immediately after time $t$ will be

$$\begin{cases} \mathbf{y}^{i \oplus j} & \text{if } u \le \kappa(y_i, y_j) \\ \mathbf{y} & \text{otherwise.} \end{cases}$$

Of course, this procedure is purely formal so far, as the set of times $t$ at which an atom $((i, j), u, t)$ arises in the Poisson random measure $N$ is everywhere dense. Let us now check that, nonetheless, this Poissonian construction is well-defined when one starts from a finite configuration. More precisely, we have to check that the set of times at which a coagulation can be observed is discrete. So pick $\mathbf{y} = (y_1, \dots) \in \mathcal{S}_f^{\downarrow}$, and suppose that $y_i = 0$ whenever $i > n$. Let $G$ be the finite set of real numbers $x$ which can be expressed in the form $x = \sum_{i \in I} y_i$, where $I$ is an arbitrary subset of $\mathbb{N}$, and set $a := \max\{\kappa(x, y) : x, y \in G\}$. As $\mathbf{y}^{i \oplus j} = \mathbf{y}$ whenever $j > n$, we only need to consider atoms $((i, j), u, t)$ with $j \le n$ and $u \le a$. This implies that the set of times $t$ at which a coagulation can be observed is included into the set of atoms of a Poisson random measure on $\mathbb{R}_+$ with intensity $a\frac{n(n-1)}{2}dt$. Since the latter is discrete a.s., the Poissonian construction is thus well-defined.

**Remark.** Consider the special case of Kingman's coalescent when $\kappa(x, y) \equiv 1$, and write $N' = \sum \delta_{((i,j),t)}$ where the summation is taken over the atoms $((i, j), u, t)$ of $N$ such that $u \leq 1$. Then $N'$ is a Poisson random measure on $\mathbb{T} \times \mathbb{R}_+$ with intensity $\# \otimes dt$, and the construction above only depends on $N'$. More precisely, this construction is merely a variation of that presented in Section 4.2.3 for exchangeable coalescents, specialized to Kingman's coalescent.

Given a Poisson random measure $N$ as above, we introduce the natural filtration induced by $N$, namely

$$\mathcal{F}_t := \sigma\left(\mathbb{1}_{\mathbb{T}\times\mathbb{R}_+\times[0,t]}N\right), \qquad t \geq 0,$$

and for every $\mathbf{x} \in \mathcal{S}_f^{\downarrow}$, we denote by $X(\mathbf{x}, \cdot) = (X(\mathbf{x}, t), t \geq 0)$ the coagulation process with values in $\mathcal{S}_f^{\downarrow}$ obtained by this procedure when the initial configuration is $X(\mathbf{x}, 0) = \mathbf{x}$.

**Proposition 5.2** *For every $\mathbf{x} \in \mathcal{S}_f^{\downarrow}$, $X(\mathbf{x}, \cdot)$ is Markovian in the filtration $(\mathcal{F}_t)_{t\geq 0}$; more precisely it is a coalescent chain with coagulation kernel $\kappa$ started from $\mathbf{x}$.*

*Proof* For every $t, t' \geq 0$, $X(\mathbf{x}, t + t')$ is constructed from $X(\mathbf{x}, t)$ and the restriction of the Poisson random measure $N$ to $\mathbb{T} \times \mathbb{R}_+ \times ]t, t']$. The superposition property of Poisson measures (cf. Lemma 2.4) easily implies that $X(\mathbf{x}, \cdot)$ enjoys the Markov property. Further, it should be plain from this construction that its jump rates coincide with that of a coalescent chain with coagulation kernel $\kappa$. □

### 5.1.2 Extension to infinite systems

Our aim in this section is to define the evolution of a coalescent process governed by a coagulation kernel $\kappa$ when there are infinitely many particles in the system, say $s_1, \ldots$, with a finite total mass $\sum_{i=1}^{\infty} s_i$. The main problem arises when

$$\sum_{1\leq i<j:s_j>0} \kappa(s_i, s_j) = \infty,$$

since then coagulations occur immediately.

This leads us to investigate continuity properties of coalescent chains starting from a finite configuration $\mathbf{s} \in \mathcal{S}_f^{\downarrow}$, as a function of $\mathbf{s}$. In this direction, the following elementary lemma will be useful later in the text.

**Lemma 5.1** *Introduce the space $\mathcal{S}_f$ of non-ranked finite numerical sequences, that is*

$$\mathcal{S}_f := \{\mathbf{x} = (x_1, \dots), x_i \geq 0 \text{ and } x_n = 0 \text{ when } n \text{ is sufficiently large}\} ,$$

*and write $\mathbf{x} \to \mathbf{x}^{\downarrow}$ for the projection from $\mathcal{S}_f$ to $\mathcal{S}_f^{\downarrow}$ which rearranges the terms of a sequence in decreasing order.*

*The map $\mathbf{x} \to \mathbf{x}^{\downarrow}$ is a contraction on $\mathcal{S}_f$, that is for every $\mathbf{x}, \mathbf{y} \in \mathcal{S}_f$ there is the inequality*

$$\sum_{i=1}^{\infty} |x_i^{\downarrow} - y_i^{\downarrow}| \leq \sum_{i=1}^{\infty} |x_i - y_i| .$$

*Proof* We will check by induction on $n$ that for arbitrary real numbers $a_1 \geq a_2 \geq \dots \geq a_n \geq 0$ and $b_1 \geq b_2 \geq \dots \geq b_n \geq 0$, and every permutation $\sigma$ of $[n] = \{1, \dots, n\}$,

$$\sum_{i=1}^{n} |a_i - b_i| \leq \sum_{i=1}^{n} |a_i - b_{\sigma(i)}| . \tag{5.3}$$

Without loss of generality, we may (and will) further assume that $a_1 \geq b_1$.

Let $p \in \mathbb{N}$ and assume that (5.3) holds for all integers $n \leq p$. Consider real numbers $a_1 \geq a_2 \geq \dots \geq a_{p+1} \geq 0$ and $b_1 \geq b_2 \geq \dots \geq b_{p+1} \geq 0$, and a permutation $\sigma$ of $[p+1]$. Plainly, (5.3) holds for $n = p+1$ and $\sigma$ whenever $\sigma$ possesses at least one fixed point.

So assume that $\sigma$ has no fixed point and let $j = \sigma(1)$. Write $\tau$ for the transposition $1 \leftrightarrow j$, and $\sigma' = \tau \circ \sigma$. So $\sigma'(1) = 1$, $\sigma'(\sigma^{-1}(1)) = j$ and $\sigma(i) = \sigma'(i)$ for $i \neq 1, \sigma^{-1}(1)$. Since $a_1 \geq b_1 \geq b_{\sigma(1)}$, we have

$$\begin{aligned} |a_1 - b_j| + |a_{\sigma^{-1}(1)} - b_1| &= |a_1 - b_1| + |b_1 - b_j| + |a_{\sigma^{-1}(1)} - b_1| \\ &\geq |a_1 - b_1| + |a_{\sigma^{-1}(1)} - b_j| , \end{aligned}$$

and therefore

$$\sum_{i=1}^{p+1} |a_i - b_{\sigma'(i)}| \leq \sum_{i=1}^{p+1} |a_i - b_{\sigma(i)}| .$$

Since $\sigma'$ has at least one fixed point,

$$\sum_{i=1}^{p+1} |a_i - b_i| \leq \sum_{i=1}^{p+1} |a_i - b_{\sigma'(i)}| ,$$

which completes the proof. □

We now introduce some spaces of sequences and paths. First, we set

$$\mathcal{S}_1^{\downarrow} := \left\{ \mathbf{x} = (x_1, \dots) \in \mathcal{S}^{\downarrow} : \sum_{i=1}^{\infty} x_i < \infty \right\} ;$$

note that the space of mass-partitions $\mathcal{P}_{\mathrm{m}} = \{\mathbf{x} = (x_1, \dots) \in \mathcal{S}^{\downarrow} : \sum_{i=1}^{\infty} x_i \leq 1\}$ can be thought of as the unit ball of $\mathcal{S}_1^{\downarrow}$. We endow $\mathcal{S}_1^{\downarrow}$ with the usual distance

$$\|\mathbf{x} - \mathbf{y}\|_1 := \sum_{i=1}^{\infty} |x_i - y_i| .$$

Then $\mathcal{S}_1^{\downarrow}$ is a Polish set and $\mathcal{S}_f^{\downarrow}$ is an everywhere dense subset of $\mathcal{S}_1^{\downarrow}$. We may also represent a configuration $\mathbf{x} = (x_1, \dots) \in \mathcal{S}_1^{\downarrow}$ by the finite measure on $\mathbb{R}_+$

$$\nu_{\mathbf{x}} := \sum_{i=1}^{\infty} x_i \delta_{x_i} ;$$

then it follows readily from Scheffé's lemma that convergence in $\mathcal{S}_1^{\downarrow}$ is equivalent to the weak convergence of finite measures on $\mathbb{R}_+$.

We observe a few elementary properties. First, the coagulation operators are contractions on $\mathcal{S}_1^{\downarrow}$, that is for every $\mathbf{x}, \mathbf{y} \in \mathcal{S}_1^{\downarrow}$ and indices $1 \leq i < j$, one has

$$\|\mathbf{x}^{i\oplus j} - \mathbf{y}^{i\oplus j}\|_1 \leq \|\mathbf{x} - \mathbf{y}\|_1 . \tag{5.4}$$

More precisely, this follows immediately from the triangle inequality and Lemma 5.1. In the same vein, we also note that Lemma 5.1 implies that for every $\mathbf{x}, \mathbf{y} \in \mathcal{S}_1^{\downarrow}$ and indices $1 \leq i < j$, one has

$$\|\mathbf{x} - \mathbf{y}^{i\oplus j}\|_1 \leq \|\mathbf{x} - \mathbf{y}\|_1 + |x_i - y_i - y_j| + x_j - |x_i - y_i| - |x_j - y_j| ,$$

so by the triangle inequality

$$\|\mathbf{x} - \mathbf{y}^{i\oplus j}\|_1 \leq \|\mathbf{x} - \mathbf{y}\|_1 + 2y_j . \tag{5.5}$$

Next consider for every $t \geq 0$ the space $\mathbb{D}_t^{\downarrow}$ of càdlàg paths $\omega : [0, t] \to \mathcal{S}_1^{\downarrow}$, endowed with the uniform distance

$$\|\omega - \omega'\|_{1,\infty} := \sup_{u \in [0,t]} \|\omega(u) - \omega'(u)\|_1 ,$$

so $\mathbb{D}_t^{\downarrow}$ is a complete metric space (specialists will note that we are using a distance that is stronger than the usual Skorokhod's distance, see for example Billingsley [57] or Jacod and Shiryaev [127]).

Now recall from Proposition 5.2 that we can construct simultaneously (i.e. couple) coalescent chains in $\mathcal{S}_f^{\downarrow}$ with the same coagulation kernel but different initial configurations. More precisely, given a Poisson random measure $N$ with intensity $\# \otimes du \otimes dt$ and an arbitrary $\mathbf{x} \in \mathcal{S}_f^{\downarrow}$, we write $X(\mathbf{x}, \cdot) = (X(\mathbf{x}, t),$

$t \geq 0$) for the coalescent chain with coagulation kernel $\kappa$ started from $\mathbf{x}$ which is built from $N$. We now state the main result of this section, which is due to Fournier [104] and claims that whenever the coagulation kernel $\kappa$ is Lipschitz-continuous, one can construct coalescent processes in $\mathcal{S}_1^{\downarrow}$ with coagulation kernel $\kappa$, by approximation from finite systems.

**Theorem 5.1** *Suppose that the coagulation kernel is locally Lipschitz-continuous, that is for every $a > 0$, there exists $c_a < \infty$ such that*

$$|\kappa(x, y) - \kappa(x', y')| \leq c_a(|x - x'| + |y - y'|), \qquad x, y, x', y' \in [0, a].$$

*For every $\mathbf{x} \in \mathcal{S}_1^{\downarrow}$, there exists a unique càdlàg process $X(\mathbf{x}, \cdot)$ with values in $\mathcal{S}_1^{\downarrow}$, called* **coalescent process with coagulation kernel** $\kappa$ *started from $\mathbf{x}$, such that the following holds. For every sequence of finite configurations $(\mathbf{x}^{(k)}, k \in \mathbb{N})$ which converges to $\mathbf{x}$ in $\mathcal{S}_1^{\downarrow}$, and every $t \geq 0$,*

$$\lim_{k\to\infty} \mathbb{E}\left( \sup_{0\leq u\leq t} \|X(\mathbf{x}, u) - X(\mathbf{x}^{(k)}, u)\|_1 \right) = 0.$$

We implicitly suppose from now on that the assumptions of Theorem 5.1 hold. The proof relies on a couple of lemmas. To start, we fix $\mathbf{x}, \mathbf{y} \in \mathcal{S}_f$, and set for every $t \geq 0$

$$f(\mathbf{x}, \mathbf{y}, t) = \mathbb{E}\left( \sup_{0\leq u\leq t} \sum_{i=1}^{\infty} |X_i(\mathbf{x}, u) - X_i(\mathbf{y}, u)| \right).$$

**Lemma 5.2** *Let $\mathbf{x}, \mathbf{y}$ be two finite mass-partitions, that is two configurations in $\mathcal{S}_f^{\downarrow}$ such that $\sum_{i=1}^{\infty} x_i \leq 1$ and $\sum_{i=1}^{\infty} y_i \leq 1$. There exists a pair of finite constants $c$ and $c'$, depending on $\kappa$ but not on $\mathbf{x}$ and $\mathbf{y}$, such that for every $t \geq 0$ and every integer $n$ for which $x_n = y_n = 0$*

$$f(\mathbf{x}, \mathbf{y}, t) \leq (1 + ct)\|\mathbf{x} - \mathbf{y}\|_1 + c'n^4t^2.$$

*Proof* Let $\Lambda_1$ denote the event that during the time interval $[0, t]$, either no coagulation occurs for $X(\mathbf{x}, \cdot)$ and $X(\mathbf{y}, \cdot)$, or a single coagulation occurs for both at the same time and involving the same pair of indices (so, if $t' \leq t$ denotes the instant when this coagulation occurs, then $X(\mathbf{x}, t') = \mathbf{x}^{i\oplus j}$ and $X(\mathbf{y}, t') = \mathbf{y}^{i\oplus j}$ for some $1 \leq i < j \leq n$). It follows from (5.4) that

$$\sup_{0\leq u\leq t} \|X(\mathbf{x}, u) - X(\mathbf{y}, u)\|_1 = \|\mathbf{x} - \mathbf{y}\|_1 \qquad \text{on } \Lambda_1.$$

Next, recall from Section 5.1.1 the construction of coagulation chains in terms of the Poisson random measure $N$ and that both initial configurations

$\mathbf{x}$ and $\mathbf{y}$ have less than $n$ particles. Observe also that for every $t \geq 0$, the largest particle in $X(\mathbf{x}, t)$ and $X(\mathbf{y}, t)$ cannot exceed 1, and set $a := \max_{0 \leq x, y \leq 1} \kappa(x, y)$. The probability of the event $\Lambda_2$ that $N$ has at least two atoms on $\{(i, j) : 1 \leq i < j < n\} \times [0, a] \times [0, t]$ is

$$1 - \exp\left(-at\frac{n(n-1)}{2}\right) - at\frac{n(n-1)}{2} \exp\left(-ta\frac{n(n-1)}{2}\right) \leq t^2 a^2 n^4 .$$

On this event, we use the obvious bound

$$\|X(\mathbf{x}, u) - X(\mathbf{y}, u)\|_1 \leq \sum_{i=1}^{\infty} (X_i(\mathbf{x}, u) + X_i(\mathbf{y}, u)) \leq 2 ,$$

where the second inequality stems from the assumptions that $\sum_{i=1}^{\infty} x_i \leq 1$ and $\sum_{i=1}^{\infty} y_i \leq 1$ and the property of conservation of the total mass for coalescent chains. This yields

$$\mathbb{E}\left(\sup_{0 \leq u \leq t} \|X(\mathbf{x}, u) - X(\mathbf{y}, u)\|_1, \Lambda_2\right) \leq c' n^4 t^2 .$$

Then, for every $1 \leq k < \ell \leq n$, we consider the event $\Lambda_{k,\ell}$ where $N$ has exactly one atom $((k, \ell), u, t')$ on $\{(i, j) : 1 \leq i < j < n\} \times [0, a] \times [0, t]$, such that $\kappa(x_k, x_\ell) < u \leq \kappa(y_k, y_\ell)$. Elementary properties of Poisson random measures (which were recalled in Section 2.2.2) easily yield that

$$\mathbb{P}(\Lambda_{k,\ell}) \leq t|\kappa(y_k, y_\ell) - \kappa(x_k, x_\ell)| \leq tc_1(|x_k - y_k| + |x_\ell - y_\ell|) ,$$

where $c_1$ denotes the Lipschitz constant of the coagulation kernel $\kappa$ on $[0, 1] \times [0, 1]$. Further, on this event, the $k$-th and $\ell$-th particles coagulate at time $t'$ in the chain $X(\mathbf{y}, \cdot)$ but not in the chain $X(\mathbf{x}, \cdot)$. It follows from (5.5) that

$$\sup_{0 \leq u \leq t} \|X(\mathbf{x}, u) - X(\mathbf{y}, u)\|_1 \leq 2y_\ell + \|\mathbf{x} - \mathbf{y}\|_1 \qquad \text{on } \Lambda_{k,\ell} .$$

This yields the upper-bound

$$\mathbb{E}\left(\sup_{0 \leq u \leq t} \|X(\mathbf{x}, u) - X(\mathbf{y}, u)\|_1 - \|\mathbf{x} - \mathbf{y}\|_1, \Lambda_{k,\ell}\right)$$
$$\leq 2tc_1 y_\ell(|x_k - y_k| + |x_\ell - y_\ell|) .$$

Similarly, we consider the event $\Lambda'_{k,\ell}$ where $N$ has exactly one atom $((k, \ell), u, t')$ on $\{(i, j) : 1 \leq i < j < n\} \times [0, c_1] \times [0, t]$, such that $\kappa(y_k, y_\ell) < u \leq \kappa(x_k, x_\ell)$. We get the upper-bound

$$\mathbb{E}\left(\sup_{0 \leq u \leq t} \|X(\mathbf{x}, u) - X(\mathbf{y}, u)\|_1 - \|\mathbf{x} - \mathbf{y}\|_1, \Lambda'_{k,\ell}\right)$$
$$\leq 2tc_1 x_\ell(|x_k - y_k| + |x_\ell - y_\ell|) .$$

We can now conclude the proof. The events considered above cover the entire probability space, so adding the upper-bounds yields

$$\mathbb{E}\left(\sup_{0\le u\le t}\|X(\mathbf{x},u)-X(\mathbf{y},u)\|_1\right)-\|\mathbf{x}-\mathbf{y}\|_1$$

$$\le c'n^4t^2+2tc_1\sum_{1\le k\le\ell}(x_\ell+y_\ell)(|x_k-y_k|+|x_\ell-y_\ell|)$$

$$\le c'n^4t^2+4tc_1\sum_{k,\ell=1}^{\infty}(x_\ell+y_\ell)|x_k-y_k|$$

$$\le c'n^4t^2+ct\sum_{k=1}^{\infty}|x_k-y_k|,$$

where the inequality at the third line stems from the fact that $x_\ell+y_\ell\le x_k+y_k$ for $k<\ell$. □

The next step in the proof of Theorem 5.1 is a sharper bound which derives from Lemma 5.2 and Gronwall's Lemma.

**Lemma 5.3** *Let* $\mathbf{x},\mathbf{y}$ *be two finite mass-partitions, that is two configurations in* $\mathcal{S}_f^{\downarrow}$ *such that* $\sum_{i=1}^{\infty}x_i\le 1$ *and* $\sum_{i=1}^{\infty}y_i\le 1$. *There exists a finite constant* $c$ *depending on* $\kappa$ *but not on* $\mathbf{x}$ *and* $\mathbf{y}$, *such that for every* $t\ge 0$

$$f(\mathbf{x},\mathbf{y},t)\le\|\mathbf{x}-\mathbf{y}\|_1 e^{ct}.$$

*Proof* We start with the following elementary observation. Let $\omega,\omega':\mathbb{R}_+\to\mathcal{S}_1^{\downarrow}$ be two càdlàg paths. Pick $t'\ge t\ge 0$ and write $\omega(t)=\mathbf{s}$, $\omega'(t)=\mathbf{s}'$. Then there is the inequality

$$\sup_{0\le u\le t'}\|\omega(u)-\omega'(u)\|_1\le\sup_{0\le u\le t}\|\omega(u)-\omega'(u)\|_1+\sup_{t\le u\le t'}\|\omega(u)-\omega'(u)\|_1-\|\mathbf{s}-\mathbf{s}'\|_1.$$

Next, pick an integer $n$ such that $x_n=y_n=0$ and recall that for every $u\ge 0$

$$X_n(\mathbf{x},u)=X_n(\mathbf{y},u)=0\quad\text{and}\quad\|X(\mathbf{y},u)\|_1\le 1\,,\ \|X(\mathbf{y},u)\|_1\le 1\,.$$

Combining the inequality above, Lemma 5.2 and the Markov property, we see that there exist two finite constants $c$ and $c'$, independent of $\mathbf{x}$ and $\mathbf{y}$, such that for all integers $\ell\le k$:

$$f(\mathbf{x},\mathbf{y},\ell t/k)\le(1+c/k)f(\mathbf{x},\mathbf{y},(\ell-1)t/k)+c'n^4k^{-2}.$$

Then summation for $\ell = 1$ to $k$ yields

$$f(\mathbf{x}, \mathbf{y}, t) \leq \|\mathbf{x}-\mathbf{y}\|_1 + ck^{-1} \sum_{\ell=0}^{k-1} f(\mathbf{x}, \mathbf{y}, \ell t/k) + c'n^4 k^{-1}.$$

Since $t \to f(\mathbf{x}, \mathbf{y}, t)$ increases, we have

$$f(\mathbf{x}, \mathbf{y}, t) \leq \|\mathbf{x}-\mathbf{y}\|_1 + c\int_0^t f(\mathbf{x}, \mathbf{y}, u)du + c'n^4 k^{-1},$$

and then letting $k \to \infty$,

$$f(\mathbf{x}, \mathbf{y}, t) \leq \|\mathbf{x}-\mathbf{y}\|_1 + c\int_0^t f(\mathbf{x}, \mathbf{y}, u)du.$$

We conclude by an application of Gronwall's lemma. □

Establishing Theorem 5.1 is now straightforward. Without loss of generality, we may assume that $\|\mathbf{x}^{(k)}\|_1 \leq 1$ for every $k \in \mathbb{N}$ (otherwise we can rescale space and time and work with $\mathbf{y}^{(k)} = a\mathbf{x}^{(k)}$ for some sufficiently small $a > 0$). Then Lemma 5.3 shows that for every $t \geq 0$, $\left(X(\mathbf{x}^{(k)}, \cdot), k \in \mathbb{N}\right)$ is a Cauchy sequence in the space of càdlàg processes $Y = (Y(u), u \leq t)$ with values in $\mathcal{S}_1^{\downarrow}$ and such that $\mathbb{E}\left(\sup_{0\leq u\leq t} \|Y\|_1\right) < \infty$, endowed with the distance

$$d(Y, Y') := \mathbb{E}\left(\sup_{0\leq u\leq t} \|Y(u) - Y'(u)\|\right).$$

The latter is a complete space, so the Cauchy sequence converges.

We now derive some consequences of Theorem 5.1; the first deals with the Markov property.

**Corollary 5.1** *Assume that the hypotheses of Theorem 5.1 are fulfilled. Then the family of processes* $\left(X(\mathbf{x}, \cdot), \mathbf{x} \in \mathcal{S}_1^{\downarrow}\right)$ *is Markovian, with a Feller semigroup: for every continuous and bounded functional* $F : \mathcal{S}_1^{\downarrow} \to \mathbb{R}$, *the map*

$$\mathrm{P}_t F : \mathbf{x} \to \mathbb{E}\left(F(X(\mathbf{x}, t))\right)$$

*is continuous on* $\mathcal{S}_1^{\downarrow}$ *for every* $t \geq 0$, *and*

$$\lim_{t\to 0} \mathrm{P}_t F(\mathbf{x}) = F(\mathbf{x}).$$

*Proof* Lemma 5.3 shows that the map $\mathbf{y} \to \mathrm{P}_t F(\mathbf{y})$ is uniformly continuous on the space $\left\{\mathbf{y} \in \mathcal{S}_f^{\downarrow} : \sum_{i=1}^{\infty} s_i \leq 1\right\}$, and it follows that its extension to $\mathcal{S}_1^{\downarrow}$ is continuous. Moreover, the right-continuity at 0 of the paths of $X(\mathbf{x}, \cdot)$ ensures by dominated convergence that $\lim_{t\to 0+} \mathrm{P}_t F(\mathbf{x}) = F(\mathbf{x})$.

Recall that $(\mathcal{F}_t)_{t\geq 0}$ for the natural filtration induced by the Poisson random measure $N$, and let $(\mathbf{x}^{(k)}, k \in \mathbb{N})$ denote a sequence in $\mathcal{S}_f^{\downarrow}$ which converges to $\mathbf{x}$ in $\mathcal{S}_1^{\downarrow}$. The Markov property of the coalescent chain $X(\mathbf{x}^{(k)}, \cdot)$ in Lemma 5.2 reads for every $t, u \geq 0$

$$\mathbb{E}\left(F(X(\mathbf{x}^{(k)}, t+u)) \mid \mathcal{F}_u\right) = \mathrm{P}_t F(X(\mathbf{x}^{(k)}, u)).$$

Taking the limit as $k \to \infty$ yields that $X(\mathbf{x}, \cdot)$ is Markovian in the filtration $(\mathcal{F}_t)_{t\geq 0}$ with semigroup $(\mathrm{P}_t, t \geq 0)$. □

Then we turn our attention to the infinitesimal generator $\mathsf{G}$ of the coalescent process with coagulation kernel $\kappa$.

**Corollary 5.2** *Assume that the hypotheses of Theorem 5.1 are fulfilled and that* $\kappa(0, 0) = 0$.

(i) *Fix* $n \in \mathbb{N}$ *and consider a symmetric function* $f : \mathbb{R}_+^n \to \mathbb{R}$ *which is of class* $\mathcal{C}^1$. *Then the functional* $F : \mathcal{S}_1^{\downarrow} \to \mathbb{R}$ *given by* $F(\mathbf{x}) = f(x_1, \ldots, x_n)$ *belongs to the domain of* $\mathsf{G}$, *and more precisely,*

$$\mathsf{G}F(\mathbf{x}) = \sum_{1\leq i<j} (F(\mathbf{x}^{i\oplus j}) - F(\mathbf{x}))\kappa(x_i, x_j),$$

*where the series in the right-hand side is absolutely convergent.*

(ii) *Consider a function* $f : \mathbb{R}_+ \to \mathbb{R}$ *which is of class* $\mathcal{C}^1$ *with* $f(0) = 0$. *Then the linear functional* $\mathbf{f} : \mathcal{S}_1^{\downarrow} \to \mathbb{R}$ *given by* $\mathbf{f}(\mathbf{x}) = \sum_{i=1}^{\infty} f(x_i)$ *belongs to the domain of* $\mathsf{G}$, *and more precisely,*

$$\mathsf{G}\mathbf{f}(\mathbf{x}) = \sum_{1\leq i<j} (f(x_i + x_j) - f(x_i) - f(x_j))\kappa(x_i, x_j),$$

*where the series in the right-hand side is absolutely convergent.*

**Remark.** When $\kappa(0, 0) > 0$, it is easy to check that for every $t > 0$, the configuration $X(\mathbf{x}, t)$ is finite a.s. (see Theorem 4.1 and its proof for Kingman's coalescent $\kappa(x, y) \equiv 1$). Thus in this situation, the infinitesimal generator is simply given by (5.1).

*Proof* It is convenient to use the characterization of the infinitesimal generator in terms of martingales; see for instance in Ethier and Kurtz [99]. Recall that $(\mathcal{F}_t)_{t\geq 0}$ denotes the natural filtration induced by the Poisson random measure $N$, and that coalescent chains $X(\mathbf{x}, \cdot)$ are Markovian in this filtration. For the sake of conciseness, we shall only establish the first claim; the proof of the second is similar. Further, without loss of generality, we shall focus on configurations $\mathbf{x} \in \mathcal{S}_1^{\downarrow}$ such that $\sum_{i=1}^{\infty} x_i \leq 1$.

To start with, we note that the assumption $\kappa(0,0) = 0$ and the Lipschitz-regularity of $\kappa$ yield the bound $\kappa(x, y) \leq c(x+y)$ where $c$ is some finite constant. On the other hand, since $F(\mathbf{x}) = f(x_1, \ldots, x_n)$ with $f$ symmetric and of class $\mathcal{C}^1$, we have $F(\mathbf{x}^{i\oplus j}) - F(\mathbf{x}) \leq c' x_j$, where $c'$ is again a finite constant. As

$$\sum_{1 \leq i < j} x_j(x_i + x_j) \leq 2 \sum_{1 \leq i < j} x_i x_j \leq 1 \,,$$

the series

$$\sum_{1 \leq i < j} (F(\mathbf{x}^{i\oplus j}) - F(\mathbf{x}))\kappa(x_i, x_j) := \mathsf{G}F(\mathbf{x})$$

is absolutely convergent. A similar argument now shows that the functional $\mathsf{G}F$ is uniformly continuous on $\left\{\mathbf{x} \in \mathcal{S}_1^{\downarrow} : \sum_{i=1}^{\infty} x_i \leq 1\right\}$.

We know from (5.1) that for every finite configuration $\mathbf{y} \in \mathcal{S}_f^{\downarrow}$, the process

$$F(X(\mathbf{y}, t)) - \int_0^t \mathsf{G}F(X(\mathbf{y}, u))du \,, \qquad t \geq 0 \,,$$

is an $(\mathcal{F}_t)$-martingale. Approaching $\mathbf{x} \in \mathcal{S}_1^{\downarrow}$ by a sequence of finite configuration and using the regularity of the functional $\mathsf{G}F$, we deduce from Theorem 5.1 that

$$F(X(\mathbf{x}, t)) - \int_0^t \mathsf{G}F(X(\mathbf{x}, u))du \,, \qquad t \geq 0 \,,$$

is an $(\mathcal{F}_t)$-martingale for any initial configuration $\mathbf{x} \in \mathcal{S}_1^{\downarrow}$. □

## 5.2 Hydrodynamic behavior and Smoluchowski's equations

We now turn our attention to a different type of asymptotics for stochastic coalescents, namely the so-called hydrodynamic behavior. For the sake of simplicity, we will assume here that the mass of a particle is an integer; for instance a particle may be a polymer and then we view the number of atoms forming this polymer as its mass. Further, we shall focus on monodisperse initial configurations $\mathbf{x}^{[k]} = (1, \ldots, 1, 0, \ldots)$ comprising $k$ particles each of unit mass, and write $X^{[k]}(\cdot) := X(\mathbf{x}^{[k]}, \cdot)$ for the stochastic coalescent with a given coagulation kernel $\kappa$ and started from the monodisperse configuration $\mathbf{x}^{[k]}$. We are interested in the quantity

$$n_t^{[k]}(x) := k^{-1} \# \left\{ i \in \mathbb{N} : X_i^{[k]}(t/k) = x \right\} , \qquad x \in \mathbb{N}$$

when $k$ is large. Note that both the number of particles of given size $x$ and the time parameter have been rescaled by the same factor $1/k$. We may imagine a volume of size $k$ containing particles which evolve according to the dynamics of stochastic coalescence, and then think of $n_t^{[k]}(x)$ as the *concentration* of particles with mass $x$ at time $t/k$. Roughly, the time-rescaling implies that the average rate of coagulation of particles with masses $x$ and $y$ per unit volume is of order $\frac{1}{2}n_t^{[k]}(x)n_t^{[k]}(y)\kappa(x, y)$, where the factor $1/2$ takes into account the obvious symmetry.

The initial concentration $n_0^{[k]}(x) = \mathbb{1}_{\{x=1\}}$ being the same for all $k$, we may expect that for every $t \geq 0$, when the volume $k$ tends to infinity, the particle concentration at time $t/k$, $(n_t^{[k]}(x), x \in \mathbb{N})$ should have a deterministic limit $(n_t(x), x \in \mathbb{N})$ and that the latter should solve

$$\frac{dn_t(x)}{dt} = -n_t(x)\sum_{y=1}^{\infty} n_t(y)\kappa(x, y) + \frac{1}{2}\sum_{y=1}^{x-1} n_t(y)n_t(x-y)\kappa(y, x-y) . \quad (5.6)$$

Indeed, the first term in the right-hand side accounts for the removal of particles of mass $x$ by coagulation with another particle (of mass $y$), and the second for the creation of a particle of mass $x$ by the coagulation of two particles with respective masses $y$ and $x - y$. The evolution equation (5.6) is known as *Smoluchowski's coagulation equation*; it has received a considerable interest in the literature. Probabilist readers are referred to the remarkable survey by Aldous [6] for much more on this topic; further classical references include Drake [84] and Dubovski [85].

In this section, we shall focus on such hydrodynamic issues for certain coagulation kernels; our main purpose is to present a rather elementary account of typical results in this vein. We shall first investigate the case of the multiplicative kernel with monodisperse initial configurations using the connection with the random graph model of Erdős and Rényi [94], and then treat sub-multiplicative kernels by a coupling argument. Sharper and more general results have been obtained by Jeon [129] and Norris [175, 176] using more sophisticated techniques.

### 5.2.1 The multiplicative kernel

In this section, we consider the *multiplicative* kernel, that is we assume henceforth that the coagulation kernel is given by

$$\kappa(x, y) = xy .$$

Recall that we focus on the case of monodisperse initial configurations, and for every integer $k$, we write $X^{[k]} = (X^{[k]}(t), t \geq 0)$ for the multiplicative coalescent started from $k$ particles, each of unit mass, that is $\mathbf{x}^{[k]} = (1, \ldots, 1, 0, \ldots)$. This special situation has an illuminating interpretation in terms of the random graph model of Erdős and Rényi [94] that we now present (see also for example Bollobás [60] for a more accessible reference).

Let us start by introducing some basic notions on graphs. We view $[k] = \{1, \ldots, k\}$ as a set of $k$ vertices. An *edge* is a pair $\{i, j\}$ of distinct vertices; we stress that edges are not oriented, that is $\{i, j\}$ and $\{j, i\}$ denote the same edge. We write $\mathcal{E}^{[k]}$ for the set of edges of $[k]$. Each edge can be either *open* or *closed*, and a *graph* $G$ on $[k]$ is a collection of open edges. Then we say that two distinct edges are adjacent if they share a common vertex, and then that two vertices $i, j \in [k]$ are *connected* in the graph $G$ and write $i \sim^G j$, if either $i = j$ or there exists a sequence of distinct vertices $v_1, \ldots, v_m \in [k]$ such that $v_1 = i$, $v_m = j$ and each edge $\{v_\ell, v_{\ell+1}\}$ is open for $\ell = 1, \ldots, m-1$. The relation $i \sim^G j$ is an equivalence relation that induces a partition of $[k]$ into connected components; we denote the latter by $\Pi^{[k]}(G)$. The size of a connected component of $G$ is the number of vertices in that component.

We next describe the Erdős-Rényi random graph model. We attach to each edge $e \in \mathcal{E}^{[k]}$ an exponential variable $\mathbf{e}_e$ with unit parameter, such that the family $(\mathbf{e}_e, e \in \mathcal{E}^{[k]})$ consists of independent variables. We then define an increasing family of random graphs $(\mathcal{G}^{[k]}(t), t \geq 0)$ by deciding that an edge $e$ is open at time $t \geq 0$ if and only if $\mathbf{e}_e \leq t$. For simplicity, we now write $\Pi^{[k]}(t) := \Pi^{[k]}(\mathcal{G}^{[k]}(t))$ for the partition of $[k]$ into connected components at time $t$. It follows immediately from the absence of memory of the exponential law that the process $\Pi^{[k]}(\cdot)$ is Markovian. More precisely, because there are $\#C \times \#C'$ edges that may connect two disjoint components $C$ and $C'$, and the minimum of $n$ independent standard exponential variables has the exponential law with parameter $n$, any pair of components of $\Pi^{[k]}$ coalesces at a rate given by the product of their sizes, independently of the other pairs of components. Turning our attention to the sizes of the connected components, the preceding observations readily yield the following basic representation.

**Lemma 5.4** *The process of the ranked sequence of the sizes of the connected components in the Erdős-Rényi random graph model $\mathcal{G}^{[k]}(\cdot)$ is a version of the multiplicative coalescent chain $X^{[k]}(\cdot)$ started from the monodisperse initial configuration.*

For every $i \in [k]$ and $t \geq 0$, let $C_i^{[k]}(t)$ denote the connected component of $\mathcal{G}^{[k]}(t)$ which contains a given vertex $i$. The key step for understanding the hydrodynamic behavior of $X^{[k]}(t)$ is provided by the following limit theorem.

**Proposition 5.3** *For $t \leq 1$, $\#C_i^{[k]}(t/k)$ converges in distribution as $k \to \infty$ to the Borel law with parameter $t$, that is the probability measure on $\mathbb{N}$ with mass-distribution*

$$\frac{(t\ell)^{\ell-1}\mathrm{e}^{-t\ell}}{\ell!}, \qquad \ell \in \mathbb{N}.$$

Let us first provide an informal argument using certain Galton-Watson branching processes (see for example [15] or [123] for background). We consider the vertex $i$ as the ancestor of a population; we set $P_0 = \{i\}$. The vertices $j \notin P_0$ such that the edge $\{i, j\}$ is open at time $t/k$ form the population $P_1$ at the first generation. The second generation consists of the vertices $\ell \notin P_0 \cup P_1$, for which there exists $j \in P_1$ such that the edge $\{j, \ell\}$ is open. We define recursively by the same procedure the population $P_n$ at any generation $n$. In other words, the population $P_n$ is the set of vertices at distance $n$ from $i$, when the random graph is endowed with its natural metric. Of course, we may have $P_n = \emptyset$ for some $n \geq 1$, and then $P_{n'} = \emptyset$ for every $n' \geq n$; this necessarily happens for $n \geq k-1$. In this setting, the component $C_i^{[k]}(t/k)$ of the random graph at time $t/k$ that contains the vertex $i$, is the reunion of the populations $P_n$, and its cardinal is given by

$$\#C_i^{[k]}(t/k) = \sum_{n=0}^{\infty} \#P_n \,.$$

There are $k-1$ edges that contain the vertex $i$, and at time $t/$ each edge is open with probability $1-\mathrm{e}^{-t}$, independently of the other edges. Thus the size of the population at the first generation, $\#P_1$, has a binomial distribution with parameter $(k-1, 1-\mathrm{e}^{-t/k})$. The asymptotic

$$\binom{k-1}{\ell}\left(1-\mathrm{e}^{-t/k}\right)^{\ell}\exp\left(-\frac{t}{k}(k-\ell-1)\right) \sim \frac{t^{\ell}\mathrm{e}^{-t}}{\ell!}, \qquad \text{as } k \to \infty$$

for each $\ell \in \mathbb{N}$, shows that when $k \to \infty$, the binomial$(k-1, 1-\mathrm{e}^{-t/k})$ distribution converges weakly to the Poisson law with parameter $t$. More generally, one may expect that when $k \to \infty$, the numbers of individuals in the population $P_{n+1}$ which are connected to given individuals in the population $P_n$ should be described by independent Poisson$(t)$ variables, that is this population model should converge weakly to a Galton-Watson process with Poisson$(t)$ reproduction law. Since (sub-)critical Galton-Watson processes generate a finite total population a.s., this suggests that for $t \leq 1$, $\#C_i^{[k]}(t/k)$ should converge weakly towards the total population in a Galton-Watson process with a single progenitor and Poisson$(t)$ reproduction law. Finally, the latter is given by the Borel$(t)$ distribution with parameter $t$, as we shall see now.

**Lemma 5.5** *Consider a (sub)-critical probability measure* $\varrho$ *on* $\mathbb{Z}_+$, *that is* $\sum_{n=0}^{\infty} n\varrho(n) \le 1$. *Let* $T$ *denote the total population of a Galton-Watson process with reproduction law* $\varrho$ *and started from* $a \in \mathbb{N}$ *progenitors. Then there is the identity*

$$\mathbb{P}(T = \ell) = \frac{a}{\ell}\mathbb{P}(S_\ell = \ell - a), \qquad \ell = a, a+1, \ldots,$$

*where* $S = (S_\ell, \ell \in \mathbb{Z}_+)$ *is a random walk with step distribution* $\varrho$ *started from* $S_0 = 0$. *In particular:*

(i) *If* $\varrho$ *is binomial*$(k, b)$ *for* $k \in \mathbb{N}$ *and* $kb \le 1$, *then*

$$\mathbb{P}(T = \ell) = \frac{a}{\ell}\binom{\ell k}{\ell - a} b^{\ell - a}(1-b)^{\ell k - \ell + a}.$$

(ii) *If* $\varrho$ *is Poisson*$(t)$ *for* $0 < t \le 1$, *then*

$$\mathbb{P}(T = \ell) = \frac{a}{\ell}\mathrm{e}^{-t\ell}\frac{(t\ell)^{\ell - a}}{(\ell - a)!}.$$

*Proof* We refer to the original paper of Dwass [90] for the easy proof of the general identity. In the special case (i), $S_\ell$ has the binomial$(k\ell, b)$ distribution and in (ii), the Poisson$(\ell t)$ law. This yields the explicit formulas. □

In order to give a rigorous proof of Proposition 5.3, we first establish the following estimate.

**Lemma 5.6** *Fix* $t > 0$ *and* $i \in [k]$. *There exists a random variable* $T_{k,t}$ *with values in* $\overline{\mathbb{N}}$, *which is distributed as the total population of a Galton-Watson process started with a single progenitor and binomial*$(k, 1 - \mathrm{e}^{-t})$ *reproduction law, such that*

$$\#C_i^{[k]}(t) \le T_{k,t} \qquad a.s.$$

*and*

$$\mathbb{P}(\#C_i^{[k]}(t) = T_{k,t}) \ge \mathbb{E}(\exp(-tT_{k,t}^2)).$$

*Proof* We investigate the connected component $C_i^{[k]}(t)$ containing $i$ using a variation of the well-known depth-first-search algorithm in combinatorics. For simplicity, denote by $c := \#C_i^{[k]}(t)$ the size of this component. We define recursively an increasing family of sets of *observed* vertices $(O_n : n = 0, \ldots, c)$ and a sequence of *explored* vertices $(j_n : n = 1, \ldots, c)$ as follows. Roughly, a new vertex of the connected component $C_i^{[k]}(t)$ is explored at each step, and has necessarily been observed before, that is $j_n \in O_{n-1}$. More precisely, we start from $O_0 = \{i\}$. Suppose that $j_1, \dots, j_{n-1}$ and $O_{n-1}$ have been constructed for some integer $1 \le n < c$ and that the inclusion $\{j_1, \dots, j_{n-1}\} \subset O_{n-1}$ is strict. Then we define the newly explored vertex as $j_n := \min O_{n-1} \backslash \{j_1, \dots, j_{n-1}\}$; in particular $j_1 = i$. The newly observed vertices at the $n$-th step (i.e. the vertices in $O_n \backslash O_{n-1}$) are formed by the subset of vertices $\ell \in C_i^{[k]}(t) \backslash O_{n-1}$ such that the edge $\{j_n, \ell\}$ is open at time $t$; equivalently we set

$$O_n = O_{n-1} \cup \left\{ \ell \in [k] : \mathbf{e}_{j_n, \ell} \le t \right\} .$$

The process of observation-exploration stops after $c$ steps when the sets of observed and explored vertices coincide, that is when the whole connected component $C_i^{[k]}(t)$ has been explored.

For example, consider the case when $i = 1$ and the random graph restricted to the connected component $C_1^{[k]}(t)$ is given by

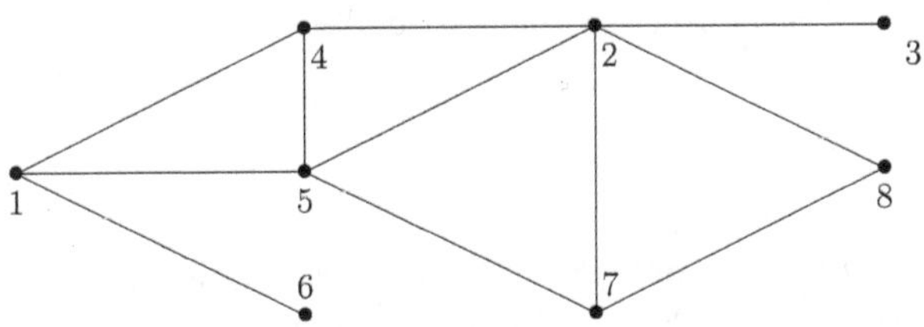


Then $j_1 = 1$ and the newly observed vertices at the first step are $4, 5, 6$. Next, $j_2 = 4$ and the sole newly observed vertex at the second step is 2. Then $j_3 = \min\{2, 5, 6\} = 2$ and the newly observed vertices at the third steps are $3, 7, 8$. All the vertices have been observed after three steps, and the rest of the sequence of explored vertices is $j_4 = 3$, $j_5 = 5$, $j_6 = 6$, $j_7 = 7$, $j_8 = 8$. In short, this can be summarized by the following picture:

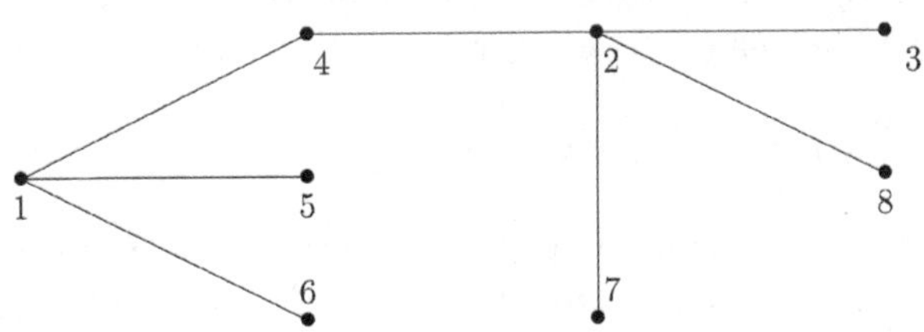


For every $n \le c$, it is convenient to view $j_n$ as the $n$-th individual in a population, and then to think of $O_n \backslash O_{n-1}$ as the set of its children. Denote $\xi_n := \#(O_n \backslash O_{n-1})$ for the number of children of the $n$-th individual, and $\mathcal{F}_n$ for the sigma-field generated by $(j_\ell, O_{\ell-1})_{1 \le \ell \le n}$. As the exponential variables attached to different edges are independent, we see that the conditional distribution of $\xi_n$ given $\mathcal{F}_n$ is the binomial law with parameter $(k-\#O_{n-1}, 1-\mathrm{e}^{-t})$ for every $n \leq c$. Recall that the sum of two independent binomial variables with respective parameters $(a, b)$ and $(a', b)$ has the binomial distribution with parameter $(a+a', b)$. This enables us to construct (on some enlarged probability space) a Galton-Watson process started from a single progenitor with reproduction law being the binomial distribution with parameter $(k, 1-\mathrm{e}^{-t})$, and to label individuals in this branching process such that the following holds. For every $n \leq c$, the number of children of the $n$-th individual in the Galton-Watson process, say $\xi'_n$, is given by

$$\xi'_n = \xi_n + \tilde{\xi}_n ,$$

where conditionally on $\mathcal{F}_n$, the variable $\tilde{\xi}_n$ is independent of $\xi_n$ and has the binomial law with parameter $(\#O_{n-1}, 1-\mathrm{e}^{-t})$.

Plainly, the total population $T_{k,t}$ of this Galton-Watson process cannot be less than $c = \#C_i^{[k]}(t)$, which is our first claim in the statement. Further, if we set

$$\tilde{T} := \sum_{n=1}^{c} \tilde{\xi}_n ,$$

then $T_{k,t} = \#C_i^{[k]}(t)$ if and only if $\tilde{T} = 0$. Observe that the conditional distribution of $\tilde{T}$ given $\mathcal{F}_\infty$ is the binomial distribution with parameter $(\sum_{n=1}^{c} \#O_{n-1}, 1-\mathrm{e}^{-t})$. Since

$$\sum_{n=1}^{c} \#O_{n-1} \leq c^2 \leq T_{k,t}^2 ,$$

this yields the second claim. □

We are now able to establish Proposition 5.3.

*Proof* For $t \leq 1$, the total population in a Galton-Watson process with a single progenitor and binomial$(k, 1-\mathrm{e}^{-t/k})$ reproduction law, has mass distribution

$$\frac{1}{\ell}\binom{k\ell}{\ell-1}\left(1-\mathrm{e}^{-t/k}\right)^{\ell-1}\exp\left(-\frac{t}{k}(\ell k-\ell+1)\right), \qquad \ell \in \mathbb{N};$$

see Lemma 5.5(i). An application of the Stirling formula shows that when $k \to \infty$, this quantity converges to the mass-distribution of the Borel law with parameter $t$. Further, it is plain that then

$$\lim_{k\to\infty} \mathbb{E}\left(\exp\left(-\frac{t}{k}T_{k,t/k}^2\right)\right) = 1 ,$$

and thus Lemma 5.6 yields Proposition 5.3. □

More generally, Proposition 5.3 can easily be extended to a joint limit theorem for the sizes of a pair of typical components in the random graph model.

**Corollary 5.3** *Pick two distinct integers $i$ and $j$ and let $t \leq 1$. Then as $k \to \infty$, the pair $(\#C_i^{[k]}(t/k), \#C_j^{[k]}(t/k))$ converges in distribution to a pair of independent Borel$(t)$ variables.*

*Proof* Clearly, the conditional probability, given the size of the component $C_i^{[k]}(t)$ which contains $i$, that $i$ and $j$ are connected at time $t$ is $(\#C_i^{[k]}(t) - 1)/(k-1)$. As a consequence, for every $\ell \in \mathbb{N}$,

$$\lim_{k\to\infty} \mathbb{P}(j \in C_i^{[k]}(t/k), \#C_i^{[k]}(t/k) = \ell) = 0\,.$$

On the other hand, as the edges are open independently of each other and with the same probability, we see that conditionally on $C_i^{[k]}(t)$, the restriction of the random graph to $[k]\backslash C_i^{[k]}(t)$ defines an Erdős-Rényi random graph on $[k]\backslash C_i^{[k]}(t)$ with the same parameter $t$. As a consequence, the conditional distribution of $\#C_j^{[k]}(t)$ given $\#C_i^{[k]}(t) = \ell$ and $j \notin C_i^{[k]}(t)$ is that of $\#C_1^{[k-\ell]}(t)$. It now follows from Proposition 5.3 that

$$\lim_{k\to\infty} \mathbb{P}(\#C_i^{[k]}(t/k) = \ell, \#C_j^{[k]}(t/k) = \ell') = \frac{(t\ell)^{\ell-1}\mathrm{e}^{-t\ell}}{\ell!} \times \frac{(t\ell')^{\ell'-1}\mathrm{e}^{-t\ell'}}{\ell'!}\,,$$
$$\ell, \ell' \in \mathbb{N}\,,$$

which proves our claim. □

We may now analyze the hydrodynamic behavior of the multiplicative coalescent started from monodisperse configurations. Recall the notation

$$n_t^{[k]}(\ell) := k^{-1}\#\left\{i \in \mathbb{N} : X_i^{[k]}(t/k) = \ell\right\}\,,$$

and, from Lemma 5.4, that this quantity can also be expressed as

$$n_t^{[k]}(\ell) = k^{-1}\#\left\{i \in \mathbb{N} : \#C_i^{[k]}(t/k) = \ell\right\}\,.$$

Further, introduce the deterministic quantity

$$n_t(\ell) := \frac{(t\ell)^{\ell-1}\mathrm{e}^{-t\ell}}{\ell(\ell!)}\,, \qquad t \geq 0 \text{ and } \ell \in \mathbb{N}\,. \tag{5.7}$$

In other words, $(\ell n_t(\ell), \ell \in \mathbb{N})$ is the mass-distribution of the Borel$(t)$ law.

**Proposition 5.4** *For every $t \leq 1$ and $\ell \in \mathbb{N}$,*

$$\lim_{k\to\infty} n_t^{[k]}(\ell) = n_t(\ell) \qquad \text{in } L^2(\mathbb{P}).$$

*Moreover, $(n_t(\ell), \ell \in \mathbb{N}$ and $0 \leq t \leq 1)$ is a solution to Smoluchowski's coagulation equation (5.6) for the multiplicative kernel $\kappa(x, y) = xy$ started from $n_0(\ell) = \mathbb{1}_{\{\ell=1\}}$.*

*Proof* We use the moment method. For every $i \in [k]$, write for simplicity $\beta_i = C_i^{[k]}(t/k)$ for the connected component of the random graph at time $t/k$ which contains the vertex $i$. We have for every $\ell \in \mathbb{N}$

$$\mathbb{E}(n_t^{[k]}(\ell)) = \frac{1}{k\ell}\mathbb{E}\left(\sum_{i=1}^{k} \mathbb{1}_{\{\#\beta_i=\ell\}}\right) = \frac{1}{\ell}\mathbb{P}(\#\beta_1 = \ell),$$

where the second equality follows from the obvious fact that the $k$-tuple $(\#\beta_1, \ldots, \#\beta_k)$ is exchangeable. Since $\#\beta_1 = \#C_1^{[k]}(t/k)$, Proposition 5.3 yields

$$\lim_{k\to\infty} \mathbb{E}(n_t^{[k]}(\ell)) = \frac{(t\ell)^{\ell-1}\mathrm{e}^{-t\ell}}{\ell(\ell!)}.$$

A similar calculation for the second moment yields

$$\begin{aligned}
\mathbb{E}((n_t^{[k]}(\ell))^2) &= \frac{1}{k^2\ell^2}\mathbb{E}\left(\sum_{i,j=1}^{k} \mathbb{1}_{\{\#\beta_i=\ell\}}\mathbb{1}_{\{\#\beta_j=\ell\}}\right) \\
&= \frac{1}{k\ell^2}\mathbb{P}(\#\beta_1 = \ell) + \frac{k-1}{k\ell^2}\mathbb{P}(\#\beta_1 = \#\beta_2 = \ell).
\end{aligned}$$

Proposition 5.3 and Corollary 5.3 now yield

$$\lim_{k\to\infty} \mathbb{E}((n_t^{[k]}(\ell))^2) = \left(\frac{(t\ell)^{\ell-1}\mathrm{e}^{-t\ell}}{\ell(\ell!)}\right)^2,$$

which completes the proof of $L^2(\mathbb{P})$-convergence.

Finally direct calculations enable us to check that $n_t(x)$ as defined in (5.7) does solve Smoluchowski's equation (5.6) for $0 \leq t \leq 1$. Indeed, observe first that

$$\frac{dn_t(x)}{dt} = \frac{(x-1)(tx)^{x-2}\mathrm{e}^{-tx}}{x!} - xn_t(x), \qquad x \in \mathbb{N}.$$

Next, recall that $xn_t(x)$ is the mass distribution at $x$ of the Borel$(t)$ law. On the one hand, we thus have

$$n_t(x)\sum_{y=1}^{\infty} n_t(y)xy = xn_t(x).$$

On the other hand, we can view

$$\sum_{y=1}^{x-1} n_t(y)n_t(x-y)y(x-y)$$

as the mass distribution at $x$ of the sum of two independent Borel($t$) variables. The latter corresponds to the total population generated by a Galton-Watson process with Poisson($t$) reproduction law and two progenitors. Hence, according to Lemma 5.5(ii), we have

$$\frac{1}{2}\sum_{y=1}^{x-1} n_t(y)n_t(x-y)y(x-y) = \frac{(x-1)(tx)^{x-2}\mathrm{e}^{-tx}}{x!},$$

and we conclude that $n_t(x)$ is a solution to (5.6) for $0 \leq t \leq 1$. □

The fact that $(n_t(\ell), 0 \leq t \leq 1)$ provides a solution to Smoluchowski's coagulation equation for the multiplicative kernel was discovered first by McLeod [158]. We also refer to the proof of the forthcoming Theorem 5.2 for a more instructive approach that circumvents explicit computations and rather emphasizes the connections with the dynamics of the stochastic coalescents. We stress that the family $(n_t(\ell), t \geq 0$ and $\ell \in \mathbb{N})$ fails to solve Smoluchowski's equation (5.6) for $t > 1$, since then $\sum_{\ell=1}^{\infty} \ell n_t(\ell) < 1$. This is due to the fact that Galton-Watson processes with Poisson($t$) reproduction laws are super-critical for $t > 1$, and thus the total populations they generate are infinite with positive probability (in other words the Borel law with parameter $t > 1$ is defective). This phenomenon is referred to as *gelation* in the literature on coagulation equations, and interpreted as the emergence of particles with infinite sizes called gel. This corresponds to the apparition of a giant component in the random graph model; see Erdős and Rényi [94], Bollobás [60], Aldous [5], ... for much more on this issue.

### 5.2.2 Sub-multiplicative kernels

Our purpose in this section is to investigate the hydrodynamic behavior of stochastic coalescents with *sub-multiplicative* kernel $\kappa$, that is such that

$$\kappa(x,y) \leq xy, \qquad x, y \geq 0,$$

using a simple coupling with multiplicative coalescents. Intuitively, the sub-multiplicativity assumption implies that there are fewer coagulations in a coalescent chain with kernel $\kappa$ than in the multiplicative coalescent. Thus, one should be able to construct the former by canceling certain coagulations which occur for the latter. Again we focus on monodisperse initial configurations; our analysis relies on a representation of coupled coalescent chains which extends the model of Erdős and Rényi that has been described in the preceding section.

Introduce a Poisson random measure $\mathtt{M}^{[k]}$ on $\mathbb{R}_+ \times [0,1] \times \mathcal{E}^{[k]}$ with intensity $ds \otimes du \otimes \#$, where $\#$ is viewed as the counting measure on the set $\mathcal{E}^{[k]}$ of edges of $[k]$. Plainly, we can enumerate the atoms of $\mathtt{M}^{[k]}$ in the form $((t_n, u_n, e_n), n \in \mathbb{N})$ where $0 < t_1 < t_2 < \cdots$ is an increasing sequence of times, $u_1, u_2, \ldots \in [0,1]$ and $e_1, e_2, \ldots \in \mathcal{E}^{[k]}$, and the exponent $[k]$ has been dropped for notational convenience. In other words, we have

$$\mathtt{M}^{[k]} = \sum_{n=1}^{\infty} \delta_{(t_n, u_n, e_n)} .$$

We shall use the Poisson measure $\mathtt{M}^{[k]}$ to construct simultaneously a couple of processes $(\mathcal{G}^{[k]}(t), t \geq 0)$ and $(\mathcal{H}^{[k]}(t), t \geq 0)$ of random graphs on $[k]$. Each random graph consists of a set of open edges, where the definition of 'open' depends both on the time $t$ and the model ($\mathcal{G}^{[k]}$ or $\mathcal{H}^{[k]}$). In both cases, all edges are closed at the initial time.

The first process is just a version of the random graph model of Erdős and Rényi. Specifically, we decide that each edge $e \in \mathcal{E}^{[k]}$ becomes (and then remains) open for $\mathcal{G}^{[k]}(\cdot)$ at the first instant

$$\mathbf{e}_e := \inf\left\{t \geq 0 : \mathtt{M}^{[k]}([0,t] \times [0,1] \times \{e\}) = 1\right\}$$

when the Poisson random measure $\mathtt{M}^{[k]}$ has an atom on the fiber $[0,t] \times [0,1] \times \{e\}$. Observe that $\mathbf{e}_e$ is a standard exponential variable, and that to different edges in $\mathcal{E}^{[k]}$ correspond independent variables (by the superposition property of Poisson random measures), so the present formulation merely rephrases that of Erdős and Rényi.

The second process depends more significantly on the Poisson random measure $\mathtt{M}^{[k]}$. We define recursively a step process $\mathcal{H}^{[k]}(\cdot)$ with values in the set graphs on $[k]$ as follows. If an edge $e \in \mathcal{E}^{[k]}$ is open at some time $t$, then it remains open at any $t' \geq t$, and all edges are closed at the initial time $t = 0$. Assume that $\mathcal{H}^{[k]}(t_{n-1})$ has been defined for some integer $n$ and set $\mathcal{H}^{[k]}(t) = \mathcal{H}^{[k]}(t_{n-1})$ for every $t_{n-1} \leq t < t_n$ (as usual, we use the convention that $t_0 = 0$). If the edge $e_n = \{v_n, v'_n\}$ was closed at time $t_{n-1}$, then it becomes open at time $t_n$ if and only if $u_n \leq \kappa(c, c')/(cc')$, where $c$ (respectively, $c'$) denotes the size of the connected component in $\mathcal{H}^{[k]}(t_{n-1})$ containing the vertex $v_n$ (respectively, $v'_n$). This defines $\mathcal{H}^{[k]}(t_n)$, and thus the graph process $\mathcal{H}^{[k]}(\cdot)$ by induction. Note that $\mathcal{H}^{[k]}(t) \subseteq \mathcal{G}^{[k]}(t)$ for every $t \geq 0$.

**Lemma 5.7** *For every $t \geq 0$, write $Y^{[k]}(t)$ for the sequence of the sizes of the connected components for $\mathcal{H}^{[k]}(t)$, ranked in decreasing order. Then the process $Y^{[k]}(\cdot)$ is a version of the coalescent chain with coagulation kernel $\kappa$ and started from the monodisperse initial configuration $\mathbf{x}^{[k]} = (1, , \ldots 1, 0, \ldots)$ given by k particles with unit mass.*

*Proof* It follows immediately from the construction of $\mathcal{H}^{[k]}(t)$ and standard properties of Poisson random measure that the process $(\mathcal{H}^{[k]}(t), t \geq 0)$ is Markovian. More precisely, given $\mathcal{H}^{[k]}(t) = H$ and an edge $e = \{v, v'\} \in \mathcal{E}^{[k]}$ which is closed at time $t$ (i.e. $e \notin H$), the rate of jump from $H$ to $H' := H \cup \{e\}$ is $\kappa(c, c')/(cc')$, where $c$ and $c'$ are the sizes of the connected components of $H$ containing respectively $v$ and $v'$. All the other jump rates from $H$ to a graph $H'$ which is not obtained from $H$ by opening a single edge $e \notin H$, are zero.

Now for every graph $H$, write $|H|^{\downarrow} \in \mathcal{S}_f^{\downarrow}$ for the sequence of the sizes of the connected components of $H$ ranked in decreasing order; note that the map $H \to |H|^{\downarrow}$ is not injective. Conditionally on $\mathcal{H}^{[k]}(t) = H$, the process $s \to |\mathcal{H}^{[k]}(t+s)|^{\downarrow}$ stays equal to $|H|^{\downarrow}$ until the first time that two distinct connected components of $H$ become connected. Observe that opening edges between pairs of vertices in the same connected component does not affect $|H|^{\downarrow}$, nor the rate at which a given edge between two distinct connected components becomes open. Moreover, given two distinct connected components of $H$, say $C$ and $C'$, with respective sizes $c$ and $c'$, there are $cc'$ graphs which can be obtained from $H$ by opening a single edge between $C$ and $C'$. We deduce that

$$\begin{aligned}
&\lim_{s\to 0} \frac{1}{s} \mathbb{P}\left(C \cup C' \text{ is a connected component of } \mathcal{H}^{[k]}(t+s) \mid \mathcal{H}^{[k]}(t) = H\right) \\
&\quad = cc' \lim_{s\to 0} \frac{1}{s} \mathbb{P}\left(\mathrm{M}^{[k]}\left([t, t+s] \times [0, \kappa(c, c')/(cc')] \times \{e\}\right) = 1\right) \\
&\quad = \kappa(c, c'),
\end{aligned}$$

where at the second line, $e$ denotes a generic edge between $C$ and $C'$ (i.e. $e = \{v, v'\}$ for some $v \in C$ and $v' \in C'$).

Elementary properties of independent exponential variables now imply that the first jump time of the process $s \to |\mathcal{H}(t+s)|^{\downarrow}$ has an exponential distribution with parameter $\frac{1}{2} \sum \kappa(\#C, \#C')$, where the sum is taken over the set of pairs $(C, C')$ of distinct connected components of $H$ and the factor $\frac{1}{2}$ accounts for the obvious symmetry. More precisely, immediately after the jump, $|\mathcal{H}(t+s)|^{\downarrow}$ results from the coagulation of a pair of distinct atoms (atoms are the sizes of the connected components of $H$), say $\{a, a'\}$, which are picked independently of the jump time and with probability proportional to $\kappa(a, a')$. This shows that the process $|\mathcal{H}(t)|^{\downarrow}$ is a stochastic coalescent chain with coagulation kernel $\kappa$. □

We next introduce some notation for certain random point measures which will naturally appear later in the text. In this direction, it is convenient to view $\mathbb{N}$ as a countable set of vertices and write $\mathcal{E}$ for the set of its edges, that is pairs $\{i, j\} = \{j, i\}$ of distinct vertices. The natural distance between two distinct edges is 1 if they are adjacent (i.e. they share a common vertex) and 2 otherwise. We work with the metric space

$$F := [0, 1]^2 \times \mathcal{E};$$

recall from Section 1.1.2 that $F_{\mathrm{p}}$ then denotes the Polish space of finite point measures on $F$.

Just as in Lemma 5.6 (for $t = 1/k$), we consider a subcritical Galton-Watson process started with a single progenitor and binomial$(k, 1 - \mathrm{e}^{-1/k})$ reproduction law. We assign distinct integer labels to the individuals of this branching process, using 1 for the ancestor (the choice of the labels for the other individuals is arbitrary). We write $\mathcal{T}_k$ for genealogical tree, that is the set of edges between parents and children, and we work conditionally on the latter. Next, we introduce a random point measure $\mathtt{N}^{[k]}$ on $F_{\mathrm{p}}$ distributed as a Poisson random measure on $[0, 1]^2 \times \mathcal{E}$ with intensity $k^{-1} ds \otimes du \otimes \#$ and conditioned by the event that $\mathtt{N}^{[k]}$ has at least one atom on the fiber $[0, 1]^2 \times \{e\}$ if $e \in \mathcal{T}_k$, and has no atoms on this fiber if $e \in \mathcal{E} \backslash \mathcal{T}_k$.

Next, recall that $C_1^{[k]}(t)$ denotes the connected component in the Erdős-Rényi graph $\mathcal{G}^{[k]}(t)$ that contains the vertex 1, and write $E_1^{[k]}(t)$ for the corresponding set of edges in $\mathcal{G}^{[k]}(t)$ which connect pairs of vertices in the component $C_1^{[k]}(t)$. We henceforth focus on time $t = 1/k$ and define a random point measure $\hat{\mathtt{M}}_1^{[k]}$ on $F$ as the restriction to $[0, 1]^2 \times E_1^{[k]}(1/k)$ of the image of $\mathtt{M}^{[k]}$ by the dilation $(s, u, e) \to (ks, u, e)$. It should be plain that conditionally on $E_1^{[k]}(1/k)$, $\hat{\mathtt{M}}_1^{[k]}$ has the law of a Poisson random measure with intensity $k^{-1} ds \otimes du \otimes \#$ and conditioned by the event of having at least one atom on the fiber $[0, 1]^2 \times \{e\}$ for each $e \in E_1^{[k]}(1/k)$. We may now state the following easy extension of Lemma 5.6.

**Lemma 5.8** *We can construct a Galton-Watson branching process with a single progenitor and binomial$(k, 1 - \mathrm{e}^{-1/k})$ reproduction law, and then a random point measure $\mathtt{N}^{[k]}$ in $F_{\mathrm{p}}$ as above, such that*

$$\# C_1^{[k]}(1/k) \leq T_k \qquad a.s.$$

*and*

$$\mathbb{P}(\hat{\mathrm{M}}_1^{[k]} = \mathrm{N}^{[k]}) \geq \mathbb{E}\left(\exp\left(-\frac{2}{k}T_k^2\right)\right),$$

*where $T_k$ denotes the total population of this subcritical Galton-Watson process.*

*Proof* We use the notation of the proof of Lemma 5.6 with $i = 1$ and $t = 1/k$. Recall that for $n \leq \#C_1^{[k]}(1/k)$, the $n$-th individual is the vertex $j_n$ that is explored at the $n$-th step (in particular $j_1 = 1$ is viewed as the ancestor), and its children are given by the newly observed vertices, namely $O_n \backslash O_{n-1}$.

Write $\mathcal{T}_1^{[k]}$ for the corresponding genealogical tree, that is the set of edges of the form $\{j_n, \ell\}$ for $\ell \in O_n \backslash O_{n-1}$ and $n \leq \#C_i^{[k]}(1/k)$. Plainly, $\mathcal{T}_1^{[k]} \subseteq E_1^{[k]}(1/k)$, and the inclusion can be strict.[1] In particular, the conditional law given $\mathcal{T}_1^{[k]}$ of the restriction of $\hat{\mathrm{M}}_1^{[k]}$ to $[0, 1]^2 \times \mathcal{T}_1^{[k]}$ is that of a Poisson measure with intensity $k^{-1} ds \otimes du \otimes \#$ and conditioned by the event of having at least one atom on the fiber $[0, 1]^2 \times \{e\}$ for each $e \in \mathcal{T}_1^{[k]}$.

Lemma 5.6 enables us to construct a Galton-Watson process with a single progenitor and offspring distribution given by the binomial$(k, 1 - \mathrm{e}^{-1/k})$ law, such that its genealogical tree $\mathcal{T}^{[k]}$ contains $\mathcal{T}_1^{[k]}$ a.s. We then define conditionally on $\mathcal{T}^{[k]}$ and $\mathcal{T}_1^{[k]}$, a random point measure $\mathrm{N}^{[k]}$ as follows. First, the point measures $\hat{\mathrm{M}}^{[k]}$ and $\mathrm{N}^{[k]}$ coincide on $[0, 1]^2 \times \mathcal{T}_1^{[k]}$. Second, on each fiber $[0, 1]^2 \times \{e\}$, $\mathrm{N}^{[k]}$ has no atoms if $e \in \mathcal{E} \backslash \mathcal{T}^{[k]}$, and is distributed according to an independent Poisson random measure with intensity $k^{-1} ds \otimes du \otimes \delta_e$ conditioned to have at least one atom if $e \in \mathcal{T}^{[k]} \backslash \mathcal{T}_1^{[k]}$. By the superposition property of independent Poisson measures, it is plain that $\mathrm{N}^{[k]}$ has the required distribution.

Note that $\mathrm{N}^{[k]}$ retains no information about the point measure $\mathrm{M}^{[k]}$ on $[0, 1/k] \times [0, 1] \times \left(\mathcal{E}^{[k]} \backslash \mathcal{T}_1^{[k]}\right)$. By construction, $\hat{\mathrm{M}}_1^{[k]}$ and $\mathrm{N}^{[k]}$ are identical if and only if first, the size $T_k$ of the Galton-Watson tree $\mathcal{T}^{[k]}$ does not exceed $\#C_1^{[k]}(1/k)$ and, second, $\mathrm{M}^{[k]}$ has no atoms on $[0, 1/k] \times [0, 1] \times \left(E_1^{[k]}(1/k) \backslash \mathcal{T}_1^{[k]}\right)$. The obvious bound

$$\#\left(E_1^{[k]}(1/k) \backslash \mathcal{T}_1^{[k]}\right) \leq T_k^2$$

and standard properties of Poisson random measures imply that the conditional probability that $\hat{\mathrm{M}}_1^{[k]} = \mathrm{N}^{[k]}$ given $\mathrm{N}^{[k]}$ and $\#C_1^{[k]}(1/k) = T_k$ is bounded from below by $\exp\left(-k^{-1} T_k^2\right)$. Since according to Lemma 5.6,

[1] More precisely equality holds if and only if the graph $E_1^{[k]}(1/k)$ is a tree, that is contains no loops.

$$\mathbb{P}(\#C_1^{[k]}(1/k) = T_k) \geq \mathbb{E}(\exp(-k^{-1}T_k^2)),$$

this completes the proof of our statement. □

Next, we would like to analyze the asymptotic behavior of the random point measure $\hat{\mathtt{M}}_1^{[k]}$ when $k \to \infty$. The labeling of vertices by $[k]$ induces a rather artificial problem (because the labels of vertices connected to 1 tend to $\infty$ when $k \to \infty$ ). To circumvent this difficulty, we define an equivalence relation between finite point measures on $F = [0, 1]^2 \times \mathcal{E}$ which enables us to 'forget' the labels of all vertices, except for the distinguished vertex 1 which plays the role of a root. Specifically, for every $m, m' \in F_{\mathrm{p}}$, we write $m \sim m'$ if and only if there is a permutation $\sigma$ on $\mathbb{N}$ which preserves 1 (i.e. $\sigma : \mathbb{N} \to \mathbb{N}$ is a bijection with $\sigma(n) = n$ for all sufficiently large $n$ and also for $n = 1$), such that $m'$ is the image of $m$ by the natural action of $\sigma$ on $F$, namely $(s, u, \{i, j\}) \to (s, u, \{\sigma(i), \sigma(j)\})$.

We write $\tilde{F}_{\mathrm{p}}$ for the set of equivalence classes of point measures on $F$, endowed with the distance $\mathrm{dist}(\tilde{m}, \tilde{m}') = \min \mathrm{dist}(m, m')$ where the minimum is taken over the finite point measures $m \in \tilde{m}$ and $m' \in \tilde{m}'$, so that $\tilde{F}_{\mathrm{p}}$ is again a Polish space. We should think of an equivalence class in $\tilde{F}_{\mathrm{p}}$ as a rooted structure consisting of edges (the vertex 1 is distinguished as the labels of the other vertices have been removed), where each edge has a mark given by a finite point measure on $[0, 1]^2$ and only finitely many edges have non-zero marks.

Finally, we need to introduce another random point measure on $F$, which can be viewed as the weak limit as $k \to \infty$ of the random point measures $\mathtt{N}^{[k]}$ that have been constructed before Lemma 5.8. Consider a critical Galton-Watson process started with a single progenitor and Poisson(1) reproduction law. We shall again work conditionally on the latter, and assign distinct integer labels to the individuals, using 1 for the ancestor. Write $\mathcal{T}$ for the genealogical tree of this branching process, that is the (finite) set of edges between parents and children. We next introduce a family $((t_e, u_e), e \in \mathcal{T})$ of i.i.d. variables where for each edge $e \in \mathcal{E}_1$, $(t_e, u_e)$ has the uniform distribution on $[0, 1]^2$. We now define a finite point measure on $F$ by

$$\mathtt{N} := \sum_{e \in \mathcal{T}} \delta_{(t_e, u_e, e)},$$

and write $\tilde{\mathtt{N}}$ for the equivalence class of $\mathtt{N}$ which enables us to forget the labeling of the vertices except 1.

**Corollary 5.4** *Denote by* $\tilde{\text{M}}_1^{[k]}$ *the equivalence class in* $\tilde{F}_\text{p}$ *of* $\hat{\text{M}}_1^{[k]}$*. Then, in the notation above,* $\tilde{\text{M}}_1^{[k]}$ *converges in distribution on* $\tilde{F}_\text{p}$ *as* $k \to \infty$ *to* $\tilde{\text{N}}$.

*Proof* Recall the description of the random point measure $\text{N}^{[k]}$ before Lemma 5.8. On the one hand, it is immediately checked that a Poisson random measure on $[0, 1]^2$ with intensity $k^{-1} ds \otimes du$ conditioned to have at least one atom, converges in law as $k \to \infty$ to the Dirac point mass at a random point which is uniformly distributed on $[0, 1]^2$. On the other hand, we know that the Poisson(1) distribution arises as the limit of the binomial$(k, 1 - \mathrm{e}^{-1/k})$ law when $k \to \infty$. It follows readily that if we write $\tilde{\text{N}}^{[k]}$ for the equivalence class in $\tilde{F}_\text{p}$ of $\text{N}^{[k]}$, then $\tilde{\text{N}}^{[k]}$ converges in law as $k \to \infty$ to $\tilde{\text{N}}$.

Next, recall that the total population $T_k$ in a Galton-Watson process with a single progenitor and binomial$(k, 1 - \mathrm{e}^{-1/k})$ reproduction law converges weakly to the Borel(1) distribution. We now apply Lemma 5.8 and see that

$$\lim_{k \to \infty} \mathbb{P}(\hat{\text{M}}_1^{[k]} = \text{N}^{[k]}) = 1 .$$

The combination with the observation above shows that $\tilde{\text{M}}_1^{[k]}$ converges in distribution to $\tilde{\text{N}}$ as $k \to \infty$. □

We have now all the technical tools to analyze the hydrodynamic behavior of coalescent chains with sub-multiplicative coagulation kernels; we just need some more notation. First, motivated by the construction of a version of the coalescent chain with coagulation kernel $\kappa$ in Lemma 5.7, we associate a coagulation cluster to fairly general finite point measures in $F_\text{p}$ as follows. Assume that $m \in F_\text{p}$ has at most one atom on the fiber $\{t\} \times [0, 1] \times \mathcal{E}$ for each $0 \leq t \leq 1$, which enables us to numerate these atoms in increasing order of their first coordinate. For every $0 \leq t \leq 1$, we associate to $m$ a graph $H(t, m)$ by the same recursive construction as for the random graph $\mathcal{H}^{[k]}(t)$, using $m$ in place of $\text{M}^{[k]}$. We then write $C(t, m)$ for the connected component of $H(t, m)$ that contains the root, that is the vertex 1. It is immediately seen that the size $\#C(t, m)$ of this component is the same as $\#C(t', m')$ for every $m'$ in the equivalence class $\tilde{m}$ of $m$.

Finally, we specialize this to the random point measure $\text{N}$ which was introduced before Corollary 5.4, and set for every $t \leq 1$

$$n_t(\ell) := \ell^{-1} \mathbb{P}(\#C(t, \text{N}) = \ell) , \qquad \ell \in \mathbb{N}. \tag{5.8}$$

We stress that the random variable $\#C(t, \text{N})$ is finite a.s., since by construction it is bounded from above by the total population generated by a critical Galton-Watson process (more precisely with Poisson(1) reproduction law). In particular, it holds that

$$\sum_{\ell=1}^{\infty} \ell n_t(\ell) = 1\,.$$

It is easily seen that in the multiplicative case when $\kappa(x, y) = xy$, then $C(t, \mathtt{N})$ is distributed as a Galton-Watson tree with a single progenitor and Poisson$(t)$ reproduction law. As a consequence, $\#C(t, \mathtt{N})$ has the Borel$(t)$ law and (5.8) agrees with (5.7).

**Theorem 5.2** *Let $X^{[k]}$ be a coalescent chain started from the monodisperse initial configuration with k particles each of unit mass, and with a sub-multiplicative coagulation kernel $\kappa$, that is $\kappa(x, y) \le xy$. Set*

$$n_t^{[k]}(\ell) := k^{-1}\#\left\{i \in \mathbb{N} : X_i^{[k]}(t/k) = \ell\right\}, \qquad \ell \in \mathbb{N}.$$

*Then for every $t \le 1$ and $\ell \in \mathbb{N}$, we have in the notation (5.8) that*

$$\lim_{k\to\infty} n_t^{[k]}(\ell) = n_t(\ell) \qquad \text{in } L^2(\mathbb{P}). \tag{5.9}$$

*Moreover, $(n_t(\ell), \ell \in \mathbb{N}$ and $0 \le t \le 1)$ is a solution of Smoluchowski's coagulation equation (5.6) started from $n_0(\ell) = \mathbb{1}_{\{\ell=1\}}$.*

*Proof* For every $i \in [k]$, write $\Gamma_i^{[k]}(t)$ for the connected component containing the vertex $i$ in the random graph $\mathcal{H}^{[k]}(t)$ and $\gamma_i^{[k]}(t) := \#\Gamma_i^{[k]}(t)$ for its size. By construction, there is the identity $\Gamma_1^{[k]}(t/k) = C(t, \hat{\mathtt{M}}_1^{[k]})$. Further, it is readily checked that for every fixed $t \in [0, 1]$ the map $m \to \#C(t, m)$, which associates the size of the connected component of the graph $H(t, m)$ that contains 1 to a finite point measure $m \in F_{\mathrm{p}}$ having at most one atom on every fiber $\{s\} \times [0, 1] \times \mathcal{E}$, is continuous. Since the point measure $\mathtt{N}$ fulfills the preceding property a.s., we deduce from Corollary 5.4 that $\gamma_1^{[k]}(t/k)$ converges in distribution as $k \to \infty$ to $\#C(t, \mathtt{N})$. More generally, we claim that for every fixed $i \neq j$, there is the convergence in distribution

$$(\gamma_i^{[k]}(t/k), \gamma_j^{[k]}(t/k)) \Longrightarrow (\xi, \xi') \qquad \text{as } k \to \infty\,,$$

where $(\xi, \xi')$ denotes a pair of independent copies of $\#C(t, \mathtt{N})$. Indeed, it is seen from an obvious symmetry that the law of $\gamma_i^{[k]}(t/k)$ does not depend on $i$, and then the asymptotic independence follows from the argument as in the proof of Corollary 5.3.

Next, Lemma 5.7 enables us to write

$$n_t^{[k]}(\ell) = \frac{1}{k\ell} \sum_{i=1}^{k} \mathbb{1}_{\{\gamma_i^{[k]}(t/k)=\ell\}}\,,$$

and moment calculations similar to that for Proposition 5.4 establish the convergence (5.9).

We now turn our attention to Smoluchowski's coagulation equation. In this direction, recall that the infinitesimal generator G of a stochastic coalescent chain is given for linear functionals by (5.2). In the present framework, we set $\mathbf{f}(\mathbf{x}) = \sum_{i=1}^{\infty} f(x_i)$ where $f(x) = k^{-1}\mathbb{1}_{\{x=\ell\}}$, so

$$\mathbf{f}(X^{[k]}(t/k)) := k^{-1}\sum_{i=1}^{\infty}\mathbb{1}_{\{X_i^{[k]}(t/k)=\ell\}} = n_t^{[k]}(\ell),$$

and then, for every $\mathbf{s} \in \mathcal{S}_f^{\downarrow}$,

$$\begin{aligned}\mathsf{G}\mathbf{f}(\mathbf{s}) = & \frac{1}{2k}\sum_{i,j=1}^{\infty}\left(\mathbb{1}_{\{s_i+s_j=\ell\}} - \mathbb{1}_{\{s_i=\ell\}} - \mathbb{1}_{\{s_j=\ell\}}\right)\kappa(s_i, s_j) \\ & - \frac{1}{2k}\sum_{i=1}^{\infty}\left(\mathbb{1}_{\{2s_i=\ell\}} - 2\mathbb{1}_{\{s_i=\ell\}}\right)\kappa(s_i, s_i).\end{aligned}$$

This leads us to introduce for $u \geq 0$

$$g_u^{[k]} := \frac{1}{2}\sum_{j=1}^{\ell-1} n_u^{[k]}(j) n_u^{[k]}(\ell - j)\kappa(j, \ell - j) - \frac{\kappa(\ell, \ell)}{2k}\left(n_u^{[k]}(\ell/2) - 2n_u^{[k]}(\ell)\right)$$

and

$$\tilde{g}_u^{[k]} := n_u^{[k]}(\ell)\sum_{j=1}^{\infty} n_u^{[k]}(j)\kappa(j, \ell).$$

Then the process

$$\mathbf{f}(X^{[k]}(t/k)) - \int_0^t \left(g_u^{[k]} - \tilde{g}_u^{[k]}\right) du, \qquad t \geq 0, \tag{5.10}$$

is a martingale.

It follows from (5.9) that for every $0 \leq u \leq 1$

$$\lim_{k\to\infty} g_u^{[k]} = \frac{1}{2}\sum_{j=1}^{\ell-1} n_u(j) n_u(\ell - j)\kappa(j, \ell - j) \quad \text{in } L^1(\mathbb{P}).$$

Let us now check that

$$\lim_{k\to\infty} \tilde{g}_u^{[k]} = n_u(\ell)\sum_{j=1}^{\infty} n_u(j)\kappa(j, \ell) \quad \text{in } L^1(\mathbb{P}). \tag{5.11}$$

It is seen from the very definition of $n_u^{[k]}(j)$ that

$$\sum_{j=1}^{\infty} j n_u^{[k]}(j) = 1,$$

and since $\sum_{j=1}^{\infty} jn_u(j) = 1$, it follows from (5.9) and Scheffé's lemma (see for example [56]) that

$$\lim_{k\to\infty} \mathbb{E}\left(\sum_{j=1}^{\infty} j|n_u^{[k]}(j) - n_u(j)|\right) = 0.$$

Since $\kappa$ is sub-multiplicative, we deduce that

$$\lim_{k\to\infty} \sum_{j=1}^{\infty} n_u(j)\kappa(j,\ell) = \sum_{j=1}^{\infty} n_u(j)\kappa(j,\ell) \quad \text{in } L^1(\mathbb{P}),$$

which in turn yields (5.11) by (5.9) since $n^{[k]}(\ell) \le 1/\ell$.

Finally, taking expectations in (5.11) and then the limit as $k \to \infty$, we obtain by dominated convergence that for every $t \le 1$

$$n_t(\ell) = n_0(\ell) + \int_0^t du \left(\frac{1}{2}\sum_{j=1}^{\ell-1} n_u(j)n_u(\ell-j)\kappa(j,\ell-j) - n_u(\ell)\sum_{j=1}^{\infty} n_u(j)\kappa(j,\ell)\right).$$

This establishes (5.6). □

## 5.3 The additive coalescence

This section is devoted to the additive coalescent, that is the stochastic coalescent with coagulation kernel

$$\kappa(x,y) = x+y, \qquad x,y \ge 0.$$

We shall first present a couple of simple properties, and then a natural connection with certain random trees and forests due to Pitman [184]. This yields a simple construction of additive coalescent chains started from monodisperse initial configurations. In turn, the latter enables us to investigate the limit behavior of processes started from a monodisperse configuration which is given by a large number of small identical particles with a fixed total mass.

### 5.3.1 Some basic properties

We start by stressing a remarkable property of additive coalescent chains, which is similar to the one observed for Kingman's coalescent in Theorem 4.1. More precisely, consider $X = (X(t), t \ge 0)$ an additive coalescent chain started from a finite configuration $\mathbf{x} = (x_1, \ldots, x_k, 0, \ldots) \in \mathcal{S}_f^{\downarrow}$ with $k$ particles (i.e. $x_k > 0$). Write $m := x_1 + \cdots + x_k$ for the total mass of the system, and for every $n = 0, \ldots, k$, $T_n$ for the instant of the $n$-th coagulation, with the convention that $T_0 = 0$ and $T_k = \infty$. Finally, we denote by $Y(n) = X(T(n))$, for $n = 0, \ldots, k-1$, the successive states of the chain.

**Proposition 5.5** *In the notation above, the following assertions hold:*

(i) *The sequences* $(T_n : n = 0, \ldots, k-1)$ *and* $(Y(n) : n = 0, \ldots, k-1)$ *are independent.*

(ii) *The sequence* $T_1 - T_0, T_2 - T_1, \ldots, T_{k-1} - T_{k-2}$ *of the waiting times between two coagulations is a sequence of independent exponential variables with respective parameters* $m(k-1), m(k-2), \ldots, m$.

(iii) *The successive states* $(Y(n) : n = 0, \ldots, k-1)$ *form a Markov sequence (i.e. a Markov chain in discrete time) on* $\mathcal{S}_f^{\downarrow}$ *with the following transition probabilities:*

$$\mathbb{P}(Y(n+1) = \mathbf{y}^{i \oplus j} \mid Y(n) = \mathbf{y}) = \frac{y_i + y_j}{m(k-n-1)}, \qquad 1 \le i < j \le k-n,$$

*where* $\mathbf{y} = (y_1, \ldots, y_{k-n}, 0, \ldots)$ *is a generic finite configuration in* $\mathcal{S}_f^{\downarrow}$ *with* $k-n$ *particles and total mass* $y_1 + \cdots + y_{k-n} = m$, *and* $\mathbf{y}^{i\oplus j}$ *stands for the configuration obtained from* $\mathbf{y}$ *by merging its i-th and j-th terms.*

*Proof* We simply observe that the first coagulation time $T_1$ has an exponential distribution with parameter

$$\sum_{1 \le i < j \le k} (x_i + x_j) = \frac{1}{2} \left( \sum_{1=i,j}^{k} (x_i + x_j) - 2 \sum_{i=1}^{k} x_i \right) = m(k-1),$$

a quantity which only depends on the total mass $m$ and the number $k$ of particles in the system. The statement then follows from the Markov property and the dynamics of coalescent chains. □

Proposition 5.5 enables us to specify the semigroup of the additive coalescent.

**Corollary 5.5** *Fix* $t > 0$ *and* $1 \le n \le k$. *In the notation above, one has:*

(i) *If* $\#(t)$ *denotes the number of particles at time* $t$, *then* $\#(t) - 1$ *has the binomial distribution with parameter* $(k-1, \mathrm{e}^{-mt})$. *Equivalently,*

$$\mathbb{P}\left(T_{k-n} \le t < T_{k-n+1}\right) = \binom{k-1}{n-1} \left(1 - \mathrm{e}^{-mt}\right)^{k-n} \mathrm{e}^{-mt(n-1)}.$$

(ii) *Consider a partition $\pi$ of $[k]$ into $n$ (non-empty) blocks, say $B_1, \dots, B_n$, and write $\Lambda_\pi(t)$ for the event that the configuration $X(t)$ of the coalescent chain at time $t$ has $n$ atoms, each resulting from the coagulation of the family of particles $\{x_i : i \in B_j\}$ for $j = 1, \dots, n$. Then*

$$\mathbb{P}(\Lambda_\pi(t)) = (1 - \mathrm{e}^{-mt})^{k-n} \mathrm{e}^{-mt(n-1)} m^{n-k} \prod_{j=1}^{n} \mathbf{x}_{B_j}^{\#B_j - 1},$$

*where $\#B_j$ stands for the cardinal of the block $B_j$ and $\mathbf{x}_{B_j} = \sum_{i \in B_j} x_j$.*

*Proof* (i) By the absence of memory of the exponential laws, Proposition 5.5 enables us to think of the sequence $T_1, \dots, T_{k-1}$ as the ordered statistics of $k-1$ i.i.d. exponential variables with parameter $m$, say $\mathbf{e}_1, \dots, \mathbf{e}_{k-1}$. In particular, we have $T_{k-n} \leq t < T_{k-n+1}$ if and only if $\#\{1 \leq i \leq k-1 : \mathbf{e}_i \leq t\} = k-n$, which yields the stated formula.

(ii) Write $\Lambda'_\pi(n)$ for the event that the state $Y(k-n)$ of the coalescent chain after $k-n$ coagulations has $n$ atoms resulting from the merging of $\{x_i : i \in B_j\}$ for $j = 1, \dots, n$. Then, by Proposition 5.5(i), we have

$$\mathbb{P}(\Lambda_\pi(t)) = \mathbb{P}(T_{k-n} \leq t < T_{k-n+1}, \Lambda'_\pi(n)) = \mathbb{P}(T_{k-n} \leq t < T_{k-n+1}) \mathbb{P}(\Lambda'_\pi(n)).$$

The first term of the product in the right-hand side has been computed in (i), so we focus on the second.

By Proposition 5.5(iii), the probability that the first coagulation involves two particles with labels in the block $B_j$ is

$$\sum_{i,i' \in B_j, i \neq i'} \frac{x_i + x_{i'}}{m(k-1)} = \frac{\mathbf{x}_{B_j}(\#B_j - 1)}{m(k-1)}.$$

Then consider an arbitrary sequence $(\ell_1, \dots, \ell_{k-n})$ in $\{1, \dots, n\}$ such that for every $j = 1, \dots, n$, $\#\{i \leq k-n : \ell_i = j\} = \#B_j - 1$. It then follows from the Markov property of the sequence $Y$ that the probability of the event that the $i$-th coagulation only involves particles formed from initial particles with labels in $B_{\ell_i}$ for all $i = 1, \dots, k-n$, is

$$\frac{(n-1)!}{m^{k-n}(k-1)!} \prod_{j=1}^{n} \left( \mathbf{x}_{B_j}^{\#B_j - 1} (\#B_j - 1)! \right).$$

The number of such sequences $(\ell_1, \dots, \ell_{k-n})$ is

$$\binom{k-n}{\#B_1 - 1, \dots, \#B_n - 1} = \frac{(k-n)!}{(\#B_1 - 1)! \dots (\#B_n - 1)!}.$$

After the cancellation of a few terms, this gives

$$\mathbb{P}(\Lambda'_\pi(n)) = \frac{1}{m^{k-n}\binom{k-1}{n-1}} \prod_{j=1}^{n} \mathbf{x}_{B_j}^{\#B_j - 1},$$

and the stated formula follows. □

### 5.3.2 Coagulation of trees in a random forest

In this section, we develop the remarkable connections between the additive coalescent and random trees which have been observed by Pitman [184]. We start by introducing further notions on graphs, forests, and trees. In this direction, it will be convenient to adopt a point of view which is slightly different from that in Section 5.2, in the sense that we worked there with unordered edges, whereas it will be important here to enumerate edges of graphs. In the present section, we shall only consider graphs with edges labeled by integers, even if we shall often omit to mention this explicitly in order to avoid repetitions.

Specifically, we consider $[k] = \{1, \dots, k\}$ as a set of $k$ labeled vertices and call a finite sequence $G = (e_1, \dots, e_n)$ of distinct edges a graph with labeled edges. As the cardinal of the set of edges is $m := k(k-1)/2$, there are $m!/(m-n)!$ graphs with $n$ labeled edges. We call a sequence of $n \geq 2$ distinct edges, say $(e_1, \dots, e_n)$, such that $e_\ell$ and $e_{\ell+1}$ are adjacent for every $\ell = 1, \dots, n$ (with the convention that $e_{n+1} = e_1$) a loop. A graph is said to be connected if and only if there is a single connected component, and then is called a *tree* if and only if it contains no loop (i.e. one cannot construct a loop using edges in $G$). More generally, we call *forest* a graph which contains no loop; the connected components of a forest then consist of a family of trees. Observe that a forest with $n$ trees on a set of $k$ vertices has $k-n$ edges. Next, we enrich these combinatorial structures by distinguishing a vertex, called the *root*, in each tree, so there are $\ell$ different possible roots in each tree with size (i.e. number of vertices) $\ell$. This enables us to enumerate the trees in a forest according to the increasing order of their roots.

For every integer $n \leq k$, we write $\mathcal{F}(n, k)$ for the set of graphs with labeled vertices on $[k]$ consisting of a forest of $n$ rooted trees. A generic forest $\varphi \in \mathcal{F}(n, k)$ can thus be constructed as follows. We first pick $n$ distinct vertices, say $r_1 < \cdots < r_n$ in $[k]$, which serve as the roots of the $n$ tree components. Then we choose a partition $E_1, \dots, E_n$ of $[k]\backslash\{r_1, \dots, r_n\}$ (some of the blocks $E_i$ may be empty), and a tree structure with unlabeled edges on each set of vertices $E_i \cup \{r_i\}$ for $i = 1, \dots, n$. Finally, we label the $k-n$ edges.

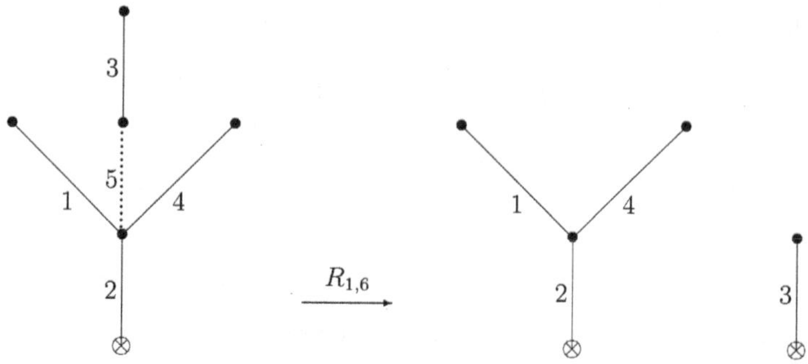


Destruction of the last edge in a rooted tree with 6 vertices ⊗ = root of a tree ;
• = ordinary vertex

For every $1 \leq n < k$, we now describe a natural map $R_{n,k} : \mathcal{F}(n, k) \to \mathcal{F}(n+1, k)$, which is the key to the remarkable connections between trees, forests and the additive coalescent. For each $\varphi \in \mathcal{F}(n, k)$, we remove the last edge (i.e. with label $k - n$), say $\{i, j\}$. This yields a forest with $n+1$ trees, all of them but one having a root. We then choose the unique vertex among $\{i, j\}$ which belongs to this unique non-rooted tree as the root of the latter. The resulting forest with $n+1$ rooted trees is denoted by $R_{n,k}(\varphi)$; see the picture above.

The following lemma provides the key to the relation with additive coalescence.

**Lemma 5.9** *For every $1 \leq n < k$, the map $R_{n,k} : \mathcal{F}(n, k) \to \mathcal{F}(n+1, k)$ is surjective. More precisely, for every $\tilde{\varphi} \in \mathcal{F}(n+1, k)$, we have*

$$\#\left\{\varphi \in \mathcal{F}(n, k) : R_{n,k}(\varphi) = \tilde{\varphi}\right\} = nk\,.$$

*Proof* That $R_{n,k}$ is surjective should be obvious. More precisely, take $\tilde{\varphi} \in \mathcal{F}(n+1, k)$. In order to construct a generic $\varphi \in R_{n,k}^{-1}(\tilde{\varphi})$, we first pick a vertex, say, $i$. We write $T$ for the tree component of $\tilde{\varphi}$ which contains $i$ and $\rho(i)$ for its root. Next we pick another tree $T' \neq T$ among tree components of $\tilde{\varphi}$ and write $j$ for its root. Then we add $\{i, j\}$ as the $(k-n)$-th edge of the graph. This new edge connects $T$ and $T'$, and yields a new tree denoted by $T \star T'$ which we root at $\rho(i)$. It should be plain from a picture that this construction induces a forest with labeled edges and $n$ rooted trees in $R_{n,k}^{-1}(\tilde{\varphi})$, that any $\varphi \in R_{n,k}^{-1}(\tilde{\varphi})$ can be obtained by this procedure, and that different choices of the vertex $i$ and the tree $T'$ yields different elements of $R_{n,k}^{-1}(\tilde{\varphi})$. Since there are $k \times n$ possible choices for $i$ and $T'$, this completes the proof of the lemma. □

The recursive deletion of edges is often referred to as the destruction of a graph. For any forest $\varphi$, say with $n$ trees, we write $|\varphi|^{\downarrow} = (s_1, \ldots, s_n, 0, \ldots) \in \mathcal{S}_f^{\downarrow}$ for the sequence of the sizes of its tree components ranked in decreasing order, and completed with an infinite sequence of 0s. Starting from a generic tree $\tau \in \mathcal{F}(1, k)$, Lemma 5.9 leads us to introduce the sequence of forests defined by

$$\varphi_\tau(1) = \tau \,, \ \varphi_\tau(n+1) = R_{n,k}(\varphi_\tau(n)) \quad \text{for } n = 1, \ldots, k-1 \,.$$

We then consider $|\varphi_\tau(n)|^{\downarrow}$, the ranked sequence of the sizes of the tree components of the forest with $n$ trees obtained from $\tau$ after deletion of its $(n-1)$ last edges. Recall the notation in Proposition 5.5

**Proposition 5.6** *Endow the space $\mathcal{F}(1, k)$ of rooted trees with labeled edges on $[k]$ with the uniform probability measure. Then the sequence of random variables*

$$\tau \to |\varphi_\tau(n)|^{\downarrow} \,, \qquad n = 1, \ldots, k \,,$$

*has the same distribution as the sequence*

$$Y(k-n) \,, \qquad n = 1, \ldots, k \,,$$

*where for $\ell = 0, \ldots, k-1$, $Y(\ell)$ is the state of an additive coalescent chain started from the monodisperse configuration with k particles each of unit mass, after the $\ell$-th coagulation.*

*Proof* Lemma 5.9 yields by induction that for every $n = 1, \ldots, k$, the random variable $\tau \to \varphi_\tau(n)$ is uniformly distributed on $\mathcal{F}(n, k)$. Suppose now $k \geq 2$, fix a forest $\tilde{\varphi} \in \mathcal{F}(n, k)$, and work conditionally on $\varphi_\tau(n) = \tilde{\varphi}$. The law of $\tau \to \varphi_\tau(n-1)$ is then the uniform probability measure on the set of $k(n-1)$ forests in $\mathcal{F}(n, k)$ that can be obtained from $\tilde{\varphi}$ by the adjunction of an edge with label $k-n$ between a vertex $i \in [k]$ and the root $j$ of one of the $n-1$ tree components of $\tilde{\varphi}$ that do not contain $i$.

Write $\tau_1, \ldots, \tau_n$ for the tree components of $\tilde{\varphi}$, listed as usual in increasing order of their roots. For every pair $\{\ell, \ell'\}$ of distinct integers in $[n]$, the probability that the vertex $i$ is picked in $\tau_\ell$ and the root $j$ in $\tau_{\ell'}$ is thus $(k(n-1))^{-1}|\tau_\ell|$, where $|\tau|$ denotes the size (i.e. number of vertices) of a generic tree $\tau$. Hence the probability that $\varphi_\tau(n-1)$ results from the merging of $\tau_\ell$ and $\tau_{\ell'}$, that is that $i$ is picked in $\tau_\ell$ and $j$ as the root of $\tau_{\ell'}$, or conversely that $i$ is picked in $\tau_{\ell'}$ and $j$ as the root of $\tau_\ell$ (recall that these two cases yield different trees as the root is that of $\tau_\ell$ in the first case, and that of $\tau_{\ell'}$ in the second), is equal to

$$\frac{|\tau_\ell|+|\tau_{\ell'}|}{k(n-1)}.$$

Now let $\mathcal{G}_\ell$ denote the sigma field generated by the random variable $\tau \to \varphi_\tau(\ell)$, so $(\mathcal{G}_\ell)_{\ell=1,\dots,k}$ is a reversed filtration. The calculation above shows that for every $n = 0, \dots, k-2$ and every finite configuration $\mathbf{s} = (s_1, \dots, s_{k-n}, 0, \dots)$ with $s_i \in \mathbb{N}$ and $\sum_{i=1}^{k-n} s_i = k$, on the event that $|\varphi_\tau(k-n)|^\downarrow = \mathbf{s}$,

$$\mathbb{P}\left(|\varphi_\tau(k-n-1)|^\downarrow = \mathbf{s}^{\ell\oplus\ell'} \mid \mathcal{G}_{k-n}\right) = \frac{s_\ell + s_{\ell'}}{k(k-n-1)}, \qquad 1 \le \ell < \ell' \le n.$$

The comparison with Proposition 5.5 completes the proof. □

It is easy to derive from Proposition 5.6 a simple construction of additive coalescent chains started from a monodisperse initial configuration, which mirrors that given in Lemma 5.4 for the multiplicative kernel in terms of the random graph model.

**Corollary 5.6** *Let $\tau^{[k]}$ be a random rooted tree with labeled edges on $[k]$, which follows the uniform distribution on $\mathcal{F}(1,k)$. For every $1 \le i \le k-1$, assign to the i-th edge $e_i$ of $\tau^{[k]}$ an independent standard exponential variable $\mathbf{e}_i$. For every $t \ge 0$, consider the random graph on $[k]$ such that each edge $e_i$ is open if and only if $\mathbf{e}_i \le t$, and denote by $X^{[k]}(t)$ the ranked sequence of the sizes (i.e. number of vertices) of its connected components. Then the process $(X^{[k]}(t)), t \ge 0)$ is an additive coalescent chain started from the monodisperse configuration $(1, \dots, 1, 0, \dots)$.*

*Proof* This is an immediate consequence of Propositions 5.5 and 5.6, and elementary properties of independent exponential variables. □

Proposition 5.6 also reduces the computation of probabilities related to the additive coalescent chain started from a monodisperse configuration to simple combinatorial arguments. In this direction, we first note that Lemma 5.9 yields the identity

$$\#\mathcal{F}(n,k) = nk\#\mathcal{F}(n+1,k),$$

and since $\#\mathcal{F}(k,k)=1$, we obtain by induction that

$$\#\mathcal{F}(n,k)=k^{k-n}(k-1)!/(n-1)!. \tag{5.12}$$

Further, since there are $(k-n)!$ ways to enumerate the $k-n$ edges of a forest in $\mathcal{F}(n,k)$, ignoring the labeling of vertices, we see that there are $\binom{k-1}{n-1}k^{k-n}$ forests with $k$ vertices and $n$ rooted trees. In particular there are $k^{k-1}$ rooted trees with $k$ vertices, which is a well-known formula due to Cayley.

We are now able to specify the one-dimensional statistics for the additive coalescent chain started from a monodisperse configuration. In this direction, the simpler expressions are obtained when particles are enumerated in random uniform order.

**Corollary 5.7** *The notation is the same as in Proposition 5.6. For every integer $n\le k$, write $\tilde{Y}(k-n)$ for the sequence obtained from $Y(k-n)$ by an independent uniform permutation of its $n$ particles, that is*

$$\tilde{Y}(k-n)=\left(Y_{\varsigma(1)}(k-n),\ldots,Y_{\varsigma(n)}(k-n)\right)$$

*where $\varsigma$ is a random uniform permutation of $[n]$ which is independent of $Y(k-n)$.*

*Then for all integers $k_1,\ldots,k_n$ with $k_1+\cdots+k_n=k$, we have*

$$\mathbb{P}(\tilde{Y}(k-n)=(k_1,\ldots,k_n))=\frac{(k-n)!}{nk^{k-n-1}}\prod_{i=1}^{n}\frac{k_i^{k_i-1}}{k_i!}.$$

*Proof* Recall that the tree components of forests are listed in increasing order of their roots. Consider a permutation $\sigma$ on $[n]$ and a forest $\varphi\in\mathcal{F}(n,k)$ with tree components $\tau_1,\ldots,\tau_n$, such that the permuted sequence $(|\tau_{\sigma(1)}|,\ldots,|\tau_{\sigma(n)}|)$ of the sizes of those trees is $(k_1,\ldots,k_n)$. Call *composition* (or ordered partition) of a finite set $E$ an ordered sequence of non-empty sets $E_1,\ldots,E_n$ which are pairwise disjoint and such that $\bigcup_{i=1,\ldots,n}E_i=E$. If we denote for each $i=1,\ldots,n$ by $E_i$ the set of vertices of $\tau_{\sigma(i)}$, then we obtain a composition of $[k]$ into blocks $E_1,\ldots,E_n$ with respective cardinals $k_1,\ldots,k_n$. Furthermore, ignoring the labeling of edges, $\varphi$ induces a rooted-tree structure on each block $E_i$.

Conversely, let us first pick a composition $(E_1,\ldots,E_n)$ of $[k]$ such that $\#E_i=k_i$ for each $i=1,\ldots,n$; there are

$$\binom{k}{k_1,\ldots,k_n}=\frac{k!}{k_1!\ldots k_n!}$$

possibilities. Next, we attach a rooted tree structure to $E_i$ without labeling the edges. By Cayley's formula, there are $k_i^{k_i-1}$ choices for each $i$. Finally, we label all the $k-n$ edges; there are $(k-n)!$ possible ways. This yields a unique forest $\varphi$ with $n$ trees, say $\tau_1, \ldots, \tau_n$, and a unique permutation $\sigma$ on $[n]$ such that $E_i$ is the set of vertices of the tree $\tau_{\sigma(i)}$ for every $i = 1, \ldots, n$.

We conclude that the total number of pairs $(\sigma, \varphi)$, where $\sigma$ is a permutation of $[n]$ and $\varphi \in \mathcal{F}(n, k)$ a forest such that $|\tau_{\sigma(i)}| = k_i$ for every $i = 1, \ldots, n$, is

$$k!(k-n)! \prod_{i=1}^{n} \frac{k_i^{k_i-1}}{k_i!}.$$

On the other hand, it follows from (5.12) that the total number of pairs $(\sigma, \varphi)$, where $\sigma$ is a permutation of $[n]$ and $\varphi \in \mathcal{F}(n, k)$, is $nk^{k-n}(k-1)!$. The statement now follows from Proposition 5.6 and the fact that $\varphi_\tau(n)$ has the uniform distribution on $\mathcal{F}(n, k)$. □

We next present the observation, due to Pavlov [177], that the distribution of the $k$-tuple $\tilde{Y}(k-n)$ in Corollary 5.7 can be conveniently expressed in terms of $k$ independent Borel(1) variables. In this direction, we first consider a compound Poisson process with drift $S = (S_t, t \geq 0)$ defined by

$$S_t := t - N_t, \qquad t \geq 0,$$

where $N = (N_t, t \geq 0)$ is a standard Poisson process. Note that $S$ has no positive jumps, and that for every integer $k$, since $k - S_k = N_k$ is a Poisson variable with parameter $k$,

$$\mathbb{P}(S_k = z) = \frac{e^{-k} k^{k-z}}{(k-z)!}, \qquad z \in \mathbb{Z}, z \leq k.$$

Next, for every $x \geq 0$, consider the first-passage time

$$\varsigma_x := \min \{k \in \mathbb{N} : S_k > x\}.$$

The absence of positive jumps for $S$ implies that $(\varsigma_n, n \in \mathbb{Z}_+)$ is a renewal process, that is the increments $\xi_1 = \varsigma_1$, $\xi_2 = \varsigma_2 - \varsigma_1, \ldots$ are i.i.d. variables. Moreover there is the well-known identity

$$\mathbb{P}(\varsigma_n = \ell) = \frac{n}{\ell} \mathbb{P}(S_\ell = n) = n \frac{e^{-\ell} \ell^{\ell-n-1}}{(\ell-n)!}, \qquad n \leq \ell; \tag{5.13}$$

see for example Corollary VII.3 in [29]. In particular, $\varsigma_1$, and thus each increment $\xi_i$ has the standard Borel law (i.e. with parameter 1)

$$\mathbb{P}(\xi_i = \ell) = \frac{e^{-\ell} \ell^{\ell-1}}{\ell!}, \qquad \ell \in \mathbb{N}.$$

We can now state the following.

**Corollary 5.8** *Fix two integers $n \le k$, consider $n$ independent variables $\xi_1, \dots, \xi_n$ with the standard Borel law, and set $\varsigma_n := \xi_1 + \cdots + \xi_n$. Then, in the notation of Proposition 5.6, $\tilde{Y}(k-n)$ has the same distribution as $(\xi_1, \dots, \xi_n)$ conditioned on $\varsigma_n = k$.*

*Proof* Let $k_1, \dots, k_n$ be $n$ integers such that $k_1 + \cdots + k_n = k$. Since the variables $\xi_i$ are independent, we have

$$\mathbb{P}(\xi_1 = k_1, \dots, \xi_n = k_n) = \mathrm{e}^{-k} \prod_{i=1}^{n} \frac{k_i^{k_i - 1}}{k_i!}.$$

By (5.13), we conclude that

$$\mathbb{P}(\xi_1 = k_1, \dots, \xi_k = k_n \mid \varsigma_n = k) = \frac{(k-n)!}{nk^{k-n-1}} \prod_{i=1}^{n} \frac{k_i^{k_i - 1}}{k_i!}.$$

The comparison with Corollary 5.7 completes the proof. □

We now turn our attention to certain asymptotics which are motivated by the study of the hydrodynamic behavior of coalescent chains in Section 5.2. We use the notation $X^{[k]} = (X^{[k]}(t), t \ge 0)$ for an additive coalescent chain started from the monodisperse configuration $(1, \dots, 1, 0, \dots)$ which consists in $k$ atoms, each of unit mass. For every $t \ge 0$, we write $\#^{[k]}(t)$ for the number of atoms in the finite configuration $X^{[k]}(t)$, and then $\xi_1^{[k]}(t)$ and $\xi_2^{[k]}(t)$ for the masses of a pair of atoms picked uniformly at random and without replacement at time $t$. This means that the conditional distribution of $(\xi_1^{[k]}(t), \xi_2^{[k]}(t))$ given $X^{[k]}(t) = (x_1, \dots, x_n, 0, \dots)$ (where $n = \#^{[k]}(t)$ is the number of atoms at time $t$) is that of $(x_{u_1}, x_{u_2})$, where $(u_1, u_2)$ is uniformly distributed on $\{(i,j) : i, j = 1, \dots, n \text{ and } i \ne j\}$. Results obtained previously in this section easily yield the following limit theorem.

**Lemma 5.10** *When $k \to \infty$, we have for every $t \ge 0$ that:*

(i) *$k^{-1}\#^{[k]}(t/k)$ converges in $L^2(\mathbb{P})$ to $\mathrm{e}^{-t}$,*
(i) *$(\xi_1^{[k]}(t/k), \xi_2^{[k]}(t/k))$ converges in distribution to a pair of independent variables with the Borel$(1 - \mathrm{e}^{-t})$ law.*

*Proof* (i) Recall from Corollary 5.5 that $\#^{[k]}(t/k) - 1$ has the binomial$(k - 1, \mathrm{e}^{-t})$ distribution. Thus the law of large numbers for Bernoulli variables implies our first claim.

(ii) Consider a sequence $\xi_1, \dots$ of i.i.d. standard Borel variables, and for every $n \in \mathbb{N}$, set $\varsigma_n := \xi_1 + \cdots + \xi_n$. It is easily seen either directly from

(5.13) or from classical large deviation arguments that for every $b \in ]0,1[$, the conditional distribution of the first two steps $(\xi_1, \xi_2)$ given $\varsigma_n = k$ converges weakly as $k \to \infty$ and $n/k \to b$ to a pair of independent variables, each with the Borel$(1-b)$ distribution. Our statement then follows from (i) and Corollary 5.8. $\square$

Lemma 5.10 and an easy variation of the argument used for the proof of Proposition 5.4 yield the following limit theorem (details are left to the reader):

**Corollary 5.9** *Let $X^{[k]} = (X^{[k]}(t), t \geq 0)$ denote the additive coalescent chain started from the monodisperse configuration $\mathbf{x}^{[k]}$ with k atoms, each of unit mass. Set for every $t \geq 0$ and $\ell \in \mathbb{N}$*

$$n_t^{[k]}(\ell) := k^{-1}\#\left\{i \in \mathbb{N} : X_i^{[k]}(t/k) = \ell\right\}.$$

*Then*

$$\lim_{k\to\infty} n_t^{[k]}(\ell) = \mathrm{e}^{-t}\frac{((1-\mathrm{e}^{-t})\ell)^{\ell-1}\exp(-(1-\mathrm{e}^{-t})\ell)}{\ell!} \qquad \text{in } L^2(\mathbb{P}).$$

Observe that the limit in Corollary 5.9 coincides with the mass distribution of the Borel$(1-\mathrm{e}^{-t})$ law up to the factor $\mathrm{e}^{-t}$, and provides a solution to Smoluchowski's coagulation equation (5.6) for the additive kernel $\kappa(x,y) = x+y$. This has been discovered originally by Golovin [117], and can be checked directly by elementary calculations. We stress that this solution is valid for all $t \geq 0$ and that

$$\sum_{\ell=1}^{\infty} \ell \mathrm{e}^{-t}\frac{((1-\mathrm{e}^{-t})\ell)^{\ell-1}\exp(-(1-\mathrm{e}^{-t})\ell)}{\ell!} = 1,$$

since the expectation of a Borel$(b)$ variable is $1/(1-b)$ for every $b \in ]0,1[$. In other words, the total mass is a preserved quantity, and there is no such phenomenon of gelation as in the case of the multiplicative kernel.

### 5.3.3 The standard additive coalescent

The purpose of this section is to present another remarkable limit theorem due to Evans and Pitman [100], which, roughly speaking, specifies the asymptotic behavior of an additive coalescent chain started from monodisperse dust. That is we consider, for a large integer $k$, the initial configuration given by $k$ particles each with mass $1/k$, so the total mass of the system is always 1. Typically, a phase transition occurs at time $\frac{1}{2}\ln k + O(1)$, in the sense that there are no macroscopic particles (i.e. of mass of order 1) in the system at times much smaller than $\frac{1}{2}\ln k$, while at times much larger than $\frac{1}{2}\ln k$, there is essentially a single macroscopic particle of mass close to 1 and all the remaining particles are much smaller. In other words, the entire macroscopic evolution of the system can be observed at times $\frac{1}{2}\ln k + O(1)$. Here is the precise statement.

**Theorem 5.3** *For each integer k, let $X^{(k)} = (X^{(k)}(t), t \geq 0)$ denote the additive coalescent chain started from the monodisperse configuration with k atoms, each of mass $1/k$. Then as $k \to \infty$, the process*

$$X^{(k)}\left(t + \frac{1}{2}\ln k\right), \qquad t \geq -\frac{1}{2}\ln k$$

*converges in distribution on the space of càdlàg paths with values in the set of mass-partitions $\mathcal{P}_{\mathrm{m}}$ and endowed with the topology of uniform convergence on bounded time-intervals. The limit is the unique additive coalescent process $X = (X(t), t \in \mathbb{R})$ parametrized by real times, which is called the* **standard additive coalescent**, *such that for each $t \in \mathbb{R}$, $X(t)$ has the law of the ranked sequence $\mathsf{a}_1 \geq \mathsf{a}_2 \geq \cdots$ of the atoms of a Poisson random measure on $]0, \infty[$ with intensity*

$$\frac{\mathrm{e}^{-t}}{\sqrt{2\pi a^3}} da, \qquad a > 0,$$

*and conditioned on $\sum_{i=1}^{\infty} \mathsf{a}_i = 1$, where this conditioning is taken in the sense of Proposition 2.4.*

The proof of Theorem 5.3 relies on the following lemma.

**Lemma 5.11** *For all integers $n < k$, write $Y^{(k)}(k-n)$ for the state of the additive coalescent started from the monodisperse initial configuration with k atoms each of unit mass, after $k-n$ coagulations. For every $b > 0$, if $k \to \infty$ and $n/\sqrt{k} \to b$, then $k^{-1}Y^{(k)}(k-n)$ converges in distribution on $\mathcal{S}_1^{\downarrow}$ to the ranked sequence $(\mathsf{a}_1, \mathsf{a}_2, \ldots)$ of the atoms of a Poisson random measure on $]0, \infty[$ with intensity*

$$\frac{b}{\sqrt{2\pi a^3}} da, \qquad a > 0,$$

*and conditioned on $\sum_{i=1}^{\infty} \mathsf{a}_i = 1$, where this conditioning is taken in the sense of Proposition 2.4.*

Let us first give a heuristic argument for Lemma 5.11 based on Corollary 5.8. Recall that $Y^{(k)}(k-n)$ can be viewed as the ranked sequence of the increments $\xi_1 = \varsigma_1 - \varsigma_0, \ldots, \xi_n = \varsigma_n - \varsigma_{n-1}$ of the first-passage time process of the random walk $S_i = i - N_i$ conditioned on $\varsigma_n = k$, where $(N_t : t \geq 0)$ is a standard Poisson process. By the invariance principle, the (unconditioned) process

$$\left(\frac{S_{kt}}{\sqrt{k}}, t \geq 0\right)$$

converges in law as $k \to \infty$ to a standard Brownian motion, say $B = (B_t, t \geq 0)$. It follows that the renormalized first passage process

$$\left(k^{-1}\varsigma_{x\sqrt{k}}, x \geq 0\right)$$

converges in distribution as $k \to \infty$ to the first-passage process of $B$,

$$\varsigma_x^B := \inf\{t \geq 0 : B_t > x\}, \qquad x \geq 0.$$

Therefore it is natural to expect that if we take $n \sim b\sqrt{k}$, then the ranked sequence of the renormalized increments $(k^{-1}\xi_1, \ldots, k^{-1}\xi_k)$ converges in law as $k \to \infty$ to the ranked sequence of the jumps of $(\varsigma_x^B, 0 \leq x \leq b)$. On the other hand, it is well known (see for example [192] on its pages 107–8) that $\varsigma^B$ is a stable subordinator with index 1/2, and more precisely with Lévy measure $(2\pi a^3)^{-1/2}da$. In particular, the point measure

$$\sum_{0<x\leq b} \delta_{\varsigma_x^B - \varsigma_{x-}^B}$$

is a Poisson random measure on $\mathbb{R}_+$ with intensity $b(2\pi a^3)^{-1/2}da$. We thus recover Lemma 5.11 by conditioning on $\varsigma_n = k$ and $\varsigma_b^B = 1$. However, this argument is only informal as it involves a singular conditioning for the Brownian limit. Let us now give a rigorous proof.

*Proof* It follows from Stirling's formula that the Borel probability function fulfills

$$\frac{\mathrm{e}^{-k}k^{k-1}}{k!} \sim (2\pi k^3)^{-1/2}, \qquad k \to \infty.$$

This enables us to apply Corollary 2.2 with $\alpha = 1/2$ and $c = 1/\sqrt{2\pi}$, and we get that $k^{-1}Y^{(k)}(k-n)$ converges in distribution on $\mathcal{P}_{\mathrm{m}}$ to the limit in the statement. Finally, an application of Scheffé's lemma enables us to strengthen the convergence to convergence in distribution on $\mathcal{S}_1^{\downarrow}$. □

We may now tackle the proof of Theorem 5.3; recall the notation there.

*Proof* Using Proposition 5.5, we see that the $(k-n)$-th coagulation in the process $X^{(k)}(\cdot)$ occurs at time $T^{(k)}_{k-n}$ which is distributed as

$$\sum_{i=n}^{k-1} \frac{1}{i}\mathbf{e}_i\,,$$

where $(\mathbf{e}_i, i \in \mathbb{N})$ is a sequence of i.i.d. standard exponential variables. In particular, if $k, n \to \infty$ with $n \sim b\sqrt{k}$ for some $b > 0$, then

$$\mathbb{E}(T^{(k)}_{k-n}) = \frac{1}{2}\ln k - \ln b + o(1) \quad \text{and} \quad \mathrm{Var}(T^{(k)}_{k-n}) = o(1)\,,$$

and therefore

$$T^{(k)}_{k-n} - \frac{1}{2}\ln k \to -\ln b \qquad \text{in probability.}$$

This shows that if we denote by $\#^{(k)}(t)$ the number of particles in the configuration $X^{(k)}(t)$, then for every $t \in \mathbb{R}$,

$$\lim_{k\to\infty} \frac{1}{\sqrt{k}}\#^{(k)}\left(\frac{1}{2}\ln k + t\right) = \mathrm{e}^{-t} \qquad \text{in probability.}$$

Alternatively, this can also be deduced from Corollary 5.5.

On the other hand, we see from an elementary change of scale that $t \to kX^{(k)}(kt)$ is an additive coalescent chain started from the monodisperse initial configuration with $k$ atoms each of unit mass. Combining the estimate above with Lemma 5.11 establishes the one-dimensional convergence in Theorem 5.3. Clearly, the additive kernel $\kappa(x, y) = x + y$ fulfills the Lipschitz condition of Theorem 5.1. By an application of the Markov property of coalescent chains, we then derive convergence in distribution on the space of càdlàg paths with values in $\mathcal{S}_1^{\downarrow}$ on some bounded time-interval and endowed with the topology of uniform convergence, from the one-dimensional convergence. □

It is remarkable that the time-parameter of the standard additive coalescent varies in the whole real line $\mathbb{R}$ and not just $[0, \infty[$; such process is called *eternal*. Plainly, one can construct further eternal processes by an elementary time-shift, that is of the type $X(t_0 + \cdot)$ for some $t_0$; and it is then natural to wonder whether there exist different eternal versions of the additive coalescent. In the framework of general Markov processes, this question is equivalent to specifying the so-called *entrance boundary* at time $-\infty$. Intuitively, any (non-degenerate) eternal additive coalescent must start at time $-\infty$ from dust, in the sense that $\lim_{t\to-\infty} X(t) = \mathbf{0} := (0, \dots)$. The mass-partition into dust $\mathbf{0}$ arises in particular as the limit when $k \to \infty$

of the monodisperse configuration $(1/k, \ldots, 1/k, 0, \ldots)$ with $k$ particles each of mass $1/k$. One might be tempted to believe that Theorem 5.3 might hold more generally for any sequence of initial configurations with total mass 1 that converges to dust, and then that the standard additive coalescent should appear as a universal limit of additive coalescent chains started from initial configurations close to dust. In fact this intuition is not correct, and there is actually a very rich family of eternal additive coalescents which can arise as such limits. We refer to Aldous and Pitman [10], and also to [33] and [166], for precise (and rather surprising) results in this direction.

### 5.3.4 A dual fragmentation process

The basic idea which has been used for computing statistics of an additive coalescent is that the destruction of a random uniform tree by successive deletion of its edges yields, upon time-reversal, the successive states of an additive coalescent chain started from a monodisperse configuration; see Proposition 5.6. In this section, we shall consider the destruction process forward in time, and show that under the same asymptotic regime as in Section 5.3.3, it yields a remarkable self-similar fragmentation process in the sense of Section 3.3 (in particular, recall Definition 3.5 and Proposition 3.7).

**Theorem 5.4** *Let $X = (X(r), r \in \mathbb{R})$ be a standard additive coalescent. Set $F(0) = (1, 0, \ldots)$ and*

$$F(t) = X(-\ln t), \qquad t > 0.$$

*The process $(F(t), t \geq 0)$ is a* **self-similar mass-fragmentation** *with index $1/2$. This means that $(F(t), t \geq 0)$ is Markovian, and for every proper mass partition $\mathbf{s} = (s_1, \ldots)$ (i.e. $\sum_{i=1}^{\infty} s_i = 1$), the conditional distribution of $F(t+u)$ given $F(t) = \mathbf{s}$ is the same as that of* $\text{Frag}(\mathbf{s}, \mathbf{s}^{(\cdot)})$*, where $\mathbf{s}^{(\cdot)} = (\mathbf{s}^{(i)}, i \in \mathbb{N})$ is a sequence of independent random mass-partitions and each $\mathbf{s}^{(i)}$ is distributed as $s_i F(u\sqrt{s_i})$.*

In order to establish Theorem 5.4, we first develop some elements on random destruction of graphs, trees and forests. Given a graph $G$, we attach to each edge $e$ of $G$ a standard exponential variable $\mathbf{e}_e$, such that to different edges correspond independent variables. We call $((e, \mathbf{e}_e), e \in G)$ a graph with exponentially marked edges. Later in the text, we shall often deal with random graphs, and it will be implicitly understood that the marks on the edges are i.i.d. standard exponential variables conditionally on the graph.

We think of the mark attached to an edge as a lifetime, in the sense that each edge $e$ in the graph is removed at time $\mathbf{e}_e$. In other words, we delete each edge at unit rate, independently of each other; more precisely we set for every $t \geq 0$

$$G(t) := ((e, \mathbf{e}_e - t) : e \in G \text{ and } \mathbf{e}_e > t) .$$

We denote by $\Delta(t) := \# \{e \in G : \mathbf{e}_e \leq t\}$ the number of edges which have been deleted up to time $t$, so $\Delta(t)$ has the binomial distribution with parameters $k$ and $1 - \mathrm{e}^{-t}$, where $k$ stands for the number of edges of $G$. If we agree to label the edges of $G$ in decreasing order of their marks, then it should be plain from the absence of memory of the exponential variables that $G(t)$ is the graph obtained from $G$ by removing its $\Delta(t)$ last edges and marking the remaining ones using independent standard exponential variables.

We now focus on the situation when $G = \tau^{(k)}$ is a random uniform tree with $k$ vertices. The preceding observations and Lemma 5.9 immediately yield the following.

**Lemma 5.12** *In the notation above, for every $t \geq 0$, the conditional distribution of $\tau^{(k)}(t)$ given $\Delta^{(k)}(t) = \ell - 1$ is that of a random uniform forest on $[k]$ with $\ell$ trees and exponentially marked edges.*

Next, denote by $F^{(k)}(t)$ the sequence of the sizes (i.e. the number of vertices) of the tree components of the forest $\tau^{(k)}(t)$, ranked in decreasing order and completed by an infinite sequence of 0s. So $F^{(k)}(\cdot)$ is a random process with values in $\mathcal{S}_f^{\downarrow}$, and it is easily seen from Lemma 5.12 that it enjoys a branching type property.

**Lemma 5.13** *Consider integers $k_1 \geq \cdots \geq k_\ell > 0$ with $k_1 + \cdots + k_\ell = k$, and then $\ell$ independent processes $\phi^{(k_1)}, \ldots, \phi^{(k_\ell)}$ such that for each $i = 1, \ldots, \ell$, $\phi^{(k_i)}$ has the same law as $F^{(k_i)}$. For every $u \geq 0$, let $\phi^{(k_1, \ldots, k_\ell)}(u)$ denote the configuration obtained by the rearrangement of the terms of $\phi^{(k_1)}(u), \ldots, \phi^{(k_\ell)}(u)$ in decreasing order.*

*Then the process $(F^{(k)}(t), t \geq 0)$ is a Markov chain with values in $\mathcal{S}_f^{\downarrow}$, and more precisely, conditionally on $F^{(k)}(t) = (k_1, \ldots, k_\ell, 0, \ldots)$, the shifted process $F^{(k)}(t + \cdot)$ has the same law as $\phi^{(k_1, \ldots, k_\ell)}(\cdot)$.*

*Proof* We work conditionally on $F^{(k)}(t) = (k_1, \ldots, k_\ell, 0, \ldots)$, and write $\tau_1, \ldots, \tau_\ell$ for the tree components of the forest $\tau^{(k)}(t)$. The set of vertices $V_i$ of $\tau_i$ for $i = 1, \ldots, \ell$ form a composition of $[k]$, and the ranked sequence of the sizes of the $V_i$ is $F^{(k)}(t)$. It follows from Lemma 5.12 that conditionally on

$(V_1, \ldots, V_\ell)$, $\tau_1, \ldots, \tau_\ell$ are independent uniform random trees on $V_1, \ldots, V_\ell$, respectively; furthermore we know that the marks on the edges of $\tau^{(k)}(t)$ are given by independent exponential variables. For every $u \geq 0$, $F^{(k)}(t+u)$ is the decreasing rearrangement of the terms of the sequences $F^{\tau_i}(u)$ for $i = 1, \ldots, \ell$, where $F^{\tau_i}(u)$ denotes the ranked sequence of the sizes of the tree components of the forest on $V_i$ which is obtained from $\tau_i$ by removing the edges with marks less than or equal to $u$. Plainly, the distribution of the process $F^{\tau_i}(u)$ only depends on the size of $V_i$, which establishes our claim. □

We are now able to establish Theorem 5.4.

*Proof* We shall prove the result by approximation, working with the processes $F^{(k)}$. Because edges are deleted at unit rate and independently of each other, for every $t \geq 0$, the number $\Delta^{(k)}(t/\sqrt{k})$ of vertices which have been destroyed up to time $t/\sqrt{k}$ fulfills

$$\lim_{k\to\infty} k^{-1/2}\Delta^{(k)}(t/\sqrt{k}) = t \qquad \text{in probability.}$$

It then follows from the analysis made in the preceding section that as $k \to \infty$, the sequence of processes

$$k^{-1}F^{(k)}(t/\sqrt{k}), \qquad t > 0,$$

converge in the sense of finite-dimensional distributions to $(X(-\ln t), t > 0)$, where $X$ denotes a standard additive coalescent.

As a consequence, for every $s > 0$, if we let $k, k' \to \infty$ with $k'/k \to s$, then for every $u > 0$

$$\frac{1}{k}F^{(k')}(u/\sqrt{k}) \text{ converges in distribution to } sX(-\ln(u\sqrt{s})).$$

We now easily deduce from Lemma 5.13 that the conditional distribution of $X(-\ln(t+u))$ given $X(-\ln(t)) = \mathbf{s} = (s_1, \ldots)$, where $\mathbf{s}$ is a proper mass-partition, is the same as the decreasing rearrangement of the terms of independent random sequences $\varsigma^{(1)}, \ldots$, where for each integer $i$, $\varsigma^{(i)}$ has the law of $s_i X(-\ln(u\sqrt{s_i}))$. The standard additive coalescent being Markovian, its transform by a deterministic time-reversal is still Markovian (possibly time-inhomogeneous); this completes the proof of Theorem 5.4. □

Since the one-dimensional distributions of $F(t) = X(-\ln t)$ are known from Theorem 5.3, Theorem 5.4 entirely characterizes the semigroup of the dual fragmentation process $F$. One can then apply the general results of this monograph to derive many interesting consequences. For instance, combining Theorem 5.3 and Proposition 2.4, one gets that the distribution of a size-biased sample from $X(-\ln t) = F(t)$ has density

$$\frac{t}{\sqrt{2\pi y(1-y)^3}} \exp\left(-\frac{t^2 y}{2(1-y)}\right), \qquad 0 < y < 1.$$

One can then deduce from Theorem 5.4 and Corollary 3.1 that the so-called tagged fragment process in the fragmentation $F$ has the same law as $(1/(1+\varsigma^B(t)), t > 0)$, where $\varsigma^B(\cdot)$ is the stable subordinator with index $1/2$ which arises as the first-passage process of a standard Brownian motion (cf. Section 5.3.3). Indeed, it is readily checked that $1/(1+\varsigma^B(\cdot))$ is a self-similar Markov process with index $1/2$, and, using the explicit calculation above, that its one-dimensional distributions coincide with those of the tagged fragment. Finally, one can also identify the characteristic $(\alpha, \mathtt{c}, \nu)$ of the self-similar fragmentation process $F$ (see Section 3.3). The index of self-similarity is given by $\alpha = 1/2$, the erosion coefficient by $\mathtt{c} = 0$ and the dislocation measure $\nu$ is the image of the measure

$$\left(2\pi x^3(1-x)^3\right)^{-1/2} dx, \qquad x \in [1/2, 1[,$$

by the map $x \to (x, 1-x, 0, \dots)$ from $[1/2, 1[$ to the space of mass-partitions $\mathcal{P}_{\mathrm{m}}$. We refer to [9] and [34] for the detailed arguments.

## 5.4 Comments

Stochastic coalescents form a class of simple mean-field processes, which are used in physics as random models for aggregation phenomena that occur for instance in the formation of droplets in aerosols, of colloids, polymers,... Besides the founding articles by Marcus [162] and Lushnikov [155], we refer to the survey by Aldous [6] which has generated a lot of interest for this topic among probabilists (in particular this article mentions several important questions which are still open). The problem of the general construction of coalescent processes, which started from configurations with an infinite number of particles, has been addressed first by Evans and Pitman [100] who used processes with values in the space $\mathcal{P}_\infty$ of partitions of $\mathbb{N}$. A major difficulty is to establish the regularity of the semigroup. Theorem 5.1, which is due to Fournier [104], improves earlier results in [100]. It is believed that Theorem 5.1 should hold under weaker assumptions than the Lipschitz condition on the coagulation kernel (typically, it would be interesting to know whether the conclusions remain valid when one only assumes the continuity of $\kappa$). Fournier and Löcherbach [104, 108] have obtained some results in this direction for certain kernels $\kappa$ being less regular at the origin.

Smoluchowski's coagulation equations have appeared first in [205] and have since motivated a considerable body of works. We refer to the surveys by Drake [84] and Aldous [6], and the monograph by Dubovski [85] as general references for many important results obtained in this field. Among the most interesting theoretical issues which have been considered in recent years, we mention existence and uniqueness of solutions for a large class of kernels (see [96, 150, 175, 176]), phenomena of gelation (criteria for occurrence or not of gelation, extension of solutions after the gelation time, see [16, 79, 96, 129, 175]), and self-similarity (existence and uniqueness of self-similar solutions, convergence of general solutions to self-similar ones; see [80, 107, 151, 165]). Contributions based on probabilistic techniques often rely on considerations about stochastic coalescents (see in particular [129, 175, 176]) as the infinitesimal generators of the latter bear clear connections with the coagulation equations. Most of the literature in this area concerns the case of binary coagulations (just as in the present chapter); we refer to Kolokoltsov [142] for an extension to $k$-nary coagulations. Further probabilistic tools such as particle systems, stochastic calculus and non-linear stochastic differential equations can be also very powerful; see [72, 73, 91, 92, 105, 106, 131, 211]. In particular, the recent work of Wagner [211] points out that the gelation phenomenon for Smoluchowski's coagulation equation is closely related to the explosion of a so-called mass flow model. In a different direction, we also mention that Smoluchowski's coagulation equations have been shown to describe the hydrodynamic behavior of certain spatial particle systems in which coagulation occurs upon collision, see in particular [32, 37, 110, 124, 149]. Constant, additive and multiplicative kernels and their linear combinations, have a special part in the study of Smoluchowski's coagulation equations, as explicit solutions can be obtained, often using Laplace transforms; see [35, 84, 158, 159, 206]. Interestingly, these are also cases for which there are natural probabilistic interpretations of the solutions in terms of branching, Poisson or Lévy processes [35, 74].

The multiplicative coalescent has been masterfully used by Aldous [5] to investigate the formation of the giant component in the Erdős-Rényi random graph model at phase transition. In this work, Aldous pointed out that the Hilbert space $\ell^2$ is a more natural state space for the multiplicative coalescent than $\mathcal{S}_1^{\downarrow}$; in particular, the multiplicative coalescent chain started from the monodisperse configuration with $k$ particles of each of mass $k^{-2/3}$ (observe that the total mass of the system is $k^{1/3}$) has then a weak limit as $k \to \infty$, which is called the *standard multiplicative coalescent*. The latter is a (Fellerian) Markov process with values in $\ell^2$, which can be viewed as the analog for the multiplicative kernel of the process in Section 5.3.3 for the additive kernel. We refer to Aldous and Limic [8] for the characterization of the entrance boundary of the multiplicative coalescence, which is much richer than one might have expected. We also refer to the recent work of Armendáriz [12] for an alternative construction of multiplicative coalescent chains, and a remarkable duality via time-reversal between the standard multiplicative coalescent and a certain fragmentation process which is not self-similar.

Additive coalescence has arisen as a model for formation of droplets in clouds [117] as well as of galaxies in the universe [204], in asymptotics for Knuth's parking [48, 68], so-called union-find algorithms [69], Burgers' turbulence with Brownian initial velocity [32] and gravitational clustering of fluid particles [110]. The remarkable connections between destruction of uniform random trees and additive coalescent chains started from monodisperse configurations can be extended to provide a construction of the standard additive coalescent. Roughly, uniform random trees converge weakly when the number of vertices tends to $\infty$, towards the so-called Continuum Random Tree, see Aldous [4]. Cutting the latter along its skeletons in a Poissonian way yields the self-similar fragmentation of Section 5.3.4 which is dual to the standard additive coalescent; see Aldous and Pitman [9]. This approach also applies to the construction of general eternal additive coalescents, provided that one replaces the Continuum Random Tree by different continuum random trees introduced in [64]; see Aldous and Pitman [10]. There exist also alternative and somewhat simpler constructions based on Brownian excursion [31] and bridges with exchangeable increments [33]; we refer to [166] and [201] for some developments in this field.

# References


[1] Abraham, R. and Delmas, J.-F. Fragmentation associated to Lévy processes using snake. Preprint available via:
`http://fr.arxiv.org/abs/math.PR/0511702`

[2] Abraham, R. and Serlet, L. Poisson snake and fragmentation. *Electron. J. Probab.* **7** (2002), 1-15. Available via:
`http://www.math.washington.edu/ ejpecp/viewarticle.php?id=1318&layout=abstract`

[3] Aldous, D. J. Exchangeability and related topics. *École d'été de probabilités de Saint-Flour, XIII*, Lecture Notes in Math. **1117**, pp. 1-198, Berlin, Springer, 1985.

[4] Aldous, D. J. The continuum random tree III. *Ann. Probab.* **21** (1993), 248-289.

[5] Aldous, D. J. Brownian excursions, critical random graphs and the multiplicative coalescent. *Ann. Probab.* **25** (1997), 812-854.

[6] Aldous, D. J. Deterministic and stochastic models for coalescence (aggregation, coagulation): a review of the mean-field theory for probabilists. *Bernoulli* **5** (1999), 3-48.

[7] Aldous, D. J. and Bandyopadhyay, A. A survey of max-type recursive distributional equations. *Ann. Appl. Prob.* **15** (2005), 1047-1110.

[8] Aldous, D. J. and Limic, V. The entrance boundary of the multiplicative coalescent. *Electron. J. Probab.* **3**, 1-59 (1998). Available via:
`http://www.math.washington.edu/ ejpecp/viewarticle.php?id=1227&layout=abstract`

[9] Aldous, D. J. and Pitman, J. The standard additive coalescent. *Ann. Probab.* **26** (1998), 1703-1726.

[10] Aldous, D. J. and Pitman, J. Inhomogeneous continuum random trees and the entrance boundary of the additive coalescent. *Probab. Theory Related Fields* **118** (2000), 455-482.

[11] Arlotti, L. and Banasiak, J. Strictly substochastic semigroups with application to conservative and shattering solutions to fragmentation equations with mass loss. *J. Math. Anal. Appl.* **293** (2004), 693-720.

[12] Armendáriz, I. Dual fragmentation and multiplicative coagulation; related excursion processes. Preprint (2005).

[13] Arratia, R., Barbour, A. D., and Tavaré, S. *Logarithmic Combinatorial Structures: a Probabilistic Approach.* Monographs in Mathematics, Zürich, European Mathematical Society, 2003.

[14] Asmussen, S. and Kaplan, N. Branching random walks. I-II. *Stochastic Processes Appl.* **4** (1976), 1-13 and 15-31.

[15] Athreya, K. B. and Ney, P. E. *Branching Processes.* Springer-Verlag, New York-Heidelberg, 1972.

[16] Bak, T. A. and Heilmann, O. J. Post-gelation solutions to Smoluchowski's coagulation equation. *J. Phys. A* **27** (1994), 4203-4209.

[17] Banasiak, J. Conservative and shattering solutions for some classes of fragmentation models. *Math. Models Methods Appl. Sci.* **14** (2004), 483-501.

[18] Barral, J. Moments, continuité, et analyse multifractale des cascades multiplicatives de Mandelbrot. *Probab. Theory Relat. Fields* **113** (1999), 535-569.

[19] Baryshnikov, Y. and Gnedin, A. Counting intervals in the packing process. *Ann. Appl. Probab.* **11** (2001), 863-877.

[20] Basdevant, A.-L. Ruelle's probability cascades seen as a fragmentation process. *Markov Process. Related Fields* (to appear). Available via: `http://hal.ccsd.cnrs.fr/ccsd-00003809`.

[21] Basdevant, A.-L. Fragmentation of ordered partitions and intervals. *Electron. J. Probab.* **11** (2006), 394–417. Available via: `http://www.math.washington.edu/ejpecp/`.

[22] Basedow, A. M., Ebert, K. H. and Ederer, H. J. Kinetic studies on the acidic hydrolysis of Dextran. *Macromolecules* **11** (1978), 774-781.

[23] Ben-Naim, E. and Krapivsky, P. L. Fragmentation with a steady source. *Phys. Lett. A* **275** (2000), 48-53.

[24] Berestycki, J. Ranked fragmentations. *ESAIM, Probabilités et Statitistique* **6** (2002), 157-176. Available via: `http://www.edpsciences.org/ps/OnlinePSbis.html`

[25] Berestycki, J. Multifractal spectrum of fragmentations. *J. Stat. Phys.* **113** (2003), 411-430.

[26] Berestycki, J. Exchangeable fragmentation-coalescence processes and their equilibrium measures. *Electron. J. Probab.* **9** (2004), 770-824. Available via: `http://www.math.washington.edu/ ejpecp/viewarticle.php?id=1477&layout=abstract`.

[27] Berestycki, J., Berestycki, N. and Schweinsberg, J. Small-time behavior of beta coalescents. Preprint available via: `http://fr.arxiv.org/abs/math.PR/0601032`

[28] Berestycki, J., Berestycki, N. and Schweinsberg, J. Beta-coalescents and continuous stable random trees. Preprint available via: `http://fr.arxiv.org/abs/math.PR/0602113`

[29] Bertoin, J. *Lévy Processes.* Cambridge, Cambridge University Press, 1996.

[30] Bertoin, J. Subordinators: Examples and Applications. *École d'été de Probabilités de St-Flour XXVII*, Lect. Notes in Maths **1717**, pp. 1-91, Berlin, Springer, 1999.

[31] Bertoin, J. A fragmentation process connected to Brownian motion. *Probab. Theory Relat. Fields* **117** (2000), 289-301.

[32] Bertoin, J. Clustering statistics for sticky particles with Brownian initial velocity. *J. Math. Pures Appl.* **79** (2000), 173-194.

[33] Bertoin, J. Eternal additive coalescents and certain bridges with exchangeable increments. *Ann. Probab.* **29** (2001), 344-360.

[34] Bertoin, J. Homogeneous fragmentation processes. *Probab. Theory Relat. Fields* **121** (2001), 301-318.

[35] Bertoin, J. Eternal solutions to Smoluchowski's coagulation equation with additive kernel and their probabilistic interpretations. *Ann. Appl. Probab.* **12** (2002), 547-564.

[36] Bertoin, J. Self-similar fragmentations. *Ann. Inst. Henri Poincaré, Probabilités et Statistiques* **38** (2002), 319-340.

[37] Bertoin, J. Self-attracting Poisson clouds in an expanding universe. *Comm. Math. Phys.* **232** (2002), 59-81.

[38] Bertoin, J. The asymptotic behavior of fragmentation processes. *J. Euro. Math. Soc.* **5** (2003), 395-416.

[39] Bertoin, J. On small masses in self-similar fragmentations. *Stochastic Process. Appl.* **109** (2004), 13-22.

[40] Bertoin, J. and Caballero, M.-E. Entrance from 0+ for increasing semi-stable Markov processes. *Bernoulli* **8** (2002), 195-205.

[41] Bertoin, J. and Gnedin, A. Asymptotic laws for nonconservative self-similar fragmentations. *Electron. J. Probab.* **9** (2004), 575-593. Available via: `http://www.math.washington.edu/ ejpecp/viewarticle.php?id=1463&layout=abstract`.

[42] Bertoin, J. and Goldschmidt, C. Dual random fragmentation and coagulation and an application to the genealogy of Yule processes. *Mathematics and Computer Science III: Algorithms, Trees, Combinatorics and Probabilities.* Drmota, M., Flajolet, P., Gardy, D. and Gittenberger, B. (Eds.) pp. 295-308, Basel, Birkhäuser, 2004.

[43] Bertoin, J. and Le Gall, J.-F. The Bolthausen-Sznitman coalescent and the genealogy of continuous-state branching processes. *Probab. Theory Relat. Fields* **117** (2000), 249-266.

[44] Bertoin, J. and Le Gall, J.-F. Stochastic flows associated to coalescent processes. *Probab. Theory Relat. Fields* **126** (2003), 261-288.

[45] Bertoin, J. and Le Gall, J.-F. Stochastic flows associated to coalescent processes, II: Stochastic differential equation. *Ann. Inst. Henri Poincaré, Probabilités et Statistiques* **41** (2005), 307-333.

[46] Bertoin, J. and Le Gall, J.-F. Stochastic flows associated to coalescent processes, III: Limits theorems. *Illinois J. Math.* (to appear). Available via: `http://hal.ccsd.cnrs.fr/ccsd-00005150`.

[47] Bertoin, J. and Martinez, S. Fragmentation energy. *Adv. Appl. Probab.* **37** (2005), 553-570.

[48] Bertoin, J. and Miermont, G. Asymptotics in Knuth's parking problem for caravans. *Random Structures Algorithms.* (to appear). Available via: `http://hal.ccsd.cnrs.fr/ccsd-00004229`.

[49] Bertoin, J. and Pitman, J. Two coalescents derived from the ranges of stable subordinators. *Elect. J. Probab.* **5** (2000), 1-17. Available via: `http://www.math.washington.edu /~ejpecp/ejp5contents.html`.

[50] Bertoin, J. and Rouault, A. Asymptotical behaviour of the presence probability in branching random walks and fragmentations. Unpublished. Available via: `http://hal.ccsd.cnrs.fr/ccsd-00002955`.

[51] Bertoin, J. and Rouault, A. Discretization methods for homogeneous fragmentations. *J. London Math. Soc.* **72** (2005), 91-109.

[52] Beysens, D, Campi, X, and Pefferkorn, E. *Fragmentation Phenomena*. Singapore, World Scientific, 1995.

[53] Biggins, J. D. Martingale convergence in the branching random walk. *J. Appl. Probability* **14** (1977), 25-37.

[54] Biggins, J. D. The central limit theorem for the supercritical branching random walk, and related results. *Stochastic Process. Appl.* **34** (1990), 255–274.

[55] Biggins, J. D. Uniform convergence of martingales in the branching random walk *Ann. Probab.* **20** (1992), 137-151.

[56] Billingsley, P. *Probability and Measure.* Third edition. New York, John Wiley & Sons, 1995.

[57] Billingsley, P. *Convergence of Probability Measures.* Second edition. New York, John Wiley & Sons, 1999.

[58] Bingham, N. H., Goldie, C. M., and Teugels, J. L. *Regular Variation.* Cambridge, Cambridge University Press, 1987.

[59] Birkner, M., Blath, J., Capaldo, M., Etheridge, A. M., Möhle, M., Schweinsberg, J. and Wakolbinger, A. Alpha-stable branching and beta-coalescents *Electron. J. Probab.* **10** (2005), 303-325. Available via: `http://www.math.washington.edu/ ejpecp/viewarticle.php? id=1496&layout=abstract`

[60] Bollobás, B. *Random Graphs.* 2nd edition. Cambridge, Cambridge University Press, 2001.

[61] Bolthausen, E. and Sznitman, A.-S. On Ruelle's probability cascades and an abstract cavity method. *Comm. Math. Physics* **197** (1998), 247-276.

[62] Bovier, A. and Kurkova, I. Derrida's generalised random energy models. I. Models with finitely many hierarchies. II. Models with continuous hierarchies. *Ann. Inst. H. Poincaré, Probabilités et Statistiques* **40** (2004), 439-480 and 481-495.

[63] Brennan, M. D. and Durrett, R. Splitting intervals. II. Limit laws for lengths. *Probab. Theory Related Fields* **75** (1987), 109-127.

[64] Camarri, M. and Pitman, J. Limit distributions and random trees derived from the birthday problem with unequal probabilities. *Electron. J. Probab.* **5** (2000), paper 2. Available via: `http://www.math.washington.edu/ ejpecp/viewarticle.php? id=1260&layout=abstract`

[65] Cannings, C. The latent roots of certain Markov chains arising in genetics: a new approach, I. Haploid models. *Adv. Appl. Prob.***6** (1974), 260-290.

[66] Cannings, C. The latent roots of certain Markov chains arising in genetics: a new approach, II. Further haploid models. *Adv. Appl. Prob.***7** (1975), 264-282.

[67] Chang, J. T. Recent common ancestors of all present-day individuals. *Adv. in Appl. Probab.* **31** (1999), 1002-1038.

[68] Chassaing, P. and Louchard, G. Phase transition for parking blocks, Brownian excursions and coalescence. *Random Structures Algorithms.* **21** (2002), 76-119.

[69] Chassaing, P. and Marchand, R. Merging costs for the additive Marcus-Lushnikov process, and union-find algorithms. *Annals Probab.* (to appear).

[70] Chauvin, B. Product martingales and stopping lines for branching Brownian motion. *Ann. Probab.* **19** (1991), 1195-1205.

[71] Chauvin, B., Klein, T., Markert, J.-F., and Rouault, A. Martingales and profile of binary search trees. *Electron. J. Probab.* **10** (2005), 420-435. Available via: `http://www.math.washington.edu/ ejpecp/viewarticle.php?id=1513&layout=abstract`.

[72] Deaconu, M., Fournier, N. and Tanré, E. A pure jump Markov process associated with Smoluchowski's coagulation equation. *Ann. Probab.* **30** (2002), 1763-1796.

[73] Deaconu, M., Fournier, N. and Tanré, E. Rate of convergence of a stochastic particle system for the Smoluchowski coagulation equation. *Methodol. Comput. Appl. Probab.* **5** (2003), 131-158.

[74] Deaconu, M. and Tanré, E. Smoluchovski's coagulation equation: probabilistic interpretation of solutions for constant, additive and multiplicative kernels. *Ann. Scuola Normale Sup. Pisa* **XXIX** (2000), 549-580.

[75] Dembo, A. and Zeitouni, O. *Large Deviations Techniques and Applications.* Second edition. Berlin, Springer, 1998.

[76] Derrida, B. Random-energy model: An exactly solvable model of disordered systems. *Phys. Rev. B* **24** (1981), 2613-2626.

[77] Diaconis, P., Mayer-Wolf, E., Zeitouni, O., and Zerner, M. P. W. The Poisson-Dirichlet law is the unique invariant distribution for uniform split-merge transformations. *Ann. Probab.* **32** (2004), 915-938.

[78] Dong, R., Goldschmidt, C. and Martin, J. B. Coagulation-fragmentation duality, Poisson-Dirichlet distributions and random recursive trees. Preprint available via: `http://fr.arxiv.org/abs/math.PR/0507591`.

[79] Dongen (van), P. G. J. and Ernst, M. H. On the occurrence of a gelation transition in Smoluchowski's coagulation equation. *J. Statist. Phys.* **44** (1986), 785-792.

[80] Dongen (van), P. G. J. and Ernst, M. H. Scaling solutions of Smoluchowski's coagulation equation. *J. Statist. Phys.* **50** (1988), 295-329.

[81] Donnelly, P. and Joyce, P. Consistent ordered sampling distributions: characterization and convergence. *Adv. in Appl. Probab.* **23** (1991), 229-258.

[82] Donnelly, P. and Kurtz, T. G. Particle representations for measured-valued population models. *Ann. Probab.* **27** (1999), 166-205.

[83] Donnelly, P. and Kurtz, T. G. Genealogical processes for Fleming-Viot models with selection and recombination. *Ann. Appl. Probab.* **9** (1999), 1091-1148.

[84] Drake, R. L. A general mathematical survey of the coagulation equation. In G. M. Hidy and J. R. Brock (eds.): *Topics in current aerosol research, part 2.* International Reviews in Aerosol Physics and Chemistry. pp. 201-376, Oxford, Pergammon Press, 1972.

[85] Dubovski, P. B. *Mathematical Theory of Coagulation.* Seoul, Seoul National University, 1994.

[86] Duquesne, T. and Le Gall, J.-F. Random trees, Lévy processes and spatial branching processes. *Astérisque* **281** (2002), vi+147 pp.

[87] Durrett, R. *Probability Models for DNA Sequences Evolution.* New York, Springer, 2002.

[88] Durrett, R., Granovsky, B. L. and Gueron, S. The equilibrium behavior of reversible coagulation-fragmentation processes. *J. Theoret. Probab.* **12** (1999), 447-474.

[89] Durrett, R. and Schweinsberg, J. A coalescent model for the effect of advantageous mutations on the genealogy of a population. *Stochastic Process. Appl.* **109** (2005), 13-22.

[90] Dwass, M. The total progeny in a branching process and a related random walk. *J. Appl. Probability* **6** (1969), 682-686.

[91] Eibeck, A. and Wagner, W. Stochastic particle approximations for Smoluchoski's coagulation equation. *Ann. Appl. Probab.* **11** (2001), 1137-1165.

[92] Eibeck, A. and Wagner, W. Stochastic interacting particle systems and nonlinear kinetic equations. *Ann. Appl. Probab.* **13** (2003), 845-889.

[93] Engen, S. *Stochastic Abundance Models with Emphasis on Biological Communities and Species Diversity.* London, Chapman and Hall, 1978.

[94] Erdős, P. and Rényi, A. On the evolution of random graphs. *Publ. Math. Inst. Hung. Acad. Sci.* **5** (1961), 17-61.

[95] Erlihson, M. and Granovsky, B. L. Reversible coagulation-fragmentation processes and random combinatorial structures: asymptotics for the number of groups. *Random Structures Algorithms* **25** (2004), 227-245.

[96] Escobedo, M., Laurençot, Ph., Mischler, S. and Perthame, B. Gelation and mass conservation in coagulation-fragmentation models. *J. Differential Equations* **195** (2003), 143-174.

[97] Escobedo, M., Mischler, S. and Rodriguez Ricard, M. On self-similarity and stationary problem for fragmentation and coagulation models. *Ann. Inst. H. Poincaré, Analyse Non Linéaire* **22** (2005), 99-125.

[98] Etheridge, A. M. *An Introduction to Superprocesses.* University Lecture Series vol. 20. Providence, American Mathematical Society, 2000.

[99] Ethier, S. N. and Kurtz, T. G. *Markov Processes. Characterization and Convergence.* New York, Wiley, 1986.

[100] Evans, S. N. and Pitman, J. Construction of Markovian coalescents. *Ann. Inst. H. Poincaré, Probabilités et Statistiques* **34** (1998), 339-383.

[101] Ewens, W. J. The sampling theory of selectively neutral alleles. *Theoretical Population Biology* **3** (1972), 87-112.

[102] Ferguson, T. A Bayesian analysis of some nonparametric problems. *Ann. Statist.* **1** (1973), 209-230.

[103] Filippov, A. F. On the distribution of the sizes of particles which undergo splitting. *Th. Probab. Appl.* **6** (1961), 275-293.

[104] Fournier, N. On some stochastic coalescents. *Probab. Theory Relat. Fields* (to appear).

[105] Fournier, N. and Giet, J.-S. On small particles in coagulation-fragmentation equations. *J. Stat. Phys.* **111** (2003), 1299-1329.

[106] Fournier, N. and Giet, J.-S. Convergence of the Marcus-Lushnikov process. *Methodol. Comput. Appl. Probab.* **6** (2004), 219-231.

[107] Fournier, N. and Laurençot, Ph. Existence of self-similar solutions to Smoluchowski's coagulation equation. *Comm. Math. Phys.* **256** (2005), 589-609.

[108] Fournier, N. and Löcherbach, E. Stochastic coalescence with homogeneous-like interaction rates. Preprint (2006).

[109] Fristedt, B. and Grey, L. *A Modern Approach to Probability Theory.* Probability and its Applications. Boston, Birkhäuser, 1997.

[110] Giraud, C. Gravitational clustering and additive coalescence. *Stochastic Process. Appl.* **115** (2005), 1302-1322.

[111] Gnedin, A. The representation of composition structures. *Ann. Probab.* **25** (1997), 1437-1450.

[112] Gnedin, A. On convergence and extensions of size-biased permutations. *J. Appl. Prob.* **35** (1998), 642-650.

[113] Gnedin, A. and Kerov, S. A characterization of GEM distributions. *Combin. Prob. Comp.* **10** (2001), 213-217.

[114] Gnedin, A. and Pitman, J. Regenerative composition structures. *Ann. Probab.* **33** (2005), 445-479.

[115] Gnedin, A., Pitman, J., and Yor, M. Asymptotic laws for compositions derived from transformed subordinators. *Ann. Probab.* **34** (2006). Available via: `http://front.math.ucdavis.edu/math.PR/0403438`.

[116] Goldschmidt, C. and Martin, J. B. Random recursive trees and the Bolthausen-Sznitman coalescent. *Electron. J. Probab.* **10** (2005), 718-745. Available via: `http://www.math.washington.edu/ ejpecp/viewarticle.php?id=1521&layout=abstract`.

[117] Golovin, A. M. The solution of the coagulation equation for cloud droplets in a rising air current. *Izv. Geophys. Ser.* **5** (1963), 482-487.

[118] Grimmett, G. R. and Stirzaker, D. R. *Probability and Random Processes.* 3rd edition. New York, Oxford University Press, 2001.

[119] Haas, B. Loss of mass in deterministic and random fragmentations. *Stochastic Process. Appl.* **106** (2003), 245-277.

[120] Haas, B. Regularity of the formation of dust in self-similar fragmentations. *Ann. Inst. Henri Poincaré, Probabilités et Statistiques* **40** (2004), 411-438.

[121] Haas, B. Equilibrium for fragmentation with immigration. *Ann. Appl. Probab.* **15** (2005), 1958-1996.

[122] Haas, B. and Miermont, G. The genealogy of self-similar fragmentations with negative index as a continuum random tree. *Electron. J. Probab.* **9** (2004), 57-97. Available via: `http://www.math.washington.edu/ ejpecp/viewarticle.php?id=1430&layout=abstract`.

[123] Haccou, P., Jagers, P., and Vatutin, V. A. *Branching Processes.* Cambridge, Cambridge University Press, 2005.

[124] Hammond, A. and Rezakhanlou, F. The kinetic limit of a system of coagulating Brownian particles. Preprint available via: `http://front.math.ucdavis.edu/math.PR/0408395`

[125] Hoppe, F. M. Pólya-like urns and the Ewens sampling formula. *J. Math. Biol.* **20** (1984), 91-94.

[126] Ignatov, T. A constant arising in the asymptotic theory of symmetric groups, and Poisson-Dirichlet measures. *Theor. Probab. Appl.* **27** (1982), 136-147.

[127] Jacod, J. and Shiryaev, A. N. *Limit Theorems for Stochastic Processes.* Berlin, Springer, 1987.

[128] Jagers, P. General branching processes as Markov fields. *Stochastic Process. Appl.* **32** (1989), 183-212.

[129] Jeon, I. Existence of gelling solutions for coagulation-fragmentation equations. *Comm. Math. Phys.* **194** (1998), 541-567.

[130] Jeon, I. Stochastic fragmentation and some sufficient conditions for shattering transition. *J. Korean Math. Soc* **39** (2002), 543-558.

[131] Jourdain, B. Nonlinear processes associated with the discrete Smoluchowski coagulation–fragmentation equation. *Markov Process. Related Fields* **9** (2003), 103-130.

[132] Kahane, J.-P. and Peyrière, J. Sur certaines martingales de Benoit Mandelbrot. *Adv. Math.* **22** (1976), 131-145.

[133] Kallenberg, O. Canonical representations and convergence criteria for processes with interchangeable increments. *Z. Wahrscheinlichkeitstheorie verw. Gebiete* **27** (1973), 23-36.

[134] Kerov, S. V. *Asymptotic Representation Theory of the Symmetric Group and its Applications in Analysis.* Translations of Mathematical Monographs **219**. Providence, American Mathematical Society, 2003.

[135] Kimmel, M. and Axelrod, D. E. *Branching Processes in Biology*. New York, Springer, 2002.

[136] Kingman, J. F. C. Random discrete distributions. *J. Roy. Statist. Soc. B* **37** (1975), 1-22.

[137] Kingman, J. F. C. The population structure associated with the Ewens sampling formula. *Theoretical Population Biol.* **11** (1977), 274-283.

[138] Kingman, J. F. C. *Mathematics of Genetic Diversity*. Philadelphia, Society for Industrial and Applied Mathematics, 1980.

[139] Kingman, J. F. C. The coalescent. *Stochastic Process. Appl.* **13** (1982), 235-248.

[140] Kingman, J. F. C. *Poisson Processes*. Oxford Studies in Probability **3**. Oxford, Clarendon Press, 1993.

[141] Kolmogorov, A. N. Über das logarithmisch normale Verteilungsgesetz der Dimensionen der Teilchen bei Zerstückelung. *C. R. (Doklady) Acad. Sci. URSS* **31** (1941), 99-101.

[142] Kolokoltsov, V. Hydrodynamic limit of coagulation-fragmentation type models of $k$-nary interacting particles. *J. Statist. Phys.* **115** (2004), 1621-1653.

[143] Krapivsky, P. L. and Ben-Naim, E. Scaling and multiscaling in models of fragmentation *Phys. Rev. E* **50** (1994), 3502-3507.

[144] Krapivsky, P. L., Ben-Naim, E. and Grosse, I. Stable distributions in stochastic fragmentation. *J. Phys. A* **37** (2004), 2863-2880.

[145] Krell, N. On the decay of fragments in homogeneous fragmentations. Preprint available via :
`http://hal.ccsd.cnrs.fr/ccsd-00018531`.

[146] Kyprianou, A. E. A note on branching Lévy processes. *Stochastic Process. Appl.* **82** (1999), 1-14.

[147] Lambert A. Coalescence times for the branching process. *Adv. Appl. Probab.* **35** (2003), 1071-1089.

[148] Lamperti, J. Semi-stable Markov processes. *Z. Wahrscheinlichkeitstheorie verw. Gebiete* **22** (1972), 205-225.

[149] Lang, R. and Nguyen, X.-X. Smoluchowski's theory of coagulation in colloids holds rigorously in the Boltzmann-Grad-limit. *Z. Wahrsch. Verw. Gebiete* **54** (1980), 227-280.

[150] Laurençot, Ph. On a class of continuous coagulation-fragmentation equations. *J. Differential Equations* **167** (2000), 245-274.

[151] Laurençot, Ph. and Mischler, S. Liapunov functionals for Smoluchowski's coagulation equation and convergence to self-similarity. *Monatsh. Math.* (to appear).

[152] Lépingle, D. La variation d'ordre $p$ des semi-martingales. *Z. Wahrscheinlichkeitstheorie und Verw. Gebiete* **36** (1976), 295–316.

[153] Liu, Q. On generalized multiplicative cascades. *Stochastic Process. Appl.* **86** (2000), 263-286.

[154] Liu, Q. Asymptotic properties and absolute continuity of laws stable by random weighted mean. *Stochastic Process. Appl.* **95** (2001), 83-107.

[155] Lushnikov, A. A. Certain new aspects of the coagulation theory. *Izv. Atmos. Ocean Phys.* **14** (1978), 738-743.

[156] Lyons, R. A simple path to Biggins' martingale convergence for branching random walk. In: Athreya, K. B., and Jagers, P. (editors). *Classical and Modern Branching Processes.* pp. 217-221, New York, Springer-Verlag, 1997.

[157] Lyons, R., Pemantle, R. and Peres, Y. Conceptual proofs of $L$ log $L$ criteria for mean behaviour of branching processes. *Ann. Probab.* **23** (1995), 1125-1138.

[158] McLeod, J. B. On an infinite set of nonlinear differential equations. *Quart. J. Math. Oxford* **13** (1962), 119-128.

[159] McLeod, J. B. On the scalar transport equation. *Proc. London Math. Soc. (3)* **14** (1964), 445-458.

[160] Mandelbrot, B. Multiplications aléatoires et distributions invariantes par moyenne pondérée aléatoire. *Comptes Rendus Acad. Sc.* **278** (1974), 289-292.

[161] Marchal, P. Nested regenerative sets and their associated fragmentation process. *Mathematics and Computer Science III: Algorithms, Trees, Combinatorics and Probabilities.* Drmota, M., Flajolet, P., Gardy, D. and Gittenberger, B. (Eds.), pp. 461-470. Basel, Birkhäuser, 2004.

[162] Marcus, A. H. Stochastic coalescence. *Technometrics* **10** (1968), 133-143.

[163] Matheron, G. *Random Sets and Integral Geometry.* Wiley Series in Probability and Mathematical Statistics. New York-London-Sydney, John Wiley & Sons, 1975.

[164] Mauldin, R. A. and Williams, S. C. Random constructions, asymptotic geometry and topological properties. *Trans. Amer. Math. Soc.* **295** (1986), 325-346.

[165] Menon, G. and Pego, R. L. Approach to self-similarity in Smoluchowski's coagulation equations. *Comm. Pure Appl. Math.* **57** (2004), 1197-1232.

[166] Miermont, G. Ordered additive coalescent and fragmentations associated to Lévy processes with no positive jumps. *Electron. J. Probab.* **6** (2001). Available via:
`http://www.math.washington.edu/ ejpecp/viewarticle.php?id=1289&layout=abstract`

[167] Miermont, G. Self-similar fragmentations derived from the stable tree I: splitting at heights. *Probab. Theory Relat. Fields* **127** (2003), 423-454.

[168] Miermont, G. Self-similar fragmentations derived from the stable tree II: splitting at nodes. *Probab. Theory Relat. Fields* **131** (2005), 341-375.

[169] Miermont, G. and Schweinsberg, J. Self-similar fragmentations and stable subordinators. *Séminaire de Probabilités* **XXXVII** pp. 333-359. Lecture Notes in Math. **1832**, Springer, Berlin, 2003.

[170] Möhle, M. and Sagitov, S. A classification of coalescent processes for haploid exchangeable population models. *Ann. Probab.* **29** (2001), 1547-1562.

[171] Nerman, O. On the convergence of supercritical general (C-M-J) branching processes. *Z. Wahrsch. Verw. Gebiete* **57** (1981), 365-395.

[172] Neuhauser, C. Mathematical models in population genetics. In: Balding *et al.* (eds) *Handbook of Statistical Genetics* pp. 153-177, Hoboken, Wiley, 2001.

[173] Nordborg, M. Coalescent theory. In: Balding *et al.* (eds) *Handbook of Statistical Genetics* pp. 179-212, Hoboken, Wiley, 2001.

[174] Norris, J. R. *Markov Chains*. Cambridge Series on Statistical and Probabilistic Mathematics. Cambridge, Cambridge University Press, 1998.

[175] Norris, J. R. Smoluchowski's coagulation equation: uniqueness, non-uniqueness and hydrodynamic limit for the stochastic coalescent. *Ann. Appl. Probab.* **9** (1999), 78-109.

[176] Norris, J. R. Cluster coagulation. *Comm. Math. Phys.* **209** (2000), 407-435.

[177] Pavlov, Ju. L. The asymptotic distribution of the maximum size of trees in a random forest. *Teor. Verojatnost. i Primenen.* **22**, 523-533.

[178] Perman, M. (1993). Order statistics for jumps of normalized subordinators. *Stochastic Process. Appl.* **46** (1977), 267-281.

[179] Perman, M., Pitman, J. and Yor, M. Size-biased sampling of Poisson point processes and excursions. *Probab. Theory Relat. Fields* **92** (1992), 21-39.

[180] Pitman, J. Exchangeable and partially exchangeable random partitions. *Probab. Theory Related Fields* **102** (1995), 145-158.

[181] Pitman, J. Random discrete distributions invariant under size-biazed permutation. *Adv. Appl. Probab.* **28** (1996), 525-539.

[182] Pitman, J. Partition structures derived from Brownian motion and stable subordinators. *Bernoulli* **3** (1997), 79-96.

[183] Pitman, J. Coalescents with multiple collisions. *Ann. Probab.* **27** (1999), 1870-1902.

[184] Pitman, J. Coalescent random forests. *J. Combin. Theory Ser. A* **85** (1999), 165-193.

[185] Pitman, J. Poisson-Dirichlet and GEM invariant distributions for split-and-merge transformation of an interval-partition. *Combin. Probab. Comput.* **11** (2002), 501–514.

[186] Pitman, J. *Combinatorial Stochastic Processes. École d'été de Probabilités de St-Flour*, Lect. Notes in Maths **1875**, Berlin, Springer, 2006. Available via: `http://stat-www.berkeley.edu /users/pitman/`

[187] Pitman, J. Poisson-Kingman partitions. In: *Science and Statistics: A Festschrift for Terry Speed*, D. R. Goldstein editor. Lecture Notes – Monograph Series 30, pp.1-34. Hayward, California, Institute of Mathematical Statistics, 2003.

[188] Pitman, J. and Yor, M. Arcsine laws and interval-partitions derived from a stable subordinator. *Proc. London Math. Soc. (3)* **65** (1992), 326-356.

[189] Pitman, J. and Yor, M. Random discrete distributions derived from self-similar random sets. *Electron. J. Probab.* **1** (1996), 28 pp. Available via: `http://www.math.washington.edu/ ejpecp/viewarticle.php?id=1206&layout=abstract`.

[190] Pitman, J. and Yor, M. On the relative lengths of excursions derived from a stable subordinator. *Séminaire de Probabilités XXXI*, Lecture Notes in Maths. 1655 pp. 272-286. Berlin, Springer, 1997.

[191] Pitman, J. and Yor, M. The two-parameter Poisson-Dirichlet distribution derived from a stable subordinator. *Ann. Probab.* **25** (1997), 855-900.

[192] Revuz, D. and Yor, M. *Continuous Martingales and Brownian Motion*, 3rd edn. Berlin, Springer, 1999.

[193] Rivero, V. A law of iterated logarithm for increasing self-similar Markov processes. *Stoch. Stoch. Rep.* **75** (2003), 443-472.

[194] Rösler, U. A fixed point theorem for distributions, *Stoch. Proc. Appl.* **42** (1992), 195-214.

[195] Rouault, A. Precise estimates of presence probabilities in the branching random walk. *Stochastic Process. Appl.* **44** (1993), 27-39.

[196] Ruelle, D. A mathematical reformulation of Derrida's REM and GREM. *Comm. Math. Physics* **108** (1987), 225-239.

[197] Sagitov, S. The general coalescent with asynchronous mergers of ancester lines. *J. Appl. Prob.* **36** (1999), 1116-1125.

[198] Sato, K.-I. *Lévy Processes and Infinitely Divisible Distributions.* Cambridge, Cambridge University Press, 1999.

[199] Schweinsberg, J. Coalescents with simultaneous multiple collisions. *Electron. J. Probab.* **5-12** (2000), 1-50. Available via: `http://www.math.washington.edu/ ejpecp/viewarticle.php?id=1270&layout=abstract`

[200] Schweinsberg, J. A necessary and sufficient condition for the Lambda-coalescent to come down from infinity. *Electr. Comm. Probab.* **5-1** (2000), 1-11. Available via: `http://www.math.washington.edu/ejpecp/ECP/viewarticle.php?id=1596&layout=abstract`

[201] Schweinsberg, J. Applications of the continuous-time ballot theorem to Brownian motion and related processes. *Stochastic Process. Appl.* **95** (2001), 151-176.

[202] Schweinsberg, J. Coalescent processes obtained from supercritical Galton-Watson processes. *Stochastic Process. Appl.* **106** (2003), 107-139.

[203] Schweinsberg, J. and Durrett, R. Random partitions approximating the coalescence of lineages during a selective sweep. *Ann. Appl. Probab.* **15** (2005), 1591-1651.

[204] Sheth, R. K. and Pitman, J. Coagulation and branching processes models of gravitational clustering. *Mon. Not. R. Astron. Soc.* (1997).

[205] Smoluchowski (von), M. Drei Vortrage über Diffusion, Brownsche Molekularbewegung und Koagulation von Kolloidteilchen. *Phys. Z.* **17** (1916), 557-571 and 585-599.

[206] Spouge, J. L. Solutions and critical times for the polydisperse coagulation equation when $a(x, y) = A + B(x + y) + Cxy$. *J. Phys. A: Math. Gen.* **16** (1983), 3127-3132.

[207] Tavaré, S. Ancestral inference in population genetics. *École d'été de Probabilités de St-Flour XXXI*, Lecture Notes in Math., **1837**, pp. 1-188. Springer, Berlin, 2004.

[208] Uchiyama, K. Spatial growth of a branching process of particles living in $R^d$. *Ann. Probab.* **10** (1982), 896-918.

[209] Vershik, A. M. The asymptotic distribution of factorizations of natural numbers into prime divisors. *Soviet Math. Dokl.* **34** (1986), 57-61.

[210] Vershik, A. M., and Shmidt, A. A. Limit measures arising in the asymptotic theory of symmetric groups. *Theor. Probab. Appl.* **22** (1977), 70-85.

[211] Wagner, W. Explosion phenomena in stochastic coagulation-fragmentation models. *Ann. Appl. Probab.* **15** (2005), 2081-2112.

[212] Wakeley, J. *Coalescent Theory: An Introduction.* Denver, Roberts & Company Publishers, 2006.

[213] Whittle, P. *Systems in Stochastic Equilibrium*. Chichester, Wiley, 1986.

# List of symbols

## Functions and operators

| | |
|---|---|
| $\mathbb{1}_B$ | indicator function of a set $B$ |
| $\#B$ | cardinal of a set $B$ |
| $\#\pi$ | number of non-empty blocks of a partition $\pi$ |
| $\lvert B\rvert$ | asymptotic frequency of a block $B$ |
| $\lvert I\rvert$ | length of some interval $I$ |
| $\lvert u\rvert$ | generation of the node $u$ in the genealogical tree $\mathcal{U}$ |
| $\lvert\vartheta\rvert^{\downarrow}$ | mass-partition induced by the interval decomposition of an open set $\vartheta \subseteq ]0,1[$ |
| $\lvert\pi\rvert^{\downarrow}$ | ranked sequence of the asymptotic frequencies of blocks for a partition $\pi$ |
| $\langle\mu, f\rangle$ | alternative notation for $\int f d\mu$, where $f$ is a measurable function and $\mu$ a measure |
| $\mathrm{Coag}(\pi, \pi')$ | coagulation of the blocks of a partition $\pi$ using the partition $\pi'$ |
| $\mathbb{E}$ | expectation with respect to a generic probability measure $\mathbb{P}$ |
| EPPF | exchangeable partition probability function |
| $\mathrm{Frag}(\pi, \pi^{(\cdot)})$ | fragmentation of the blocks of a partition $\pi$ using the sequence $\pi^{(\cdot)}$ |
| $\mathsf{G}$ | infinitesimal generator |
| $\kappa(q)$ | see Formula (1.17) |
| $\kappa(x, y)$ | coagulation kernel |
| $\Phi$ | see Formula (3.7) |
| $\mathbf{s}^{i\oplus j}$ | sequence obtained from $\mathbf{s} = (s_1, \ldots)$ by merging its $i$-th and $j$-th terms |

## Measures and rates

| | |
|---|---|
| $\delta_x$ | Dirac point mass at $x$ |
| $\epsilon$ | erosion rate |
| $\Lambda$ | measure on $]0, \infty[$ with $\int (1 \wedge x)\Lambda(dx) < \infty$ |
| $\mu^{\mathrm{K}}$ | binary coagulation rate corresponding to Kingman's coalescent |
| $\nu$ | in Chapters 1 and 3: dislocation measure<br>in Chapter 4: measure of multiple coagulations |

| | |
|---|---|
| $\mathbb{P}$ | generic probability measure |
| $\mathbb{P}^*$ | probability measure associated to the randomly tagged branch |
| $q(x,\cdot)$ | jump rate from the state $x$ in a Markov chain |
| $q(x)$ | total jump rate $q(x,E)$ from the state $x$ in a Markov chain |
| $\overline{q}(x,\cdot)$ | normalized probability kernel $q(x,\cdot)/q(x)$ |
| $q_\pi$ | in Chapter 3: jump rate from $\mathbf{1}_{[n]}$ to $\pi$ in a homogeneous fragmentation<br>in Chapter 4: jump rate from $\mathbf{0}_{[n]}$ to $\pi$ in a homogeneous coalescent |
| $\varrho_{\mathbf{s}}$ | law of the paint-box based on the mass-partition $\mathbf{s}$ |

## Partitions

| | |
|---|---|
| $\mathbf{0}$ | mass-partition $(0,\dots)$ |
| $\mathbf{1}$ | mass-partition $(1,0,\dots)$ |
| $\mathbf{1}_B$ | trivial partition $(B,\emptyset,\dots)$ of a block $B$ |
| $\mathbf{0}_B$ | partition of a block $B$ into singletons |
| $\pi$ | generic partition, usually in $\mathcal{P}_n$ or $\mathcal{P}_\infty$ |
| $\pi_i$ | $i$-th block of a partition $\pi$ |
| $\pi_{\mid B}$ | restriction of a partition $\pi$ to a block $B$ |
| $\mathbf{s}$ | generic element $(s_1,s_2,\dots)$ of $\mathcal{S}^\downarrow$ or $\mathcal{P}_{\mathrm{m}}$ |
| $\mathbf{s}^*$ | size-biased reordering of a proper mass-partition $\mathbf{s}$ |
| $s_0$ | mass of dust $1-\sum_{i=1}^\infty s_i$ of a mass-partition $\mathbf{s}=(s_1,s_2,\dots)$ |

## Processes and random variables

| | |
|---|---|
| $B_{t,t'}$ | flow associated to a simple exchangeable coalescent |
| $\hat{B}_{t,t'}$ | dual flow $B_{-t',-t}$ |
| $\mathrm{D}$ | process of dust (Sections 1.3.2 and 4.3.2) |
| $\mathbf{e}$ | standard exponential variable |
| GEM($\theta$) | Griffiths-Engen-McClosey variable with parameter $\theta>0$ |
| $\mathrm{K}^{[n]}$ | state sequence for Kingman coalescent on $[n]$ |
| $\mathcal{M}$ | intrinsic martingale |
| $\mathcal{M}_\infty$ | terminal value of the intrinsic martingale |
| $\Pi$ | in Chapter 3: exchangeable fragmentation<br>in Chapter 4: exchangeable coalescent |
| $\Pi_1(t)$ | block of the random partition $\Pi(t)$ which contains 1 |
| $\Pi^{\mathrm{K}}$ | Kingman coalescent |
| $\Pi^{\mathrm{BS}}$ | Bolthausen-Sznitman coalescent |
| PD($\alpha,\theta$) | Poisson-Dirichlet random mass-partition with parameter $(\alpha,\theta)$ |
| $X$ | in Chapter 1: self-similar fragmentation chain<br>in Chapters 3 and 4: process $\lvert\Pi\rvert^\downarrow$ of the ranked asymptotic frequencies of $\Pi$ |
| $\chi$ | process of the size of the tagged particle |
| $\xi(t)$ | subordinator such that $\lvert\Pi_1(t)\rvert=\exp-\xi(t)$ |

## Sets and spaces

| | |
|---|---|
| $[n]$ | set of the first $n$ integers $\{1, \ldots, n\}$ |
| $[\infty]$ | alternative notation for $\mathbb{N}$ |
| $E_{\mathrm{p}}$ | space of finite point measures on a space $E$ |
| $\mathbb{N}$ | set of positive integers $\{1, 2, \ldots\}$ |
| $\overline{\mathbb{N}}$ | $\mathbb{N} \cup \{\infty\}$ |
| $\mathcal{P}_B$ | space of partitions of a block $B$ |
| $\mathcal{P}_{\mathrm{I}}$ | space of interval-partitions |
| $\mathcal{P}_{\mathrm{m}}$ | space of mass-partitions |
| $\mathcal{P}_n$ | space of partitions of $[n]$ |
| $\mathcal{P}_\infty$ | space of partitions of $\mathbb{N}$ |
| $\mathcal{S}^\downarrow$ | space of non-increasing numerical sequences $\mathbf{s}$ with limit 0 |
| $\mathcal{S}_f^\downarrow$ | sub-space of $\mathcal{S}^\downarrow$ of sequences having only finitely many non-zero terms |
| $\mathcal{S}_1^\downarrow$ | sub-space of $\mathcal{S}^\downarrow$ formed by summable sequences |
| $\mathcal{U}$ | genealogical tree |
| $\mathbb{Z}$ | set of integers $\{\ldots, -1, 0, 1, \ldots\}$ |
| $\mathbb{Z}_+$ | set of non-negative integers $\{0, 1, \ldots\}$ |

## Some important parameters

| | |
|---|---|
| $\alpha$ | index of self-similarity |
| $\mathtt{c}$ | in Chapter 3: coefficient of erosion<br>in Chapter 4: coefficient of binary coagulations |
| $p^*$ | Malthusian exponent |
| $\underline{p}$ | see Formula (1.16) |
| $\bar{p}$ | see Lemma 1.6 |

# Index

www.ingramcontent.com/pod-product-compliance
Lightning Source LLC
LaVergne TN
LVHW102008210926
848483LV00010B/171
*9780521867283*